Bacterial Circadian Programs

Jayna L. Ditty • Shannon R. Mackey
Carl H. Johnson
Editors

Bacterial Circadian Programs

Springer

Editors
Dr. Jayna L. Ditty
University of St. Thomas
Department of Biology
2115 Summit Avenue
St. Paul, MN 55105, USA
e-mail: jlditty@stthomas.edu

Dr. Carl H. Johnson
Vanderbilt University
Department of Biological Sciences
Box 1634, Station B
Nashville, TN 37235, USA
e-mail: carl.h.johnson@vanderbilt.edu

Dr. Shannon R. Mackey
St. Ambrose University
Department of Biology
518 W. Locust St.
Davenport, IA 52803, USA
e-mail: mackeyshannonr@sau.edu

Cover illustration: The cover depicts the character in Japanese for "kai", the name of the central circadian clock gene cluster in cyanobacteria. "Kai" means cycle or rotation number in Japanese, and is therefore apropos for a gene cluster that controls circadian cycles in cyanobacteria.

ISBN 978-3-540-88430-9 e-ISBN 978-3-540-88431-6
DOI: 10.1007/978-3-540-88431-6

Library of Congress Control Number: 2008939987

Cover design: WMXDesign GmbH, Heidelberg, Germany

Printed on acid-free paper

9 8 7 6 5 4 3 2 1

springer.com

Preface

Internal biological clock systems exist in nearly all organisms, including humans, rodents, insects, plants, fungi, and bacteria. These biological (circadian) rhythms allow for each system to maintain internal time and likely provide an adaptive advantage to those organisms. The discovery of circadian rhythms in the cyanobacteria was surprising to some who believed that bacteria were too "simple" to possess the machinery necessary for generating these internal rhythms; however, investigations into the basic biology of the temporal separation of oxygen-evolving photosynthesis and oxygen-sensitive nitrogen fixation demonstrated that this diverse group of bacteria was capable of generating and maintaining internal timing.

Since the discovery of a biological clock in cyanobacteria in the 1980s, the field has exploded with new information. The cyanobacterial model system for studying circadian rhythms, *Synechococcus elongatus* PCC 7942, has allowed for a detailed genetic dissection of the bacterial clock due to the methods in molecular biology and biochemistry that are currently available. Although the majority of research has been conducted using *S. elongatus*, work in other cyanobacterial species has been instrumental to our understanding of the bacterial biological clock. In addition, examination of the various, fully sequenced cyanobacterial genomes suggest that there may be several variations upon the same theme for producing internal rhythms in prokaryotes. Through mathematical modeling and generating synthetic oscillators in other bacterial strains, in conjunction with information derived from in vivo and in vitro oscillations, the mechanism for the generation of biological rhythms in a single cell can be better elucidated.

The rapid advancement in our understanding of the bacterial circadian clock is due to many different avenues of discovery and inquiry. The success in understanding bacterial circadian programs is due, in part, to the genetically malleable *S. elongatus* PCC 7942 system and the insightful investigations of geneticists, molecular biologists, evolutionary biologists, and biochemists. What cannot be overlooked when discussing the success of this model system is that the molecular work stands on the shoulders of hundreds of years of circadian insights into the physical, physiological, and chemical basis of rhythms defined by circadian biologists outside the prokaryotic arena. Currently the *S. elongatus* system is arguably one of the best characterized circadian clock systems of any model system, even though it is one of the newest model systems to be investigated.

Thanks to the many advances in our understanding of the bacterial biological clock, this book serves as a timely review of the fundamental process of circadian timing in prokaryotes. It is also organized as a compendium of the most current data on the circadian mechanism in prokaryotes. The chapters in this book are intended to address the history and background of the cyanobacteria and initial investigations and discovery of circadian rhythms in this diverse group of microorganisms (Chaps. 1, 2, 3, 4). The molecular basis and structure of the circadian clock system are reviewed (Chaps. 5, 6, 7), as well as entrainment of the oscillator with the environment (Chap. 8) and the downstream genes and behavioral activities that are controlled by the clock (Chaps. 9, 10, 11). A demonstration of the adaptive significance of the circadian clock in cyanobacteria (Chap. 12) and the prokaryotic clock's remarkable stability are also discussed (Chap. 13). Due to the great diversity of the cyanobacteria as a group, investigations have been conducted to address the evolution of cyanobacterial clock genes and whether those genes are involved in the generation of circadian rhythms in cyanobacterial strains other than the *S. elongatus* model system (Chaps. 2, 14, 15) and mathematical models for *S. elongatus* clock function and synthetic oscillator models are included (Chaps. 16, 17).

Our hope is that this book will serve many audiences, spanning from those who are currently expanding the studies discussed within, to those who are beginning their endeavor into the wonderful world of prokaryotic clock systems. We envision this text as a comprehensive reference of past accomplishments, but hopefully also a stepping stone for future work on this amazing group of microorganisms and timing. We are grateful to each of our colleagues and friends who contributed to this work. It is our hope that you enjoy reading each chapter as much as we enjoyed putting this combined work together.

Jayna L. Ditty
Shannon R. Mackey
Carl H. Johnson

Contents

Contributors

Setsuyuki Aoki
Graduate School of Information Science, Nagoya University, Furo-cho, Chikusa-ku, Nagoya 464-8601, Japan, e-mail: aoki@is.nagoya-u.ac.jp

David Arps
Department of Biological Chemistry, University of Michigan Medical School, Ann Arbor, MI 48109-0606, USA

Stephen Atkins
Department of Biological Chemistry, University of Michigan Medical School, Ann Arbor, MI 48109-0606, USA, e-mail: sjatkins@umich.edu

Mariette R. Atkinson
Department of Biological Chemistry, University of Michigan Medical School, Ann Arbor, MI 48109-0606, USA

Mark Byrne
Department of Physics, 4000 Dauphin Street, Spring Hill College, Mobile, AL 36608, USA, e-mail: mbyrne@shc.edu

You Chen
Department of Biology and Center for Research on Biological Clocks, Texas A&M University, College Station, TX 77843, USA, e-mail: ychen@syntheticgenomics.com

Donald Court
National Cancer Institute–Frederick, Frederick, MD 21702-1201, USA, e-mail: court@ncifcrf.gov

Jayna L. Ditty
Department of Biology, The University of St. Thomas, St. Paul, MN 55105, USA, e-mail: jlditty@stthomas.edu

Volodymyr Dvornyk
School of Biological Sciences, The University of Hong Kong, Pokfulam Road, Hong Kong SAR, P.R. China, e-mail: dvornyk@hkucc.hku.hk

Martin Egli
Department of Biochemistry, School of Medicine, Vanderbilt University, Nashville, TN 37232, USA, e-mail: martin.egli@vanderbilt.edu

Daniel Forger
Department of Mathematics, University of Michigan, Ann Arbor, MI 48109-1043, USA, e-mail: forger@umich.edu

Susan S. Golden
Department of Biology and Center for Research on Biological Clocks, Texas A&M University, College Station, TX 77843, USA, e-mail: sgolden@tamu.edu

Tan-Chi Huang
Institute of Plant and Microbial Biology, Academia Sinica, Taipei, Taiwan, Republic of China

Hideo Iwasaki
Department of Electrical Engineering & Bioscience, Waseda University; and PRESTO, Japan Science & Technology Agency, 2-2 Wakamatsu-cho, Shinjuku, Tokyo 162-8480, Japan, e-mail: hideo-iwasaki@waseda.jp

Carl Hirschie Johnson
Department of Biological Sciences, Vanderbilt University, Nashville, TN 37235, USA, e-mail: carl.h.johnson@vanderbilt.edu

Rong-Fong Lin
Institute of Medical BioTechnology, Central Taiwan University of Science and Technology, TaiChung, Taiwan, Republic of China, e-mail: rflin@ctust.edu.tw

Shannon R. Mackey
Department of Biology, St. Ambrose University, Davenport, IA 52803, USA, e-mail: MackeyShannonR@sau.edu

Avraham E. Mayo
Weizmann Institute of Science, Rehovot, Israel, e-mail: avimayo@weizmann.ac.il

Irina Mihalcescu
Laboratoire de Spectrométrie Physique, Université de Grenoble–CNRS UMR5588, 38402 Saint Martin d'Hères, France, e-mail: Irina.Mihalcescu@ujf-grenoble.fr

Tetsuya Mori
Department of Biological Sciences, Vanderbilt University, VU Station B 35-1634, 2301 Vanderbilt Place, Nashville, TN 37235-1634, USA, e-mail: tetsuya.mori@vanderbilt.edu

Alexander J. Ninfa
Department of Biological Chemistry, University of Michigan Medical School, Ann Arbor, MI 48109-0606, USA, e-mail: aninfa@umich.edu

Kiyoshi Onai
Center for Gene Research, Nagoya University, Furo-cho, Chikusa-ku, Nagoya 464-8602, Japan, email: onai@gene.nagoya-u.ac.jp

Tokitaka Oyama
Kyoto University, Graduate School of Science, Department of Botany, Kitashirakawa-Oiwake-cho, Sakyo-ku, Kyoto, 606-8502, Japan, e-mail: oyama@cosmos.bot.kyoto-u.ac.jp

Nicolas Perry
Department of Biophysics, University of Michigan, Ann Arbor, MI 48109, USA, e-mail: nperryp@umich.edu

Stephen Selinsky
Department of Biological Chemistry, University of Michigan Medical School, Ann Arbor, MI 48109-0606, USA

Rachelle M. Smith
Divison of Biological Sciences, University of California San Diego, La Jolla, CA 92093, USA, e-mail: rms005@ucsd.edu

Phoebe L. Stewart
Department of Molecular Physiology and Biophysics, School of Medicine, Vanderbilt University, Nashville, TN 37232, USA

Ioannis Vakonakis
Department of Biochemistry, University of Oxford, South Parks Road, Oxford, OX1 3QU, United Kingdom, e-mail: ioannis.vakonakis@bioch.ox.ac.uk

Stanly B. Williams
Life Science Building, Department of Biology, University of Utah, Salt Lake City, UT 84112, USA, e-mail: williams@biology.utah.edu

Mark A. Woelfle
Department of Biological Sciences, Vanderbilt University, Nashville, TN 37235, USA, e-mail: mark.woelfle@vanderbilt.edu

Yao Xu
Department of Biological Sciences, Vanderbilt University, Nashville, TN 37235, USA, e-mail: yao.xu@vanderbilt.edu

Gil Zeidner
Department of Biology and Center for Research on Biological Clocks, Texas A&M University, College Station, TX 77843, USA, e-mail: gzeidner@mail.bio.tamu.edu

Chapter 1
Classic Circadian Characteristics: Historical Perspective and Properties Relative to the *Synechococcus elongatus* PCC 7942 Model

Jayna L. Ditty and Shannon R. Mackey

Abstract The purpose of this chapter is to introduce the basics of circadian biology relative to the cyanobacterial model system. It is meant to define the terms, characteristics, and rules that pertain to the study of circadian biology in the context of the cyanobacterial systems used to elucidate the mechanisms by which the prokaryotic circadian clock functions. In addition, its purpose is to serve as a conduit to the chapters in this book, which comprehensively review our most recent understanding about each of these canonical characteristics in the *Synechococcus elongatus* PCC 7942 model system as well as other cyanobacterial and prokaryotic systems.

1.1 Introduction

1.1.1 Overview

Our planet rotates about its axis every 24 h, which exposes the majority of plants and animals that inhabit the earth to sidereal fluctuations of light and temperature. This daily change in light and dark was a strong selective force (for those organisms that are subject to it) to devise physiological mechanisms with which to respond to, or better yet predict, when these daily changes were going to occur. As a result of this pressure, organisms have evolved internal timing mechanisms to anticipate the daily variations in light and temperature; this anticipatory behavior provides a selective advantage to the organism (DeCoursey 1961; Ouyang et al. 1998; Michael et al. 2003; Woelfle 2004; Johnson 2005).

J.L. Ditty(✉)
Department of Biology, The University of St. Thomas, St. Paul, MN 55105, USA,
e-mail: jlditty@stthomas.edu

S.R. Mackey
Department of Biology, St. Ambrose University, Davenport, IA 52803, USA,
e-mail: MackeyShannonR@sau.edu

J.L. Ditty et al. (eds.), *Bacterial Circadian Programs.*

This daily clock phenomenon was termed "circadian" in 1959 by Franz Halberg using the Latin terms *circa* for "about" and *dies* "day". Therefore circadian phenomenon pertain to biological activities with a frequency of one activity cycle every 24 h (Halberg et al. 1977). The purpose of this chapter is to introduce the basics of "circadiana": to define the numerous terms, characteristics, and rules that pertain to the study of circadian biology in the context of the cyanobacterial systems that have been used to elucidate the mechanism by which the prokaryotic circadian clock functions. In addition, its purpose is to serve as a conduit to the chapters in this book, which comprehensively review the most recent understanding about each of these canonical characteristics in the *Synechococcus elongatus* PCC 7942 model system as well as other cyanobacterial and prokaryotic systems.

1.1.2 Historical Perspectives

Investigations into the mechanism that organisms use to relate and respond to diurnal fluctuations in light and temperature have been undertaken at least as early as the 1700s. One of the earliest reports that correlates behavior with specific times of day came from the French astronomer Jean-Jacques d'Ortous deMairan, who made the observation that the leaves of heliotrope plants move in response to changes in light. Even more importantly, he recognized that these leaves would continue to move in the same pattern when kept in constant darkness (DD), generating the first evidence that a behavioral activity could be regulated by an internal mechanism of the plant, and not a result of environmental light and dark cues (deMairan 1729). During the same period, the Swedish botanist Carl Linneaus developed his *horologium florae* or "flower clock," which could be used to tell the time of day based upon when particular plant species would flower (Freer 2003).

The modern field of chronobiology, or the study of biological timing processes in living things, was initiated in the mid-1950s by Colin S. Pittendrigh and Jürgen Aschoff. They were instrumental in defining and organizing the principles of a circadian system that mapped the course for circadian research, and these rules still hold true to the present time (Aschoff 1960, 1981; Pittendrigh 1961, 1981). While the characteristics and principles of circadian biology were being brought to bear by early circadian biologists, a particular question of interest was whether circadian activity was a learned behavior in organisms or had a genetic basis. The work of Erwin Bünning in 1935 alluded to the answer by providing evidence that period length was heritable in bean plants (Bünning 1935); however, it was not until the early 1970s that the first evidence for a genetic basis to circadian activity was brought to light by two independent groups working in fruit flies and fungus. Ronald Konopka and Seymour Benzer isolated *Drosophila melanogaster* mutants that had altered eclosion and activity rhythms. Each of the mutations was complemented by one genetic locus, termed the *period* gene (Konopka and Benzer 1971). Soon after, Jerry Feldman and Marian Hoyle identified the *frequency* gene, which was shown to be essential for rhythms of asexual spore formation in *Neurospora crassa* (Feldman and Hoyle 1973).

The study of circadian clocks and rhythms was sequestered to eukaryotic models as historical circadian dogma dictated that nuclear structure, intercellular communication, and generation times longer than 24 h were required for rhythmic activity – characteristics that are lacking in prokaryotic cells and, at least in part, in unicellular eukaryotes (Edmunds 1983; Kippert 1987). However, in the 1980s, several lines of evidence were emerging to contradict the "eukaryocentric" circadian requirements. The cyanobacteria are a large and diverse group of microorganisms that are typically photoautotrophic and diazotrophic, and are responsible for a vast majority of the carbon and nitrogen fixation in the environment (see Chap. 2; Garrity 2001). Within several different cyanobacterial species, circadian activity in nitrogen fixation, amino acid uptake, and cell division were identified (see Chap. 3; Grobbelaar et al. 1986; Mitsui et al. 1986; Sweeney and Borgese 1989; Huang et al. 1990; Chen et al. 1991; Grobbelaar and Huang 1992; Schneegurt et al. 1994). While the physiological evidence drastically changed the manner by which scientists thought about circadian biology, a good model system for prokaryotic circadian research was lacking. Ultimately *S. elongatus* PCC 7942 became the model of choice in part because of the vast amount of molecular tools available in this strain (see Chap. 4; Golden 1987; Golden 1988; Kondo et al. 1993, 1994; Ishuira et al. 1998; Andersson 2000).

1.2 Properties of a Clock-Controlled Rhythm

Regardless of the model system one is using to understand the circadian process, the underlying mechanisms achieve a similar goal: maintain an internal, 24-h time. A circadian clock system is defined as an endogenous mechanism that allows an organism to temporally regulate biological activity as a function of the 24-h day. Such biological activities that are regulated by the circadian clock are therefore coined circadian rhythms (Pittendrigh 1981; Edmunds 1983; Dunlap et al. 2004; Koukkari and Sothern 2006). The rhythmic nature of daily activity can be described by three terms that correspond to the characteristic descriptions of a waveform: period, phase, and amplitude.

1.2.1 Period

The period of a rhythm is defined as the duration of one complete activity cycle (Fig. 1.1). Therefore, a circadian period would be an activity that completed its cycle (with a frequency of approximately 1) over a 24-h period of time (Dunlap et al. 2004; Koukkari and Sothern 2006). When measured under constant conditions (see Sect. 1.3.1) the period is defined as the free-running period (FRP),

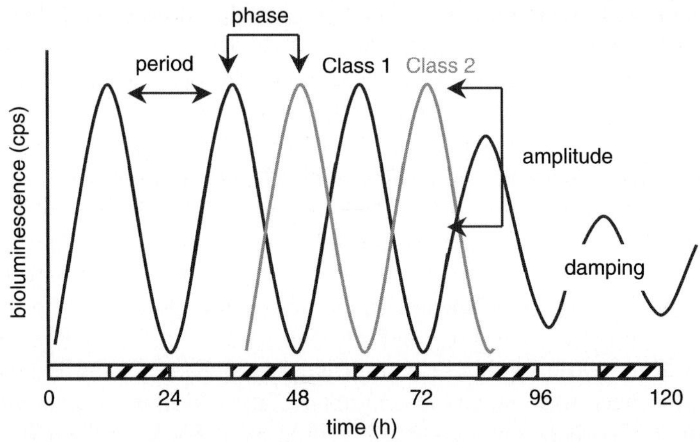

Fig. 1.1 Properties of a circadian activity rhythm as measured in *Synechococcus elongatus* PCC 7942. The traces depict levels of bioluminescence in counts per second (*cps*) over time (*h*) from two representative *S. elongatus* luciferase reporter strains maintained in constant conditions. Alternating *open* and *hatched* bars on the abscissa represent subjective day and subjective night, respectively. Period is defined here by peak-to-peak activity duration over approximately 24 h. Phase is defined here as the time when peak activity is reached. In the *S. elongatus* model, two phases are typically described: Class 1 (*black*) peaks at subjective dusk, while Class 2 (*gray*) peaks at subjective dawn. Amplitude is defined as the magnitude of the oscillation from the mean, where damping is a general trend whereby there is a decrease in rhythmic robustness over time under constant conditions

represented by τ (the Greek symbol *tau*). Observable, and therefore measurable, rhythms of a circadian timing system are not easily measured in bacteria due to their microscopic size and lack of obvious overt behaviors. Therefore, in *S. elongatus* PCC 7942, cyanobacterial promoters were engineered to produce the LuxAB luciferase proteins, as well as their necessary substrates (LuxCDE), from *Vibrio fischeri* and *V. harveyi* respectively, for bioluminescence as an easily measurable and quantifiable output (Kondo et al. 1993). While the average period for circadian rhythms in *S. elongatus* PCC 7942 is approximately 24–25 h (Kondo et al. 1993; Ishiura et al. 1998), the form or shape of the activity rhythm can vary considerably depending upon the promoter used to drive expression of *luxAB*. Waveform patterns of gene expression have been shown to be symmetrical sine curves, asymmetric, saw-tooth, or step-like in form (Liu et al. 1995).

1.2.2 Phase

The phase of an activity rhythm is defined as the instantaneous state of an oscillation within a period (Fig. 1.1; Dunlap et al. 2004; Koukkari and Sothern 2006). For example, the highest point of any rhythmic activity would be defined as the peak of activity (trough for the lowest). These peaks (or troughs) of activity can be used as reference points for determining at what point a particular activity occurs the most (or least) during a 24-h day. The majority of genes expressed in a circadian manner, as measured by random promoter:luciferase fusions in *S. elongatus* PCC 7942, were categorized into a number of different classes with the majority of the genes falling into either Class 1, where activity peaked at subjective dusk (the time in constant light, LL, that corresponds to dusk of the entraining light/dark, LD, cycle), or Class 2, where activity peaked at subjective dawn (the time in LL that corresponds to dawn of the entraining LD cycle; Liu et al. 1995).

1.2.3 Amplitude

The amplitude of a rhythm is defined as the magnitude from the mean activity level to either the peak or to the trough of activity (Fig. 1.1; Dunlap et al. 2004; Koukkari and Sothern 2006). Amplitude is an obvious requirement of a cyclic activity, but it is much more difficult to quantify and interpret than the period or phase of a rhythmic behavior, particularly in the cyanobacterial system. Typically, cultures of cyanobacterial cells (in lieu of individual cyanobacterial cells) are measured for circadian activity. Therefore, careful consideration of the number of cells being measured, the innate differences in the particular promoter driving expression of the reporter, and the level of substrate available for luciferase could each affect the measurement of the amplitude (Kondo et al. 1993; Andersson et al. 2000). Additionally, damping, or a decrease in rhythmic activity over time, can confound amplitude measurements; however, this has not been extremely problematic in the *S. elongatus* PCC 7942 model system, as robust rhythms have been measured for over two weeks in constant conditions (Golden and Canales 2003).

1.2.4 Time

The period, phase, and amplitude are all characteristics of activity patterns that are measured over time. When considering time in a circadian system, there are important distinctions that must be noted. Standard clock time is measured by mechanical or atomic clocks that are used to determine time of day with midnight placed in the middle of the dark and noon when the sun is at its highest point. Therefore, when activity is measured under standard conditions, light and dark cycles persist, and activity can be influenced by these light cues. When measured under these cues, activity is measured in zeitgeber time (ZT, "time-giver" in German), as the environmental cues (zeitgebers) of light, dark, and temperature (to name a few) are present to affect behavior (Fig. 1.2; Dunlap et al. 2004; Koukkari and Sothern 2006).

To truly measure endogenously generated circadian activities, rhythms must be measured in the absence of these environmental cues (see Sect. 1.3.1).

In contrast to ZT time, circadian time (CT) is subject only to the internal timing mechanism of the organism and is independent of ZT; as such, CT is measured under constant environmental conditions. Depending on the system, constant conditions can be maintained in either LL or DD. Under LL conditions, the CT "hour" for *S. elongatus* is calculated by dividing the FRP (approximately 25 h) by the 24-h standard time (for an approximate value of 1.04 h). When measuring circadian activity in these unnatural constant conditions, the CT subjective day starts at CT0 (subjective dawn) and refers to the time that corresponds to lights on in ZT. In contrast, subjective night begins at CT12 (subjective dusk), which corresponds to lights off (Fig. 1.2; Daan et al. 2002; Dunlap et al. 2004; Koukkari and Sothern 2006).

1.3 Defining Characteristics of a Circadian Rhythm

An important distinction must be made between whether a particular activity or behavior is merely *responding* to an environmental cue (i.e., lights on), or if the organism is *predicting* the next environmental cue, thereby generating an activity pattern that is regulated by an internal timing mechanism. Therefore, there are three tenets that describe and define a rhythmic activity that is generated internally and is truly circadian in nature. A circadian rhythm: (1) persists in the absence of environmental cues, (2) is entrained by environmental stimuli, and (3) is temperature-compensated (Pittendrigh 1981; Edmunds 1983; Dunlap et al. 2004; Koukkari and Sothern 2006).

1.3.1 Persistence under Constant Conditions

The first intrinsic characteristic of the circadian rhythm is that the pattern of activity continues in the absence of any environmental cue (Fig. 1.2). In the absence of zeitgebers such as light, dark, temperature, or humidity, rhythmic activity continues with the period that is set by the circadian clock. The time needed for one circadian oscillation to occur (either peak-to-peak or trough-to-trough) under these artificial constant conditions is therefore defined as the FRP, as the activity pattern is free from zeitgeber influence and is driven solely by circadian clock control, which demonstrates the endogenous source of the control mechanism. The range of FRPs for organisms is typically 22–25 h (Pittendrigh 1960; Aschoff 1981; Dunlap et al. 2004; Koukkari and Sothern, 2006).

Although circadian rhythms persist under constant conditions, the FRP is still sensitive to changes or differences in various zeitgebers. The FRP in LL has been shown to vary in response to changing light intensities. First described by Jürgen Aschoff and now affectionately referred to as Aschoff's rule, diurnal organisms typically display a shorter FRP, or a "faster" clock, under high light intensities

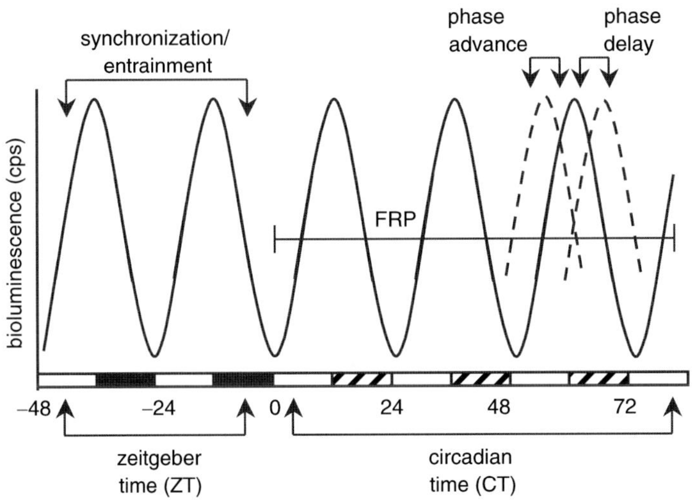

Fig. 1.2 General characteristics of a circadian rhythm. The abscissa designates time (*h*). Negative values represent zeitgeber time, in which *S. elongatus* cultures are exposed to LD cycles in order to entrain their internal clocks. Alternating *open* and *black* bars represent when cells are exposed to light and darkness, respectively. Positive values represent circadian time, when cells are in constant light (LL). Alternating *open* and *hatched* bars represent subjective day and subjective night, respectively. *FRP* is the free-running period, which is the measured persistence of the rhythm in the absence of an environmental cue. In response to zeitgeber cues provided during LL, phase shifting occurs to ensure that activity coincides with the correct time of day. The resulting phase of the activity pattern is shifted either earlier (phase advance) or later (phase delay), while the period is not altered

than at low light intensities. Conversely, nocturnal organisms exhibit the opposite response with a longer FRP under constant high light intensities than under constant low light (Aschoff 1981).

1.3.2 Entrainment by Environmental Cues

Under natural conditions, where organisms are subject to daily changes in light and dark, circadian rhythms are not free running; rather, they are entrained to local environmental cues (Fig. 1.2). Because FRPs are close to (but very rarely) 24 h in length, the circadian system must be able to reset its rhythm by zeitgebers each day to avoid falling out of phase with local standard time. If this were not the case, an organism with an FRP of 22 h would become active approximately 2 h earlier each day, such that after six days, the activity rhythm would occur a full 12 h out of phase from the natural environmental cycle. Therefore, the circadian system must be cognizant and responsive to these environmental cues such that the circadian clock, and therefore activity, is entrained to local time.

A synchronizer is an agent or signal that promotes the synchrony of multiple clocks within a population. Typically in *S. elongatus* experiments, two LD cycles are used to synchronize all clocks within the culture population. Entrainment results in the internal biological rhythm having a period that matches that of the environmental stimulus, and entrainment to a particular cue (such as light or dark) ensures two things: (1) the period of the activity rhythm is equal to that of the LD cycle, (2) the phase of the activity rhythm is appropriately stable and occurring at the correct time of day. The difference (in hours) between the phase of the clock-driven rhythm and the rhythm of the entraining stimulus is the phase angle (Aschoff 1960; Moore-Ede et al. 1982; Dunlap et al. 2004; Koukkari and Sothern 2006).

When the circadian mechanism responds to a zeitgeber, the ultimate goal is to maintain the proper phase of activity or inactivity, which ensures that daily behaviors are occurring at the proper time within a cycle. This process, known as phase shifting, is the change in the timing of the phase of a rhythm in relation to the zeitgeber information of the previous phase. Again, to ensure that activity phases are occurring at the correct time of day, activity phases can be shifted to occur earlier (advance) or later (delay) in the day (Fig. 1.2; Aschoff 1960; Johnson 1999; Dunlap et al. 2004; Koukkari and Sothern 2006).

The most important zeitgebers for the circadian clock are the daily cycles in light and dark. While the clock mechanism must be sensitive to these cues, it does not mean that the circadian clock is sensitive, or responsive, to these cues at all times of day. As has been shown for many organisms that use light as an entraining signal, light exposure in the early subjective night produces phase delay shifts whereas light in the late subjective night produce phase advance shifts; light during the subjective day produces a very small, if any, phase shift. A phase response curve (PRC) for an organism in response to a particular zeitgeber can be generated by plotting the time at which the zeitgeber signal is provided to the organism on the *x*-axis, and the magnitude (in hours) and direction of the shift (advance or delay) to that signal are plotted on the *y*-axis (advance on the positive and delay on the negative *y*-axis; Aschoff 1960; Johnson 1999; Dunlap et al. 2004; Koukkari and Sothern 2006). In general, an organism's clock is less responsive to pulses of darkness during the subjective night and pulses of light during the subjective day.

1.3.3 Temperature Compensation

For a circadian clock to be accurate in the environment, it must maintain its periodicity despite changes in daily temperature. This property, named temperature compensation, is a result of the observation that the value of the FRP changes very little over different temperatures within the physiological range of the organism. One could imagine this as an important facet of the circadian mechanism, as it would be detrimental to an organism to have its circadian clock run faster or slower on a warm or cold day, respectively (Aschoff 1960; Sweeny and Hastings 1960; Dunlap et al. 2004; Koukkari and Sothern 2006). In a typical enzymatic reaction, temperature directly influences the rate at which that reaction proceeds. The Q_{10} temperature coefficient is the measure of this phenomenon, whereby the rate of a reaction tends to increase by a factor of two or three with every 10°C increase. The Q_{10} value for the period of a circadian rhythm is typically 0.8–1.2, indicating that the rhythmic activity is insulated from changes in temperature. Temperature insulation of the circadian mechanism should not be misconstrued as temperature insensitivity. Temperature has been demonstrated to be a strong zeitgeber for many circadian systems and can be used to entrain the circadian clock to adjust the phase of the activity rhythm (Aschoff 1960; Liu et al. 1998; Dunlap et al. 2004; Koukkari and Sothern 2006). While LD cycles are the strongest environmental cues for entraining the *S. elongatus* clock, temperature has been shown to be an effective zeitgeber as well (Lin et al. 1999; Schmitz et al. 2000; Ditty et al. 2003).

1.4 Introduction to the Cyanobacterial Circadian Clock Mechanism

Underlying the properties characteristic of the circadian rhythm is the circadian mechanism itself. It is comprised of internal molecules that drive rhythmic gene expression and therefore regulate cellular processes on a 24-h time scale (Aschoff 1960; Pittendrigh 1981; Dunlap et al. 2004; Koukkari and Sothern 2006). The individual molecular players of a circadian clock are very complex, but the overall game in which they play is easily modeled with discrete, but interacting, units. This simple organization divides the circadian mechanism into three basic elements: the oscillator, an input pathway and an output pathway (Fig. 1.3).

Central to the circadian mechanism is the oscillator (also known as the pacemaker) that is responsible for maintaining and disseminating the 24-h time information. As dictated by the characteristics of the circadian rhythms it controls, it is entrainable by environmental cues. The ability of the oscillator to communicate

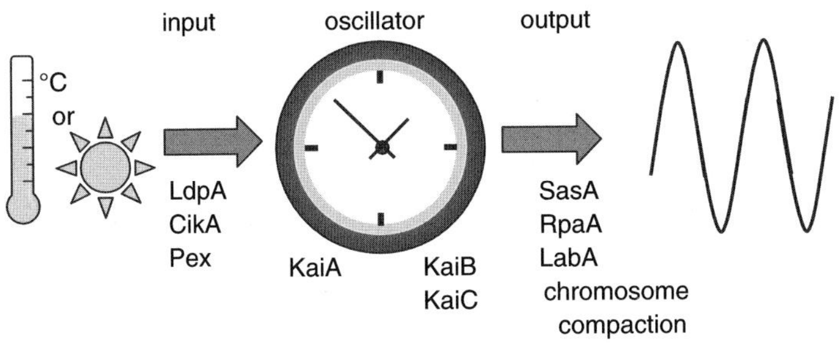

Fig. 1.3 Simple model for the *S. elongatus* PCC 7942 circadian clock, showing the three conceptual designations for the circadian clock (*input*, *oscillator*, *output*), along with known *S. elongatus* PCC 7942 genes involved in each unit. See text for brief descriptions of each gene and directives to chapters within this book that fully describe each of these processes

with environmental zeitgebers occurs via input pathways. Temporal information that is maintained by the oscillator must be transmitted to the processes that it controls, including gene expression or downstream behaviors, which are measured as the overt circadian rhythms. While this simple model separates these three components of the circadian mechanism, it is important to recognize that communication between the molecular underpinnings of the input pathways, oscillator, and output pathways is integral to the circadian process (Aschoff 1960; Pittendrigh 1981; Dunlap et al. 2004; Koukkari and Sothern 2006). The following sections are meant to introduce the vast amount of information that is currently understood about the input, oscillator, and output mechanisms in cyanobacteria. Please see the subsequent chapters of this book (indicated parenthetically) for more detailed descriptions.

1.4.1 Oscillator

The core oscillator of the *S. elongatus* PCC 7942 clock is based on the activity of three proteins named KaiA, KaiB, and KaiC (Ishiura et al. 1998). Originally isolated by random chemical mutagenesis (see Chaps. 4, 5), the *kaiABC* genes, when mutated, demonstrate period, phase, and amplitude defects, as well as arrhythmia (Kondo et al. 1994). When any, or all, of the *kai* genes are deleted, cells are viable but are rendered arrhythmic (Ishiura et al. 1998). Significant progress has been made in understanding the mechanisms by which the Kai oscillator keeps time. Insights into the biochemistry and structure of KaiC (see Chaps. 6, 7) have elucidated that hexameric structure and phosphorylation of this protein are central to the time-keeping process (Nishiwaki et al. 2000; Pattanayek et al. 2004). KaiA and KaiB seem to serve auxiliary functions for KaiC, as KaiA and KaiB have been shown to augment and inhibit KaiC phosphorylation events, respectively. It is the state of KaiC phosphorylation throughout the course of the day that is believed to be critical for circadian timing (Iwasaki et al. 2002; Williams et al. 2002; Kitayama et al. 2003; Xu et al. 2003).

In the circadian mechanism of many eukaryotic model systems, the function of the clock is regulated by transcriptional and translational feedback loops that increase and decrease key clock components at particular times of day (Cheng et al. 2001; Harmer et al. 2001; Van Gelder et al. 2003; Dunlap et al. 2004; Koukkari and Sothern 2006). While it was originally thought that these feedback loops were required to maintain robust rhythmic activity of the *kai* genes and their protein products in cyanobacteria (Ishiura et al. 1998), recent evidence suggests that this feedback may only be required for fine-tuning the system (Xu et al. 2003; Ditty et al. 2005). The most striking evidence of this phenomenon came from experimentation that demonstrated a temperature-compensated, circadian rhythm of KaiC phosphorylation in vitro using only KaiA, KaiB, KaiC, and ATP, which dismisses the absolute requirement for a regulatory feedback loop of gene expression in the cyanobacterial system (see Chap. 5; Nakajima et al. 2005).

1.4.2 Input

One of the key roles of the oscillator is to communicate with the external environment through input pathways. In *S. elongatus* PCC 7942, a few genes and their subsequent proteins have been shown to function in this capacity. In many eukaryotic model systems, bona fide photoreceptors have been identified that directly connect the perception of light to the central oscillator (Liu 2003; Millar 2003). However, the molecular mechanisms for cyanobacterial entrainment are not as clearly understood (see Chap. 8). A true photoreceptor has not yet been identified in *S. elongatus* that is dedicated to transducing external light information to the central oscillator. In fact, a canonical photoreceptor may not be required because intracellular redox state appears to trigger the input response. In *S. elongatus*, the LdpA and CikA proteins both function in this capacity as they have been shown to mediate the ability of the organism to obey Aschoff's rule and reset to zeitgeber cues, respectively (Schmitz et al. 2000; Katayama et al. 2003; Ivleva et al. 2005). The function of the input pathway is two-fold: to recognize environmental zeitgebers and to relay that information to the oscillator. The Pex protein, which has been implicated in maintaining period and phase-resetting, has also recently been shown to bind to the promoter region of *kaiA* to repress expression of this core oscillator component (Kutsuna et al. 1998; Takai et al. 2006a; Kutsuna 2007).

1.4.3 Output

The pathways through which the circadian oscillator relays temporal information to downstream activities ensure that specific cellular activities occur at the correct time of day, and a number of proteins have been identified that function in this capacity for *S. elongatus*. SasA has been shown to receive temporal information from KaiC; this protein–protein interaction then stimulates the autophosphorylation of the histidine protein kinase, SasA, and subsequent transfer of the phosphoryl group to the response regulator protein, RpaA. RpaA may act as a transcriptional regulator of gene expression (Smith and Williams 2006; Takai et al. 2006b). In addition, the LabA protein is proposed to act as a repressor of KaiC and RpaA activity based upon information from the oscillator (Taniguchi et al. 2007).

The mechanism by which two-component regulatory systems, including that of SasA/RpaA, affect gene expression has yet to be deciphered (see Chap. 9). What is known is that the *S. elongatus* circadian clock regulates gene expression globally; nearly every *S. elongatus* promoter tested via luciferase fusions oscillates in a circadian manner (Liu et al. 1995; Woelfle and Johnson 2006). One current hypothesis for mediating this global circadian gene expression is via higher-order chromosome structure (see Chap. 10). It has been demonstrated that compaction of the cyanobacterial chromosome is Kai-oscillator dependent but SasA independent. The exact flow of information from the Kai oscillator to chromosome compaction, and the manner by which chromosome compaction mediates circadian gene expression is still elusive. Chromosome compaction may alter the availability of promoter regions to transcriptional machinery, which would affect gene expression (Smith and Williams 2006; Woelfle et al. 2007).

Because measuring clock output by reporter gene expression is easy and reliable in *S. elongatus*, there are few actual clock-controlled physiological activities in this organism that have been investigated. The circadian clock has been shown to regulate the process of cell division (see Chap. 11), in that cell division is only allowed to occur at particular times of day (Sweeney and Borgese 1989; Mori et al. 1996; Mori and Johnson 2000). The physiological characteristic of temporal separation of nitrogen fixation and oxygenic photosynthesis, that has been shown to be under clock control in other cyanobacteria, is irrelevant to the *S. elongatus* PCC 7942 model system as this strain is not diazotrophic (Herrero et al. 2001). Without a need to separate the oxygen-producing process of photosynthesis from the oxygen-labile process of nitrogen fixation, the necessity for *S. elongatus* to maintain a circadian clock is not obvious. But, the *S. elongatus* clock does indeed impart a selective growth advantage to those cells whose internal period closely matches that of the environmental LD cycle (see Chap. 12). When strains with different endogenous periods were put in competition with one another, the strain with the endogenous period that most closely matched the external LD cycle prevailed (Ouyang et al 1998; Woelfle et al. 2004). The mechanism by which the circadian clock delivers this competitive edge to cells is not understood; however, there seems to be little, if any, communication between cells to maintain robust rhythmicity in *S. elongatus* (Mihalcescu et al. 2004). Measurement of bioluminescence expression from rhythmic single cells has shown that the circadian clock is indeed a property of individual cells. Therefore the stability of the clock seems to be ensured by an intracellular mechanism, as intercellular coupling of period information seems to be insignificant (see Chap. 13; Mihalcescu et al. 2004; Amdaoud et al. 2007).

1.4.4 The Periodosome

While the model for input, oscillator, and output is a good way to begin to understand the molecular basis for a circadian clock, it is a compartmentalized oversimplification. What many different lines of investigation have shown is that the clock itself functions due to a dynamic flux of input, oscillator, and output protein alterations and associations. The current model that has emerged from clock protein structure (see Chaps. 6, 7), size exclusion gel filtration chromatography, and immunoblot experiments shows circadian clock function based upon the assembly and disassembly of a large, heteromultimeric physical complex over a 24-h period. Included in this "periodosome" (Golden 2004) are the oscillator proteins KaiC, KaiA, and KaiB along with the input and output proteins CikA, LdpA, and SasA (Kageyama et al. 2002; Ivleva et al. 2005; Ivleva et al. 2006); it is the combination

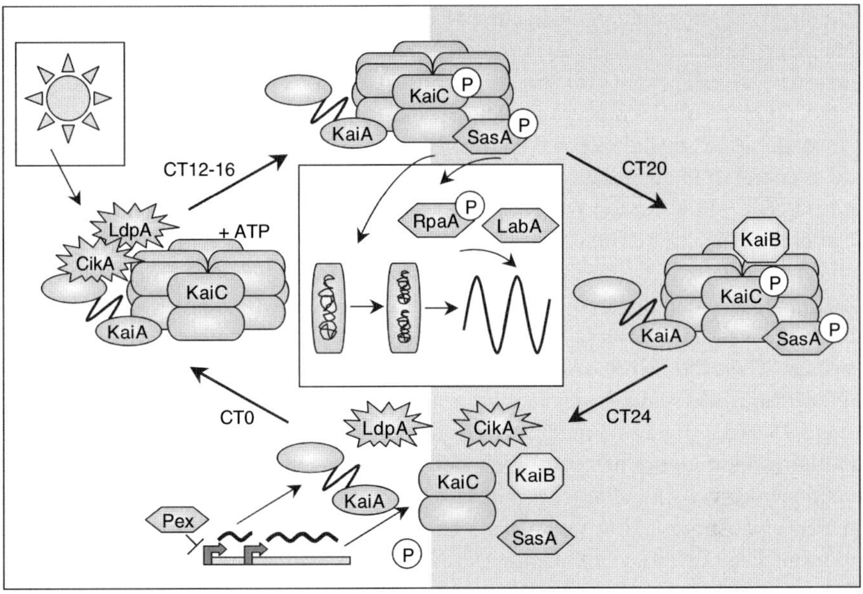

Fig. 1.4 Current model for the circadian mechanism and periodosome structure in *S. elongatus* PCC 7942. The diagram constitutes interactions and covalent (*encircled P* designates phosphorylation events) changes to known circadian proteins and the periodosome over a 24-h period. The *sun inset* designates unknown input mechanisms to proteins with known input function in this system. The *central inset* designates known and yet undiscovered output mechanisms. See the text for a brief description and directives to chapters within this book that fully describe the current model

and biochemical state of the different components at different times during the 24-h day that maintain the circadian cycle (Fig. 1.4).

1.5 Conclusions

What has become clear over the past two decades of working with the *S. elongatus* PCC 7942 model system is that there is still much to be learned. While the model depicted in Fig. 1.4 shows what is currently understood about this circadian system, it is not a complete picture. There are still gaps in our understanding of the input and output pathways that feed into and out of the Kai oscillator. In addition, it may be rash to assume that all of the intricate pieces of the Kai oscillator have already been identified.

So the question is, where does the field go from here? How do we further our understanding of the known clock components and how do we identify new components? A particularly powerful tool that can be used to better understand the *S. elongatus* system may result from looking at other cyanobacterial species and studying their circadian systems. As one of the oldest groups of microorganisms on the planet, and with the full genomic sequences of at least 40 cyanobacterial genomes, investigation into the evolutionary relationships between the structure and occurrence of clock genes and the evolutionary constraints to mutation in clock genes of various cyanobacterial species have been helpful in elucidating genes that are essential in the generation of a circadian system (see Chap. 14). While *S. elongatus* PCC 7942 was chosen as the model system because of its genetic malleability, it does have its limitations. Again, *S. elongatus* does not fix nitrogen; therefore *Synechococcus* RF-1 (the cyanobacterial strain in which circadian rhythms were first identified) continues to be studied (see Chap. 3). It would be remiss to characterize the vast and diverse Orders of cyanobacteria based upon two unicellular species. Therefore work from trichomatous and thermophilic genera have been pursued to understand the cyanobacterial circadian system more inclusively (see Chap. 15). The power of mathematical modeling can also be an important tool in divulging new components to a circadian system. Modeling the rates of clock component biochemical reactions have been helpful in elucidating the mechanism of KaiC phosphorylation (see Chap. 16) and understanding synthetic oscillators in *Escherichia coli* may help to identify previously unforeseen components or characteristics of the circadian mechanism as well (see Chap. 17).

We have a long way to go in our understanding of the circadian clock in cyanobacteria. Only time, and the continued efforts of those who love cyanobacteria and their circadian underpinnings, will tell.

References

Amdaoud M, Vallade M, Weiss-Schaber C, Mihalcescu I (2007) Cyanobacterial clock, a stable phase oscillator with negligible intercellular coupling. Proc Natl Acad Sci USA 104:7051–7056

Andersson CR, Tsinoremas NF, Shelton J, Lebedeva NV, Yarrow J, Min H, Golden SS (2000) Application of bioluminescence to the study of circadian rhythms in cyanobacteria. Methods Enzymol 305:527–542

Aschoff J (1960) Exogenous and endogenous components in circadian rhythms. Cold Spring Harb Lab Quant Biol 25:11–28

Aschoff J (ed) (1981) Handbook of behavioral neurobiology 4: biological rhythms. Plenum, New York

Bruce VG (1960) Environmental entrainment of circadian rhythms. Cold Spring Harb Lab Quant Biol 25:29–48

Bünning E (1973) The physiological clock, 3rd edn. Springer, Heidelberg

Castenholz, RW (ed) (2001) Phylum BX cyanobacteria: oxygenic, photosynthetic bacteria. In: Garrity GM (ed) Bergey's manual of systematic bacteriology: the Archaea and the deeply branching and phototrophic bacteria, vol 1. Springer, Heidelberg, pp 473–600

Chen T-H, Chen T-L, Hung L-M, Huang T-C (1991) Circadian rhythm in amino acid uptake by *Synechococcus* RF-1. Plant Physiol 97:55–59

Cheng P, Yang Y, Liu Y (2001) Interlocked feedback loops contribute to the robustness of the *Neurospora* circadian clock. Proc Natl Acad Sci USA 98:7408–7413

Daan S, Merrow M, Rönnenberg T (2002) External time–internal time. J Biol Rhythms 17:107–109

DeCoursey PJ (1961) Effect of light on the circadian activity rhythm of the flying squirrel, *Glaucomys volans*. Z Vgl Physiol 44:331–354

deMairan J-J (1729) Observation botanique. Histoire de l'Academie Royale des Sciences. Academie Royale des Sciences, Paris, pp 35–36

Ditty JD, Williams SB, Golden SS (2003) A cyanobacterial timing mechanism. Annu Rev Genet 37:513–543

Ditty JL, Canales SR, Anderson BE, Williams SB, Golden SS (2005) Stability of the *Synechococcus elongatus* PCC 7942 circadian clock under directed anti-phase expression of the *kai* genes. Microbiology 151:2605–2613

Dunlap JC, Loros JJ, DeCoursey PJ (eds) (2004) Chronobiology: biological timekeeping. Sinauer, Sunderland, Mass.

Edmunds LN Jr (1983) Chronobiology at the cellular and molecular levels: models and mechanisms for circadian timekeeping. Am J Anat 168:389–431

Feldman JF, Hoyle MN (1973) Isolation of circadian clock mutants of *Neurospora crassa*. Genetics 75:605–613

Freer S (2003) Linnaeus' philosophia botanica (translation). Oxford University Press, Oxford

Golden SS (1987) Genetic engineering of the cyanobacterial chromosome. Methods Enzymol 153:215–231

Golden SS (1988) Mutagenesis of cyanobacteria by classical and gene-transfer-based methods. Methods Enzymol 167:714–727

Golden SS (2004) Meshing the gears of the cyanobacterial circadian clock. Proc Natl Acad Sci USA 101:13697–13698

Golden SS, Canales SR (2003) Cyanobacterial circadian clocks – timing is everything. Nat Rev Microbiol 1:191–199

Grobbelaar N, Huang T-C (1992) Effect of oxygen and temperature on the induction of a circadian nitrogenase activity rhythm in *Synechococcus* RF-1. Plant Physiol 140:391–394

Grobbelaar N, Huang T-C, Lin HY, Chow TJ (1986) Dinitrogen fixing endogenous rhythm in *Synechococcus* RF-1. FEMS Microbiol Lett 37:173–177

Halberg F, Carandente F, Cornelissen G, Katinas GS (1977) Glossary of chronobiology. Chronobiologica 4:1–189

Harmer SL, Panda S, Kay SA (2001) Molecular bases of circadian rhythms. Annu Rev Cell Dev Biol 17:215–253

Herrero A, Muro-Pastor AM, Flores E (2001) Nitrogen control in cyanobacteria. J Bacteriol 183:411–425

Huang T-C, Tu J, Chow TJ, Chen T-H (1990) Circadian rhythm of the prokaryote *Synechococcus* sp. RF-1. Plant Physiol 92:531–533

Ishiura M, Kutsuna S, Aoki S, Iwasaki H, Andersson CR, Tanabe A, Golden SS, Johnson CH, Kondo T (1998) Expression of a gene cluster *kaiABC* as a circadian feedback process in cyanobacteria. Science 281:1519–1523

Ivleva NB, Bramlett MR, Lindahl PA, Golden SS (2005) LdpA: a component of the circadian clock senses redox state of the cell. EMBO J 24:1202–1210

Iwasaki H, Nishiwaki T, Kitayama Y, Nakajima M, Kondo T (2002) KaiA-stimulated KaiC phosphorylation in circadian timing loops in cyanobacteria. Proc Natl Acad Sci USA 99:15788–15793

Johnson CH (1999) Forty years of PRCs – what have we learned? Chronobiol Int 16:711–743

Johnson CH (2005) Testing the adaptive value of circadian systems. Methods Enzymol 393:818–837

Kageyama H, Kondo T, Iwasaki H (2002) Circadian formation of clock protein complexes by KaiA, KaiB, KaiC and SasA in cyanobacteria. J Biol Chem 278:2388–2395

Katayama M, Kondo T, Xiong J, Golden SS (2003) *ldpA* encodes an iron–sulfur protein involved in light-dependent modulation of the circadian period in the cyanobacterium *Synechococcus elongatus* PCC 7942. J Bacteriol 185:1415–1422

Kitayama Y, Iwasaki H, Nishiwaki T, Kondo T (2003) KaiB functions as an attenuator of KaiC phosphorylation in the cyanobacteria circadian clock system. EMBO J 22:1–8

Kippert F (1987) Endocytobiotic coordination, intracellular calcium signaling, and the origin of endogenous rhythms. Ann NY Acad Sci 503:476–495

Kondo T, Strayer CA, Kulkarni RD, Taylor W, Ishiura M, Golden SS, Johnson CH (1993) Circadian rhythms in prokaryotes: luciferase as a reporter of circadian gene expression in cyanobacteria. Proc Natl Acad Sci USA 90:5672–5676

Kondo T, Tsinoremas NF, Golden SS, Johnson CH, Kutsuna S, Ishiura M (1994) Circadian clock mutants of cyanobacteria. Science 266:1233–1236

Kondo T, Mori T, Lebedeva NV, Aoki S, Ishiura M, Golden SS (1997) Circadian rhythms in rapidly dividing cyanobacteria. Science 275:224–227

Konopka RJ, Benzer S (1971) Clock mutants of *Drosophila melanogaster*. Proc Natl Acad Sci USA 68:2112–2116

Koukkari WL, Sothern RB (2006) Introducing biological rhythms. Springer, Heidelberg

Kutsuna S, Kondo T, Aoki S, Ishiura M (1998) A period-extender gene, *pex*, that extends the period of the circadian clock in the cyanobacterium *Synechococcus* sp. strain PCC 7942. J Bacteriol 180:2167–2174

Kutsuna S, Kondo T, Ikegami H, Uzumaki T, Katayama M, Ishiura M (2007) The circadian clock-regulated gene *pex* regulates a negative *cis* element in the *kaiA* promoter region. J Bacteriol 189:7690–7696

Lin R-F, Chou H-M, Huang T-C (1999) Priority of light/dark entrainment over temperature in setting the circadian rhythms of the prokaryote *Synechococcus* RF-1. Planta 209:202–206

Liu Y (2003) Molecular mechanisms of entrainment in the *Neurospora* circadian clock. J Biol Rhythms 18:195–205

Liu Y, Tsinoremas NF, Johnson CH, Lebdeva NV, Golden SS, Ishiura M, Kondo T (1995) Circadian orchestration of gene expression in cyanobacteria. Genes Dev 9:1469–1478

Liu Y, Merrow M, Loros JJ, Dunlap JC (1998) How temperature changes reset a circadian oscillator. Science 281:825–829

Michael TP, Salome PA, Yu HJ, Spencer TR, Sharp EL, McPeek MA, Alonso JM, Ecker JR, McClung CR (2003) Enhanced fitness conferred by naturally occurring variation in the circadian clock. Science 302:1049–1053

Mihalcescu I, Hsing W, Leibler S (2004) Resilient circadian oscillator revealed in individual cyanobacteria. Nature 430:81–85

Millar AJ (2003) A suite of photoreceptors entrains the plant circadian clock. J Biol Rhythms 18:217–226

Mitsui A, Kumazawa S, Takahashi A, Ikemoto H, Cao S, Arai T (1986) Strategy by which nitrogen-fixing unicellular cyanobacteria grow photoautotrophically. Nature 323:720–722

Moore-Ede MC, Sulzman FM, Fuller CA (1982) The clocks that time us. Harvard University Press, Cambridge, Mass.

Mori T, Johnson CH (2000) Circadian control of cell division in unicellular organisms. Prog Cell Cycle Res 4:185–192

Mori T, Binder B, Johnson CH (1996) Circadian gating of cell division in cyanobacteria growing with average doubling times of less than 24 hours. Proc Natl Acad Sci USA 93:10183–10188

Nakajima M, Imai K, Ito H, Nishiwaki T, Murayama Y, Iwasaki H, Oyama T, Kondo T (2005) Reconstitution of circadian oscillation of cyanobacterial KaiC phosphorylation in vitro. Science 308:414–415

Nishiwaki T, Iwasaki H, Ishiura M, Kondo T (2000) Nucleotide binding and autophosphorylation of the clock protein KaiC as a circadian timing process of cyanobacteria. Proc Natl Acad Sci USA 97:495–499

Ouyang Y, Andersson CR, Kondo T, Golden SS, Johnson CH (1998) Resonating circadian clocks enhance fitness in cyanobacteria. Proc Natl Acad Sci USA 95:8660–8664

Pattanayek R, Wang J, Mori T, Xu Y, Johnson CH, Egli M (2004) Visualizing a circadian clock protein: crystal structure of KaiC and functional insights. Mol Cell 15:375–388

Pittendrigh CS (1960) Circadian rhythms and the circadian organization of living systems. Cold Spring Harb Lab Quant Biol 25:159–184

Pittendrigh CS (1961) On temporal organization in living systems. Harvey Lect 56:93–125
Pittendrigh CS (1981) Circadian systems: general perspective and entrainment. In: Aschoff J (ed) Handbook of behavioral neurobiology. Plenum, New York, pp 57–77
Schmitz O, Katayama M, Williams SB, Kondo T, Golden SS (2000) CikA, a bacteriophytochrome that resets the cyanobacterial circadian clock. Science 289:765–768
Schneegurt MA, Sherman DM, Nayar S, Sherman LA (1994) Oscillating behavior of carbohydrate granule formation and dinitrogen fixation in the cyanobacterium *Cyanothece* sp. strain ATCC51142. J Bacteriol 176:1586–1597
Smith RM, Williams SB (2006) Circadian rhythms in gene transcription imparted by chromosome compaction in the cyanobacterium *Synechococcus elongatus*. Proc Natl Acad Sci USA 103:8564–8569
Sweeney BM, Borgese MB (1989) A circadian rhythm in cell division in a prokaryote, the cyanobacterium *Synechococcus* WH7803. J Phycol 25:183–186
Sweeney BM, Hastings JW (1957) Characteristics of the diurnal rhythm of luminescence in *Gonyaulax polyedra*. J Cell Comp Physiol 49:115–128
Sweeney BM, Hastings JW (1960) Effects of temperature upon diurnal rhythms. Cold Spring Harb Quant Biol 25:87–104
Takai N, Ikeuchi S, Manabe K, Kutsuna S (2006a) Expression of the circadian clock-related gene *pex* in cyanobacteria increases in darkness and is required to delay the clock. J Biol Rhythms 21:235–244
Takai N, Nakajima M, Oyama T, Kito R, Sugita C, Sugita M, Kondo T, Iwasaki H (2006b) A KaiC-associating SasA-RpaA two-component regulatory system as a major circadian timing mediator in cyanobacteria. Proc Natl Acad Sci USA 103:12109–12114
Taniguchi Y, Katayama M, Ito R, Takai N, Kondo T, Oyama T (2007) *labA*: a novel gene required for negative feedback regulation of the cyanobacterial circadian clock protein KaiC. Genes Dev 21:60–70
Van Gelder RN, Herzog ED, Schwartz WJ, Taghert PH (2003) Circadian rhythms: in the loop at last. Science 300:1534–1535
Williams SB, Vakonakis I, Golden SS, LiWang AC (2002) Structure and function from the circadian clock protein KaiA of *Synechococcus elongatus*: a potential clock input mechanism. Proc Natl Acad Sci USA 99:15357–15362
Woelfle MA, Johnson CH (2006) No promoter left behind: global circadian gene expression in cyanobacteria. J Biol Rhythms 21:419–431
Woelfle MA, Ouyang Y, Phanvijhitsiri K, Johnson CH (2004) The adaptive value of circadian clocks: an experimental assessment in cyanobacteria. Curr Biol 14:1481–1486
Woelfle MA, Xu Y, Qin X, Johnson CH (2007) Circadian rhythms of superhelical status of DNA in cyanobacteria. Proc Natl Acad Sci USA 104:18819–18824
Xu Y, Mori T, Johnson CH (2003) Cyanobacterial circadian clockwork: roles of KaiA, KaiB and the *kaiBC* promoter in regulating KaiC. EMBO J 22:2117–2126

Chapter 2
Speculation and Hoopla: Is Diversity Expected in Cyanobacterial Circadian Timing Systems?

Stanly B. Williams

Abstract Cyanobacteria are an extremely diverse group of photoautotrophic prokaryotes. *Synechococcus elongatus* PCC 7942 has become the model cyanobacterium for biological research directed at understanding the circadian timing mechanism and its central role in prokaryote circadian biology. Working primarily with *S. elongatus*, genetic and biochemical experimentation over the past two decades has identified the key components (and their functions) of a fascinating circadian timing mechanism. Of course, many basic questions remain regarding cyanobacterial circadian biology. Among those questions: is there a model system that can accurately represent such a diverse group of organisms? As a first step toward addressing that question, this chapter introduces several aspects of cyanobacterial diversity and then discusses the similarities and differences among likely circadian clock protein components from 39 different species of cyanobacteria. Although sound conclusions remain elusive, the information within the chapter should at least serve as a reminder to interpret model system data within the biological context under which it was determined. Ecology and evolutionary history are always important components of understanding molecular biological data.

2.1 Introduction

During the most calamitous of times, Texans – even nonhuman ones – can be heard hollering, "Remember the Alamo!" This colloquial bellow somehow inspires them to persevere. Perhaps it is time for cyanobacteriologists to holler, "Remember Jacob and Monod!" or "Remember K 12!" This should serve as inspiration and reminder that François Jacob and Jacques Monod thought that they had answered the problematic questions of gene regulation by studying regulatory patterns from the *Escherichia coli* K 12 *lac* operon (Pardee et al. 1959). Oops! And, even with the

S.B. Williams
Life Science Building, Department of Biology, University of Utah, Salt Lake City, UT 84112, USA, e-mail: williams@biology.utah.edu

Frenchmen's faux pas in mind, contemporary researchers working with that and other enteric microorganisms still generalize their particular conclusions as being representative of "the bacteria." Say what? As you raise your glass and nod in recognition and remembrance, ask yourself if this is now happening in contemporary circadian biology research? Consider attempts to explain cyanobacterial circadian biology with three *Synechococcus elongatus* Kai proteins and a soupçon of ATP. *Caution* is the word of the day.

These current explanatory attempts are somewhat awkward, certainly premature and perhaps even totally unwarranted at every scale of relevant biological inquiry – from molecules to microbes. Nonetheless, are they also correct? The point of this chapter is to provide some of the information necessary to initiate consideration of that question. The biochemical details of a cyanobacterial circadian timing system are discussed in other chapters and these details will perhaps provide even more cautionary tales. Chapters 5, 9, and 8, respectively, give a description of function for those proteins listed in Table 2.1: KaiA, KaiB, KaiC, SasA, and CikA. As it is currently understood, these proteins make up the central circadian timing system in *S. elongatus* PCC 7942. Below, I briefly examine the awe-inspiring ecological, behavioral, and genetic diversity of the cyanobacteria. What might that tell us? Within this examination, we maintain a consideration of circadian timing and daily rhythm biology. Several broad and interesting questions surface. Can there truly be a model or representative cyanobacterium in the study of circadian rhythms? How could one type of circadian timing mechanism hope to be the functional representative of all the others found within such an amazingly diverse group of creatures? Is *S. elongatus* PCC 7942 the best candidate for model organism? We can consider the astonishing diversity of habitat, morphology, metabolism, and behavior among the cyanobacteria before attempting to answer any of these questions. As above, we must be cautious.

2.2 Some General Cyanobacteriology

The cyanobacteria are not now and never have been "blue-green algae."(For the record, they are also not "pond scum.") Cyanobacteria are a monophyletic clade of photoautotrophic prokaryotes (Tomitani et al. 2006; Shi and Falkowski 2008). They contribute significantly to global primary production and the diazotrophs among them are central players in our planetary nitrogen cycle (Zehr et al. 2001; Karl 2002; Martinez-Alonso et al. 2004; Karjalainen et al. 2007). Interestingly, the earliest cyanobacterium was likely a thermophilic, phototrophic organism incapable of nitrogen fixation (Shi and Falkowski 2008). Their accepted lineage is ancient; and fossilized cyanobacterium-like organisms have been found in old conglomeratic Apex chert [3500 million years ago (Mya); Schopf and Packer 1987; Schopf 1993; Brasier et al. 2002]. Contemporary cyanobacteria – prokaryotic, photoautotrophic, and in some cases diazotrophic – can use water as a reductant during light-driven respiration. Concomitant with this biological process of water oxidation is oxygen

Table 2.1 Some cyanobacteria and a subset of their circadian clock-related proteins. Information in this table is taken primarily from the NCBI database. Only 30 of the 39 listed organisms have had their genomes completely sequenced. Common habitat and metabolic characteristics are abbreviated as follows: *F* freshwater, *M* marine, *R* plant root-associated, *T* terrestrial; *chl* chlorophyll, *f* filaments, *H* heterocyst-forming, *N* nitrogen-fixing, *u* unicells. *Clock-related proteins* lists the putative number of amino acyl residues encoded by the relevant gene (percent sequence identity with the *S. elongatus* PCC 7942 protein is indicated parenthetically). *Chromosome size* does not include plasmids or other nonchromosomal DNA (percent G+C is given parenthetically)

Organism	Habitat; metabolic characteristics	Clock-related proteins					Chromosome size; Mbp (%C+G)
		KaiC	KaiB	KaiA	SasA	CikA	
Synechococcus elongatus PCC 7942	F; u	519	102	284	387	754	2.7 (56)
S. elongatus PCC 6301	F; u	519 (100)	102 (100)	284 (99)	399 (99)	754 (100)	2.7 (56)
Acaryochloris marina	M; N, chl *d*	522 (79)	104 (88); 272 (45)	299 (41)	390 (40)	735 (46)	6.5 (48)
Anabaena sp. strain PCC 7120	T, F; H, N f	519 (81)	108 (87); 254 (52)	102 (53)	401 (44)	676 (50)	6.4 (41)
A. variabilis ATCC 29413	F, T; H, N f	519 (81)	108 (87); 254 (52)	89 (50)	401 (44)	676 (50)	6.4 (41)
Crocosphaera watsonii WH 8501	M; N	519 (80); 504 (55)	104 (85); 94 (55); 253 (45)	309 (45)	381 (43)	758 (43)	6.2 (37)
Cyanothece sp. CCY 0110	M; N	495 (82); 504 (56)	94 (56); 98 (42); 253 (43)	282 (45)	382 (43)	764 (44)	5.9 (37)
Lyngbya sp. PCC 8106	M; N	522 (81); 575 (36); 485 (28)	104 (88); 110 (65); 103 (42); 268 (50)	278 (47)	394 (42)	641 (47)	7.0 (41)
Nostoc punctiforme PCC 73102	R; H, N f	520 (80)	104 (85); 285 (48)	193 (36)	395 (44)	683 (50)	9.0 (41)
Nodularia spumigena CCY 9414	M, brackish; N	521 (80)	104 (88); 261 (47)	101 (46)	395 (44)	–	5.3 (41)
Prochlorococcus marinus AS 9601	M; chl *b*, u	509 (75)	105 (82)	–	372 (35)	–	1.7 (31)
P. marinus CCMP 1375 (SS120)	M; chl *b*, u	501 (76)	117 (84)	–	381 (35)	–	1.8 (36)

(*continued*)

Table 2.1 (Continued)

Organism	Habitat; metabolic characteristics	Clock-related proteins					Chromosome size; Mbp (%C+G)
		KaiC	KaiB	KaiA	SasA	CikA	
P. marinus CCMP 1986 (MED4)	M; chl *b*, u	509 (75)	107 (85)	–	372 (36)	–	1.7 (31)
P. marinus MIT 9211	M; chl *b*, u	512 (77)	114 (84)	–	370 (37)	–	1.7 (38)
P. marinus MIT 9215	M; chl *b*, u	512 (73)	105 (82)	–	372 (36)	–	1.7 (31)
P. marinus MIT 9301	M; chl *b*, u	509 (75)	105 (82)	–	372 (36)	–	1.6 (31)
P. marinus MIT 9303	M; chl *b*, u	488 (79)	119 (87)	–	370 (35)	–	2.7 (50)
P. marinus MIT 9312	M; chl *b*, u	498 (75)	105 (83)	–	372 (36)	–	1.7 (31)
P. marinus MIT 9313	M; chl *b*, u	499 (78)	119 (86)	–	370 (37)	–	2.4 (51)
P. marinus MIT 9515	M; chl *b*, u	509 (73)	108 (85)	–	372 (36)	–	1.7 (31)
P. marinus NATL1A	M; chl *b*, u	500 (77)	107 (83)	–	381 (34)	–	1.9 (35)
P. marinus NATL2A	M; chl *b*, u	500 (78)	107 (83)	–	381 (34)	–	1.8 (35)
Synechococcus sp. BL107	M; u	501 (79)	120 (88)	292 (40)	410 (38)	–	2.3 (54)
Synechococcus sp. CC 9311	M; u	511 (78)	119 (89)	328 (39)	383 (37)	–	2.6 (52)
Synechococcus sp. CC 9605	M; u	512 (79)	121 (88)	294 (43)	411 (38)	–	2.5 (59)
Synechococcus sp. CC 9902	M; u	512 (78)	120 (88)	292 (40)	383 (38)	–	2.2 (54)
Synechococcus sp. JA-2-3B'	Hot spring; u	541 (79)	100 (82)	334 (36)	377 (45)	–	3.0 (59)
Synechococcus sp. JA-3-3-Ab	Hot spring; u	534 (80)	100 (82)	304 (37)	377 (45)	–	2.9 (60)
Synechococcus sp. RCC 307	M; u	517 (79)	110 (87)	293 (42)	370 (38)	–	2.2 (61)
Synechococcus sp. RS9916	M; u	512 (80)	119 (89)	294 (40)	397 (35)	–	2.7 (60)
Synechococcus sp. RS9917	M; u	519 (79)	119 (88)	302 (40)	383 (38)	–	2.6 (65)
Synechococcus sp. WH 5701	M, F; u	514 (79)	92 (87)	295 (40)	397 (40)	–	3.0 (65)
Synechococcus sp. WH 7803	M; u	512 (79)	119 (88)	295 (43)	380 (38)	–	2.4 (60)
Synechococcus sp. WH 7805	M; u	512 (80)	119 (88)	296 (44)	383 (38)	–	2.6 (58)
Synechococcus sp. WH 8102	M; u	512 (79)	104 (88)	296 (41)	383 (37)	–	2.4 (59)

Synechocystis sp. PCC 6803	F; u	519 (81); 505 (55); 568 (36)	105 (86); 108 (54); 102 (47)	299 (41)	383 (40)	750 (39)	3.6 (48)
Thermosynechococcus elongatus	Hot spring; u	518 (81)	103 (88); 267 (46)	283 (43)	380 (39)	729 (41)	2.6 (54)
Trichodesmium erythraeum	M; N, u	519 (83)	104 (86); 254 (47)	325 (41)	412 (39)	853 (46)	7.8 (34)
Gloeobacter violaceus PCC 7421	T, rock; u	–	–	–	–	–	4.7 (62)

production. In fact, this type of metabolism seems to have evolved within this lineage and the cyanobacteria are the only prokaryotes thought capable of so-called oxygenic photosynthesis. That is, they use water as a source of reductant for carbon dioxide fixation via the Calvin cycle. Over eons, this metabolic activity appears to have led to both the creation of our oxygen-enriched atmosphere and the increased oxygen tension in the upper strata of our oceans (Kasting and Siefert 2002; Anbar et al. 2007; Kaufman et al. 2007; Scott et al. 2008). These enrichment events subsequently allowed the evolution of more composite (complex?), multicellular organisms that rely upon oxygen for their obligatory aerobic respiration (Falkowski et al. 2005; Falkowski 2006; Raymond and Segre 2006).

The word ubiquitous may actually be appropriate when used to describe the cyanobacterial niche. Cyanobacteria are everywhere and generally serve as the primary producers in their particular ecosystems. Their metabolic activities (such as oxygenic photosynthesis, nitrogen fixation, and de novo vitamin and enzyme cofactor biosynthesis) which have allowed them to inhabit practically every environment have also made them common participants in symbiotic associations. The genera *Calothrix*, *Cylindrospermum*, *Fischerella*, and *Nostoc* each include species that form endophytic, epiphytic, and true symbiotic relationships with numerous plants, fungi, sponges, and protists (Janson et al. 1999; Costa et al. 2001; Thomas 2001; Gorelova and Korzhenevskaia 2002; Guevara et al. 2002; Rikkinen et al. 2002; Wong and Meeks 2002; Douglas and Raven 2003). Recent work has even shown that many sponge-related compounds with activity against cancerous human cells are actually the products of the bacterial consortia living within those sponges (Wang 2006).

Modern cyanobacteria and plant chloroplasts are considered homologous. It is widely accepted that modern plastids evolved from a free-living cyanobacterium after its sequestration by a primitive eukaryotic-like cell (Cavalier-Smith 2002; Martin et al. 2002). One unique or particular $1–2 \times 10^3$ Mya endosymbiotic event was evidently highly successful, because all extant plastids are considered monophyletic (Douglas and Raven 2003). Primary plastids, those directly descended from that first cyanobiont, still exist among the rhodophyte, chlorophyte, and glaucocystophyte algae (Douglas and Raven 2003). Evolutionary relationships among cyanobacteria and these plastids remain intriguing and are actively studied (Morden and Golden 1989; Suzuki and Bauer 1995; Tomitani et al. 1999; Cavalier-Smith 2000; Nobles et al. 2001; Cavalier-Smith 2002; Martin et al. 2002; Ting et al. 2002; Raven and Allen 2003). Note that many plant nuclear genes also have cyanobacterial origins. Approximately 18% of the genes encoded in the *Arabidopsis thaliana* genome appear to be derived from cyanobacteria (Deusch et al. 2008). Even a brief consideration of the small amount of information discussed above makes it reasonable for one to argue that the cyanobacteria have had an enormous impact on both the Earth's biogeochemistry and the subsequent biological evolution. Cyanobacteria are perhaps the most important group of prokaryotes on this planet. Oddly and inexplicably, we still know relatively little about the ecology and evolutionary history of these most fascinating and utilitarian organisms.

2.3 Briefly: Cyanobacterial Ecology and Behavior

Ecological studies of the cyanobacteria have tended to focus upon the temporal and spatial relationships of community population densities, their symbiotic relationships, and less often on the relative role these communities play in real-world carbon and nitrogen fixation (DeLong and Pace 2001; de la Torre et al. 2003; Taton et al. 2003; Walker and Pace 2007). Cyanobacteria are found in essentially every habitat where sunlight infiltrates and even in a few habitats where it does not. One extreme example is the cave-dwelling *Gloeocapsa* species that survives under light intensities as low as 1 lux (~0.02 mmol photon m^{-2} s^{-1}; Cox et al. 1981). Nothing is yet known regarding the circadian biology of this species. We could speculate that light is probably not a significant zeitgeber for this organism. That would then make its putative circadian timing system unusual and therefore interesting. Cyanobacteria have also been identified in the darkness of the human gut (Frank and Pace 2001, 2008). I have no clue.

There is some speculation that the earliest cyanobacteria were thermophilic and, of course, cyanobacteria can still be isolated from high temperature environs around the globe (Garcia-Pichel et al. 1998; Nandi and Sengupta 1998; Ohto et al. 1999; Abed et al. 2002; Nakamura et al. 2002). These thermophilic cyanobacteria have maximum growth temperatures ranging from 50°C to 74°C. *Thermosynechococcus elongatus* BP-1 was isolated from the Beppu hot spring in Japan and has an optimal growth temperature near 57°C (Yamaoka et al. 1978; Rippka et al. 1979; Stanier 1980). *T. elongatus* has all three *kai* genes, a *sasA*, and a *cikA* gene and they are all similar in size and sequence to those of the model circadian organism *S. elongatus* PCC 7942 (Table 2.1; see Chap. 15). Appropriately, the circadian timing system in *T. elongatus* is temperature-compensated, even up to 60°C (Onai et al. 2004). Mesophilic cyanobacteria, like *S. elongatus* PCC 7942, can be identified essentially everywhere and have been isolated from most dry land ecosystems, including karst and travertine regions. Also, they flourish in benthic, limnetic, lotic, and pelagic fresh- and saltwater habitats (Paerl 1996; Carmichael et al. 1997; Martinez et al. 1997; Olson et al. 1998; Ostensvik et al. 1998; Richter et al. 1998; Sano et al. 1998; Atkins et al. 2001; Cuvin-Aralar et al. 2002; Frank 2002). *S. elongatus* PCC 7942, formerly known as *Anacystis nidulans* R2 and *S. leopoliensis*, was originally isolated from a freshwater habitat (see http://www.pasteur.fr/recherche/banques/PCC/).

Many mesophiles, including *Oscillatoria agardhii*, *Aphanizomenon flos-aquae*, *Microcystis aeruginosa*, and many of those listed in Table 2.1 (such as *Trichodsmium erythraeum*) produce proteinacous gas vesicles that allow them to adjust their position in the water column. Presumably, they are adjusting their position for efficient light absorption and perhaps to avoid predation by zooplankton (Damerval et al. 1989, 1991; Walsby 1994; Beard et al. 2002; van Gremberghe et al. 2008). Nothing is known about circadian clock regulation of this buoyancy behavior. *Aphanothece halophytica*, *Dactylococcopsis salina*, *Microcoleus chthonoplastes*, and *Spirulina major*, among many other cyanobacterial species, are halotolerant if not true halophilic organisms (Gabbay-Azaria et al. 1988; Nubel et al. 2000). Cyanobacteria have also been isolated from extreme hypersaline environments (Brock 1976; Ehrlich and Dor 1985; Davis and Giordano 1996). Our laboratory has isolated several cyanobacterial species from the Great Salt Lake (Utah, USA) and we are currently determining whether the *kai* and any related clock genes are present in these isolates. Psychrophilic species, like *Nodularia harveyana*, *Phormidum frigidum*, and *Rivularia minutula*, are characteristically the predominant life forms in their low-temperature environs. These species thrive in seemingly inhospitable habitats that include the tundra, ice shelves, glacial moraines, and polar desert soils of both the Arctic and the Antarctic regions (Wharton et al. 1981; Sheath et al. 1996; Priscu et al. 1998). Cyanobacteria isolated from desert climates have an uncommon and astonishing ability to withstand multiple rounds of desiccation and subsequent rehydration (Billi and Potts 2002; Ballal and Apte 2005; Ohad et al. 2005; Rothrock and Garcia-Pichel 2005; Tamaru et al. 2005). Again, nothing is known about the role of circadian timing and biology, if any, in the survival and behavior of psychrophiles or these desiccation tolerant species. It could be especially interesting to compare the biochemistry of a psychrophile's circadian timing mechanism to that of the mechanism in the mesophilic *Synechococcus elongatus*.

Cyanobacteria have evolved mechanisms to ingeniously generate their own macroscopic, insular environments. This seems to be a general survival strategy and allows them to thrive under the harsh conditions found in complex ecosystems such as benthic, coastal tidal, hot spring, and hypersaline microbial mats, the environmentally essential (and extremely sensitive) cryptobiotic desert crusts, and even in fresh- and saltwater blooms (Grotzschel and de Beer 2002; Neilan et al. 2002; Urmeneta et al. 2003; Ohad et al. 2005; Rothrock and Garcia-Pichel 2005). Unfortunately, we know nothing about the role of circadian timing or biology in the production of these insular environments. The regulatory and metabolic signals controlling this behavior and their impact on circadian timing systems would be a fascinating area of study. The range of cyanobacterial growth rates and metabolic activities are worth mentioning. Typical aquatic *Synechococcus* species have doubling times of 6–8 h, whereas those of the *Prochlorococcus* genera tend to be around 20–30 h (Vaulot et al. 1995; Shalapyonok et al. 1998; Jacquet et al. 2001). Amazingly, it has been estimated that some cyanobacterial populations in the cold, oligotrophic, dry deserts of Antarctica may have doubling times of nearly 10,000 years (Friedmann et al. 1993; Nienow and Friedmann 1993). Carbon dating of samples from these polar regions supports one implication of this slow growth rate estimate by showing that living cyanobacterial cells may be over 1000 years old (Bonani et al. 1988). It would be interesting to compare the circadian timing systems from these incredibly slow-growing organisms to those of organisms like *S. elongatus* that having doubling times of less than a day. Clearly, many interesting survival strategies have evolved in the cyanobacteria and not all of them are particularly accommodating. *Cylindrospermopsis raciborskii*, *Hapalosiphon fontinalis*, *Hormothamnion enteromorphoides*, *Umezakia natans*, many of the species listed in Table 2.1, and most of the aforementioned genera make a wide array of cyanotoxins as secondary metabolites (Beasley et al. 1989; Yoshizawa et al. 1990; Harada et al.

1991; Harada et al. 1994; Kuiper-Goodman et al. 1999; Ito et al. 2002; Mwaura et al. 2004; Welker et al. 2005; Kellmann et al. 2006; Stewart et al. 2006). These toxic metabolites are species-specific and include alkaloids, macrolides, and short linear or cyclic peptides that can be cytotoxic, hepatotoxic, or even neurotoxic to many organisms, including mammals (Stewart et al. 2006). And, guess what? No information exists as to how toxin production might be regulated by the circadian timing systems in these organisms. Interestingly, one explanatory model that grew out of studies of the reproductive fitness supplied to strains of *S. elongatus* by the circadian clock suggests that the clock may improve fitness in the cyanobacteria by utilizing the correct timing of toxin production and toxin resistance (Ouyang et al. 1998; Mori and Johnson 2001; Gonze et al. 2002; Woelfle et al. 2004). This is often termed the "peeing on your neighbor" model.

2.4 Some Cyanobacterial Genetic Diversity

The genetic diversity among extant cyanobacteria, exemplified by comparing the mol% G+C content of their genomes, is especially noteworthy. For example, *Cyanobium* species strain PCC 6707 has nearly 70% G+C, *S. elongatus* PCC 7942 has 56% G+C, and the *Nostoc* species strain PCC 7524 genome has only 39% G+C (Tandeau de Marsac and Houmard 1987; Table 2.1). Also, genome size among the cyanobacteria ranges from 9.0 Mbp in *N. punctiforme* down to 1.6 Mbp in *Prochlorococcus marinus* MIT 9301 (Table 2.1). Another glimpse into the genetic diversity among the cyanobacteria can perhaps be gleaned from their diverse cell morphologies (Whitton and Potts 2000). The sizes and shapes of these organisms are diverse and beguiling. Many species, including those within the *Aphanacapsa*, *Chroococcus*, *Merismopedia*, *Prochlorococcus*, *Synechocystis*, and *Synechococcus* genera, grow as ovoid- or rod-shaped unicells that can range from 0.4 mm to 40 mm in diameter (Whitton and Potts 2000). Most of these unicellular species live as single cells. However, others remain in tightly grouped cell aggregates after cell division (Paerl 1996; Palinska et al. 1996). Some species appear to regulate this lifestyle choice based upon the prevailing environmental conditions (Palinska et al. 1996). The cell aggregates are often highly organized, perhaps reflecting an underlying social order (Paerl and Priscu 1998; Gorelova 2000; De Philippis et al. 2005; Fuks et al. 2005). Little is known about this social order or what factors might regulate it. *S. elongatus* forms cell aggregates containing elongated cells under starvation conditions. It is not known what role if any the circadian clock plays in this process. Other cyanobacterial species, including those from the genera *Anabaena*, *Lyngbya*, *Nostoc*, *Scytonema*, *Spirulina*, *Stigonema*, *Tolypothrix*, and *Trichodesmium*, are long, relatively thin, multicellular filaments commonly surrounded by a mucilaginous sheath. Generally, these filaments are several microns in diameter and can be several hundred microns long (Green et al. 1989; Shi et al. 1995). We isolated a filamentous, halophilic *Spirulina* species from the Great Salt Lake that was over 200 µm in length (Williams 2007). Filamentous species often regulate the formation of differentiated cells including hormagonia (motile fragment of a cyanobacterial filament), akinetes (resting cyanobacterial spores), and terminally differentiated cells called heterocysts, which develop under nitrogen-limited conditions and essentially function as anaerobic chambers for nitrogen fixation (Golden and Yoon 1998; Garrity 2001; Yoon and Golden 2001). Many filamentous species have the genes that encode a circadian timing system (Table 2.1; see Chap. 15). However, little work has been published that directly examines the role of the circadian clock in any of those regulated differentiation pathways (Campbell et al. 2007).

Given their exceptional diversity of form, faculty, and function – a diversity that has only been hinted at in this abbreviated presentation – what biological properties might define a cyanobacterium? A recent comparative genome analysis using sequences from 13 different cyanobacteria, all of which are represented in Table 2.1, revealed a highly conserved set of 323 genes (Shi and Falkowski 2008). The encoded gene products of this core set are primarily involved in photosynthesis and mRNA translation. This set makes up only about 4% of the total number of genes in *Nostoc punctiforme* but is nearly 21% of the total *Prochlorococcus marinus* MIT 9301 gene assemblage. More ecologically informative analyses have come from comparisons of the genome sequences of closely related marine *Synechococcus* and *Prochlorococcus* species. Physiological adaptations to particular oceanic niches were evident in the sequence divergence between specific strains (Rocap et al. 2002; Palenik et al. 2003; Rocap et al. 2003). As a result of their shared core gene set, cyanobacteria also contain an intracytoplasmic, thylakoid membrane used to house their photosynthesis machinery (Kaftan et al. 2002). [If an exception makes the rule, then consider *Gloeobacter violaceous* PCC 7421 (Table 2.1), whose photosystems are located within its cytoplasmic membrane (Rippka et al. 1974; Mangels et al. 2002; Inoue et al. 2004; Mimuro et al. 2005).] As alluded to above, they also utilize both photosystem II and photosystem I and, as stated above, use water as the primary reductant during oxygenic photosynthesis (Fromme et al. 2001; Zak et al. 2001; Zouni et al. 2001). Also, cyanobacteria absorb light energy for photosynthesis by synthesizing and employing the chlorophyll *a* molecule, phycobiliproteins, and accessory phycobilin pigments like phycoerythrin, allophycocyanin, and phycocyanin (Brown et al. 1989; Meyer 1994). High concentrations of these latter two pigments often make the organisms appear greenish-blue, leading to their previous and incorrect designation as "blue-green algae" and the current moniker of cyanobacteria. Of course, not all are blue-green in color and the broadly distributed prochlorophyte species (*Prochlorococci*) use both chlorophylls *a* and *b* as antenna pigments and do not make elaborate phycobilin, light-harvesting antennae (Palenik and Haselkorn 1992; Kehoe and Grossman 1994; Table 2.1). In addition to all these traits, cyanobacteria appear to have circadian timing systems (Lorne et al. 2000; Dvornyk et al. 2002, 2003; Ditty et al. 2003; Dvornyk and Nevo 2004). Support for this speculation comes from the 39 species listed in Table 2.1 and from data showing 40 different cyanobacterial species with a *kaiC* gene (see Chap. 14; Lorne et al. 2000).

2.5 Diversity in Cyanobacterial Circadian Timing Mechanisms

So, having briefly examined the awe-inspiring ecological, behavioral, and genetic diversity of the cyanobacteria, what did we learn? Unfortunately, we were not able to draw any firm conclusions from this examination. These amazing creatures are found everywhere but we simply do not have sufficient data regarding their specific biological interactions or their metabolic needs and exchanges within those habitats to correlate with known circadian timing functions. Thus, we cannot describe realistic relationships between those parameters. However, all the information presented above is worth keeping in mind while examining the data in Table 2.1. Our question remains: is it possible that a single type of circadian timing system functions within all of these organisms?

The Kai proteins are the central components of the circadian timing mechanism in *S. elongatus*. The CikA protein is a primary timing-input device (see Chaps. 8, 10) and the SasA protein is a primary timing-output device (see Chap. 9). The KaiC protein sequences listed in Table 2.1 are remarkably similar to one another. With the exception of *G. violaceus*, which evidently does not have a circadian timing system, all listed organisms have a putative KaiC protein that is approximately 500 amino acyl residues in length and has at least 73% sequence identity with the *S. elongatus* KaiC protein. Each of these proteins also has the corresponding residues that are phosphorylated in the *S. elongatus* KaiC protein. Phosphorylation is a key biochemical function for KaiC (see Chaps. 5, 6, 7). Likewise and with the same exception, all of those listed organisms have a putative KaiB protein that is approximately 100 amino acyl residues in length and has at least 56% sequence identity with the *S. elongatus* KaiB protein. Even though we do not have genetic complementation or any other functional data, it is probably safe to assume that all these KaiB and KaiC proteins have similar function. The presence of a longer and perhaps redox-sensitive KaiB protein in the nitrogen fixing strains and in the thermophilic *T. elongatus* strain is trying to tell us something. I discuss these proteins in more detail elsewhere (Williams 2007). Several species listed in Table 2.1 have additional (more than one) *kaiB* and *kaiC* genes. There are no published data from which to discuss a role for their gene products in circadian timing. *S. elongatus* has only one set of *kai* genes.

The SasA protein is an important output component of the *S. elongatus* circadian timing mechanism. It evidently interacts directly with the KaiC protein and transduces timing information from the circadian clock to downstream gene regulation pathways (see Chaps. 9, 10). Again, with the exception of *G. violaceus*, all organisms listed in Table 2.1 have a putative SasA protein that is approximately 400 amino acyl residues in length and has at least 35% sequence identity with the *S. elongatus* SasA protein. This level of identity is undoubtedly functionally significant as sensory kinases like SasA are known to have highly variable carboxy-terminal regions (Stock et al. 1995; Williams and Stewart 1999). It is their amino-terminal domains that determine input source. As mentioned above, input is evidently from interactions with their respective KaiC proteins.

Diversity in the basic circadian timing system is evidently on the input side of the process. Briefly, the *S. elongatus* CikA protein is thought to be a light-responsive redox sensor that relays information into the central timing mechanism through the KaiA protein (see Chap. 8). Structure and function information regarding KaiA can be found in Chap. 6. None of the *Prochlorococcus* species in Table 2.1 has a CikA or KaiA protein; and the listed marine *Synechococcus* species have typical KaiA proteins but seem not to have CikA. If we assume that these species all have functional circadian timing systems, based upon the likely presence of KaiB, KaiC, and SasA, then clearly the mechanism by which information flows into the timing system is different in these marine organisms than it is in the terrestrial and even freshwater organisms. Curiously, the *Crocosphaera, Cyanothece, Lyngbya,* and *Nodularia* species all have typical KaiA and CikA proteins (Table 2.1). Again, there are no functional data regarding proteins other than those from *S. elongatus* but the similarity of the CikA and KaiA protein sequence listed in Table 2.1 is highly suggestive of functional similarity. These organisms are found in marine environments also. There are no sufficiently detailed data from the ecology of these different organisms to explain the similarities or differences in their timing-input mechanisms. Perhaps small, open-ocean organisms like the *Prochlorococci* rely on environmental signals other than light-responsive redox potentials for timing-input information. Another input anomaly that appears in Table 2.1 is the length of the KaiA proteins in the three *Anabaena* and *Nostoc* species. These species have KaiA proteins that are equivalent to only the carboxyl-terminal domain of the *S. elongatus* KaiA (Williams et al. 2002). Interestingly, these organisms appear to have a typical CikA protein. No gene encoding protein resembling the amino-terminal domain of the *S. elongatus* KaiA is obvious in the genome sequence of these three nitrogen-fixing, heterocyst-forming species. Again, we know too little about the biology of these organisms to speculate about their unusual KaiA proteins.

Can there truly be a model or representative cyanobacterium in the study of circadian rhythms? That depends. If one is interested in a mechanistic, biochemical description of how the Kai proteins interact to keep time, then one system is probably as good as another. However, if one is more curious about and fascinated by the complexity of circadian biology as it pertains to a particular cyanobacterium, then each type of circadian timing system – there are at least five represented in Table 2.1 – must be understood in the context of that cyanobacterium's interactions with its particular environment. And, because much of the variability seems to be on the input side of the system, the organism's ability to sense the environment determines the molecular details of circadian timing input. Thus, understanding the ecology and evolutionary history of a particular organism, as hinted at above, is paramount to understanding that organism's circadian timing mechanism. Is *S. elongatus* PCC 7942 the best candidate for model organism? Yes, serendipitously it appears to have been a good choice. It has the simplest circadian timing system that includes all the primary parts, CikA, KaiA, KaiB, KaiC, and SasA.

Acknowledgements Research accomplished in the author's laboratory has been supported by the National Institutes of Health, the National Science Foundation, and the University of Utah. Assistance from the book editors during the preparation of this chapter is rightfully noted.

References

Abed RM, Garcia-Pichel F, Hernandez-Marine M (2002) Polyphasic characterization of benthic, moderately halophilic, moderately thermophilic cyanobacteria with very thin trichomes and the proposal of *Halomicronema excentricum* gen. nov., sp. nov. Arch Microbiol 177:361–370

Anbar AD, Duan Y, Lyons TW, Arnold GL, Kendall B, Creaser RA, Kaufman AJ, Gordon GW, Scott C, Garvin J, Buick R (2007) A whiff of oxygen before the great oxidation event? Science 317:1903–1906

Atkins R, Rose T, Brown RS, Robb M (2001) The *Microcystis* cyanobacteria bloom in the Swan River – February 2000. Water Sci Technol 43:107–114

Ballal A, Apte SK (2005) Differential expression of the two kdp operons in the nitrogen-fixing cyanobacterium *Anabaena* sp. strain L-31. Appl Environ Microbiol 71:5297–5303

Beard SJ, Hayes PK, Pfeifer F, Walsby AE (2002) The sequence of the major gas vesicle protein, GvpA, influences the width and strength of halobacterial gas vesicles. FEMS Microbiol Lett 213:149–157

Beasley VR, Dahlem AM, Cook WO, Valentine WM, Lovell RA, Hooser SB, Harada K, Suzuki M, Carmichael WW (1989) Diagnostic and clinically important aspects of cyanobacterial (blue-green algae) toxicoses. J Vet Diagn Invest 1:359–365

Billi D, Potts M (2002) Life and death of dried prokaryotes. Res Microbiol 153:7–12

Bonani G, Friedmann EI, Ocampo-Friedmann R, McKay CP, Woelfli W (1988) Preliminary report on radiocarbon dating of cryptoendolithic microorganisms. Polarforschung 58:199–200

Brasier MD, Green OR, Jephcoat AP, Kleppe AK, Van Kranendonk MJ, Lindsay JF, Steele A, Grassineau NV (2002) Questioning the evidence for Earth's oldest fossils. Nature 416:76–81

Brock TD (1976) Halophilic blue-green algae. Arch Microbiol 107:109–111

Brown SB, Holroyd JA, Vernon DI, Shim YK, Smith KM (1989) The biosynthesis of the chromophore of phycocyanin. Pathway of reduction of biliverdin to phycocyanobilin. Biochem J 261:259–263

Campbell EL, Summers ML, Christman H, Martin ME, Meeks JC (2007) Global gene expression patterns of *Nostoc punctiforme* in steady-state dinitrogen-grown heterocyst-containing cultures and at single time points during the differentiation of akinetes and hormogonia. J Bacteriol 189:5247–5256

Carmichael WW, Evans WR, Yin QQ, Bell P, Moczydlowski E (1997) Evidence for paralytic shellfish poisons in the freshwater cyanobacterium *Lyngbya wollei* (Farlow ex Gomont) comb. nov. Appl Environ Microbiol 63:3104–3110

Cavalier-Smith T (2000) Membrane heredity and early chloroplast evolution. Trends Plant Sci 5:174–182

Cavalier-Smith T (2002) Chloroplast evolution: secondary symbiogenesis and multiple losses. Curr Biol 12:R62–R64

Costa JL, Paulsrud P, Rikkinen J, Lindblad P (2001) Genetic diversity of *Nostoc* symbionts endophytically associated with two bryophyte species. Appl Environ Microbiol 67:4393–4396

Cox G, Benson D, Dwarte DM (1981) Ultrastructure of a cave wall cyanophyte, *Gloeocapsa* NS 4. Arch Microbiol 130:165–174

Cuvin-Aralar ML, Fastner J, Focken U, Becker K, Aralar EV (2002) Microcystins in natural blooms and laboratory cultured *Microcystis aeruginosa* from Laguna de Bay, Philippines. Syst Appl Microbiol 25:179–182

Damerval T, Castets AM, Guglielmi G, Houmard J, Tandeau de Marsac N (1989) Occurrence and distribution of gas vesicle genes among cyanobacteria. J Bacteriol 171:1445–1452
Damerval T, Castets AM, Houmard J, Tandeau de Marsac N (1991) Gas vesicle synthesis in the cyanobacterium *Pseudanabaena* sp.: occurrence of a single photoregulated gene. Mol Microbiol 5:657–664
Davis JS, Giordano M (1996) Biological and physical events involved in the origin, effects, and control of organic matter in solar saltworks. Int J Salt Lake Res 4:335–347
de la Torre JR, Goebel BM, Friedmann EI, Pace NR (2003) Microbial diversity of cryptoendolithic communities from the McMurdo Dry Valleys, Antarctica. Appl Environ Microbiol 69:3858–3867
De Philippis R, Faraloni C, Sili C, Vincenzini M (2005) Populations of exopolysaccharide-producing cyanobacteria and diatoms in the mucilaginous benthic aggregates of the Tyrrhenian Sea (Tuscan Archipelago). Sci Total Environ 353:360–368
DeLong EF, Pace NR (2001) Environmental diversity of bacteria and archaea. Syst Biol 50:470–478
Deusch O, Landan G, Roettger M, Gruenheit N, Kowallik KV, Allen JF, Martin W, Dagan T (2008) Genes of cyanobacterial origin in plant nuclear genomes point to a heterocyst-forming plastid ancestor. Mol Biol Evol 25:748–761
Ditty JL, Williams SB, Golden SS (2003) A cyanobacterial circadian timing mechanism. Annu Rev Genet 37:513–543
Douglas AE, Raven JA (2003) Genomes at the interface between bacteria and organelles. Phil Trans R Soc Lond B Biol Sci 358:5–18
Dvornyk V, Nevo E (2004) Evidence for multiple lateral transfers of the circadian clock cluster in filamentous heterocystic cyanobacteria Nostocaceae. J Mol Evol 58:341–347
Dvornyk V, Vinogradova O, Nevo E (2002) Long-term microclimatic stress causes rapid adaptive radiation of *kaiABC* clock gene family in a cyanobacterium, *Nostoc linckia*, from “Evolution Canyons” I and II, Israel. Proc Natl Acad Sci USA 99:2082–2087
Dvornyk V, Vinogradova O, Nevo E (2003) Origin and evolution of circadian clock genes in prokaryotes. Proc Natl Acad Sci USA 100:2495–2500
Ehrlich A, Dor I (1985) Photosynthetic microorganisms of the Gavesh sabkha. In: Friedmann GM, Krumbein WE (ed) Hypersaline ecosystems. Springer, Heidelberg, pp 296–321
Falkowski PG (2006) Evolution. Tracing oxygen’s imprint on earth’s metabolic evolution. Science 311:1724–1725
Falkowski PG, Katz ME, Milligan AJ, Fennel K, Cramer BS, Aubry MP, Berner RA, Novacek MJ, Zapol WM (2005) The rise of oxygen over the past 205 million years and the evolution of large placental mammals. Science 309:2202–2204
Frank CA (2002) Microcystin-producing cyanobacteria in recreational waters in southwestern Germany. Environ Toxicol 17:361–366
Frank DN, Pace NR (2001) Molecular-phylogenetic analyses of human gastrointestinal microbiota. Curr Opin Gastroenterol 17:52–57
Frank DN, Pace NR (2008) Gastrointestinal microbiology enters the metagenomics era. Curr Opin Gastroenterol 24:4–10
Friedmann EI, Kappen L, Meyer MA, Nienow JA (1993) Long-term productivity in the cryptoendolithic microbial community of the Ross Desert, Antarctica. Microb Ecol 25:51–69
Fromme P, Jordan P, Krauss N (2001) Structure of photosystem I. Biochim Biophys Acta 1507:5–31
Fuks D, Radic J, Radic T, Najdek M, Blazina M, Degobbis D, Smodlaka N (2005) Relationships between heterotrophic bacteria and cyanobacteria in the northern Adriatic in relation to the mucilage phenomenon. Sci Total Environ 353:178–188
Gabbay-Azaria R, Tel-Or E, Schonfeld M (1988) Glycinebetaine as an osmoregulant and compatible solute in the marine cyanobacterium *Spirulina subsalsa*. Arch Biochem Biophys 264:333–339
Garcia-Pichel F, Nubel U, Muyzer G (1998) The phylogeny of unicellular, extremely halotolerant cyanobacteria. Arch Microbiol 169:469–482

Garrity GM (ed) (2001) Bergey's manual of systematic bacteriaology. Springer, Heidelberg
Golden JW, Yoon HS (1998) Heterocyst formation in *Anabaena*. Curr Opin Microbiol 1:623–629
Gonze D, Roussel MR, Goldbeter A (2002) A model for the enhancement of fitness in cyanobacteria based on resonance of a circadian oscillator with the external light–dark cycle. J Theor Biol 214:577–597
Gorelova OA (2000) Spatial integration of partners and heteromorphism of cyanobacterium *Nostoc muscoum* CALU 304 in mixed culture with *Rauwolfia*. Mikrobiologiia 69:565–573
Gorelova OA, Korzhenevskaia TG (2002) Formation of giant and ultramicroscopic forms of *Nostoc muscorum* CALU 304 during cocultivation with *Rauwolfia* tissues. Mikrobiologiia 71:654–661
Green JW, Knoll AH, Swett K (1989) Microfossils from silicified stromatolitic carbonates of the Upper Proterozoic Limestone-Dolomite 'Series', central East Greenland. Geol Mag 126: 567–585
Grotzschel S, de Beer D (2002) Effect of oxygen concentration on photosynthesis and respiration in two hypersaline microbial mats. Microb Ecol 44:208–216
Guevara R, Armesto JJ, Caru M (2002) Genetic diversity of *Nostoc* microsymbionts from *Gunnera tinctoria* revealed by PCR-STRR fingerprinting. Microb Ecol 44:127–136
Harada K, Ogawa K, Matsuura K, Nagai H, Murata H, Suzuki M, Itezono Y, Nakayama N, Shirai M, Nakano M (1991) Isolation of two toxic heptapeptide microcystins from an axenic strain of *Microcystis aeruginosa*, K-139. Toxicon 29:479–489
Harada KI, Ohtani I, Iwamoto K, Suzuki M, Watanabe MF, Watanabe M, Terao K (1994) Isolation of cylindrospermopsin from a cyanobacterium *Umezakia natans* and its screening method. Toxicon 32:73–84
Inoue H, Tsuchiya T, Satoh S, Miyashita H, Kaneko T, Tabata S, Tanaka A, Mimuro M (2004) Unique constitution of photosystem I with a novel subunit in the cyanobacterium *Gloeobacter violaceus* PCC 7421. FEBS Lett 578:275–279
Ito E, Takai A, Kondo F, Masui H, Imanishi S, Harada K (2002) Comparison of protein phosphatase inhibitory activity and apparent toxicity of microcystins and related compounds. Toxicon 40:1017–1025
Jacquet S, Partensky F, Marie D, Casotti R, Vaulot D (2001) Cell cycle regulation by light in *Prochlorococcus* strains. Appl Environ Microbiol 67:782–790
Janson S, Wouters J, Bergman B, Carpenter EJ (1999) Host specificity in the *Richelia*–diatom symbiosis revealed by *hetR* gene sequence analysis. Environ Microbiol 1:431–438
Kaftan D, Brumfeld V, Nevo R, Scherz A, Reich Z (2002) From chloroplasts to photosystems: *in situ* scanning force microscopy on intact thylakoid membranes. EMBO J 21:6146–6153
Karjalainen M, Engstrom-Ost J, Korpinen S, Peltonen H, Paakkonen JP, Ronkkonen S, Suikkanen S, Viitasalo M (2007) Ecosystem consequences of cyanobacteria in the northern Baltic Sea. Ambio 36:195–202
Karl DM (2002) Nutrient dynamics in the deep blue sea. Trends Microbiol 10:410–418
Kasting JF, Siefert JL (2002) Life and the evolution of Earth's atmosphere. Science 296:1066–1068
Kaufman AJ, Johnston DT, Farquhar J, Masterson AL, Lyons TW, Bates S, Anbar AD, Arnold GL, Garvin J, Buick R (2007) Late Archean biospheric oxygenation and atmospheric evolution. Science 317:1900–1903
Kehoe DM, Grossman AR (1994) Complementary chromatic adaptation: photoperception to gene regulation. Semin Cell Biol 5:303–313
Kellmann R, Mills T, Neilan BA (2006) Functional modeling and phylogenetic distribution of putative cylindrospermopsin biosynthesis enzymes. J Mol Evol 62:267–280
Kuiper-Goodman T, Falconer I, Fitzgerald J (1999) Human health aspects. In: Choros I, Bartram J (ed) Toxic cyanobacteria in water: a guide to their public health consequences, monitoring and management. Spon, London, pp 113–153
Lorne J, Scheffer J, Lee A, Painter M, Miao VP (2000) Genes controlling circadian rhythm are widely distributed in cyanobacteria. FEMS Microbiol Lett 189:129–133

Mangels D, Kruip J, Berry S, Rogner M, Boekema EJ, Koenig F (2002) Photosystem I from the unusual cyanobacterium *Gloeobacter violaceus*. Photosynth Res 72:307–319

Martin W, Rujan T, Richly E, Hansen A, Cornelsen S, Lins T, Leister D, Stoebe B, Hasegawa M, Penny D (2002) Evolutionary analysis of *Arabidopsis*, cyanobacterial, and chloroplast genomes reveals plastid phylogeny and thousands of cyanobacterial genes in the nucleus. Proc Natl Acad Sci USA 99:12246–12251

Martinez A, Pibernat I, Figueras J, Garcia-Gil J (1997) Structure and composition of freshwater microbial mats from a sulfur spring ("Font Pudosa", NE Spain). Microbiologia 13:45–56

Martinez-Alonso M, Mir J, Caumette P, Gaju N, Guerrero R, Esteve I (2004) Distribution of phototrophic populations and primary production in a microbial mat from the Ebro Delta, Spain. Int Microbiol 7:19–25

Meyer TE (1994) Evolution of photosynthetic reaction centers and light harvesting chlorophyll proteins. Biosystems 33:167–175

Mimuro M, Tsuchiya T, Inoue H, Sakuragi Y, Itoh Y, Gotoh T, Miyashita H, Bryant DA, Kobayashi M (2005) The secondary electron acceptor of photosystem I in *Gloeobacter violaceus* PCC 7421 is menaquinone-4 that is synthesized by a unique but unknown pathway. FEBS Lett 579:3493–3496

Morden CW, Golden SS (1989) *psbA* genes indicate common ancestry of prochlorophytes and chloroplasts. Nature 337:382–385

Mori T, Johnson CH (2001) Circadian programming in cyanobacteria. Semin Cell Dev Biol 12:271–278

Mwaura F, Koyo AO, Zech B (2004) Cyanobacterial blooms and the presence of cyanotoxins in small high altitude tropical headwater reservoirs in Kenya. J Water Health 2:49–57

Nakamura Y, Kaneko T, Sato S, Ikeuchi M, Katoh H, Sasamoto S, Watanabe A, Iriguchi M, Kawashima K, Kimura T, Kishida Y, Kiyokawa C, Kohara M, Matsumoto M, Matsuno A, Nakazaki N, Shimpo S, Sugimoto M, Takeuchi C, Yamada M, Tabata S (2002) Complete genome structure of the thermophilic cyanobacterium *Thermosynechococcus elongatus* BP-1. DNA Res 9:123–130

Nandi R, Sengupta S (1998) Microbial production of hydrogen: an overview. Crit Rev Microbiol 24:61–84

Neilan BA, Burns BP, Relman DA, Lowe DR (2002) Molecular identification of cyanobacteria associated with stromatolites from distinct geographical locations. Astrobiology 2:271–280

Nienow JA, Friedmann EI (1993) Terrestrial lithophytic (rock) communities. In: Friedmann EI (ed) Antarctic microbiology. Wiley–Liss, New York, pp 353–412

Nobles DR, Romanovicz DK, Brown RM Jr (2001) Cellulose in cyanobacteria. Origin of vascular plant cellulose synthase? Plant Physiol 127:529–542

Nubel U, Garcia-Pichel F, Muyzer G (2000) The halotolerance and phylogeny of cyanobacteria with tightly coiled trichomes (*Spirulina turpin*) and the description of *Halospirulina tapeticola* gen. nov., sp. nov. Int J Syst Evol Microbiol 50:1265–1277

Ohad I, Nevo R, Brumfeld V, Reich Z, Tsur T, Yair M, Kaplan A (2005) Inactivation of photosynthetic electron flow during desiccation of desert biological sand crusts and *Microcoleus* sp.-enriched isolates. Photochem Photobiol Sci 4:977–982

Ohto C, Ishida C, Nakane H, Muramatsu M, Nishino T, Obata S (1999) A thermophilic cyanobacterium *Synechococcus elongatus* has three different Class I prenyltransferase genes. Plant Mol Biol 40:307–321

Olson JB, Steppe TF, Litaker RW, Paerl HW (1998) N_2-fixing microbial consortia associated with the ice cover of Lake Bonney, Antarctica. Microb Ecol 36:231–238

Onai K, Morishita M, Itoh S, Okamoto K, Ishiura M (2004) Circadian rhythms in the thermophilic cyanobacterium *Thermosynechococcus elongatus*: compensation of period length over a wide temperature range. J Bacteriol 186:4972–4977

Ostensvik O, Skulberg OM, Underdal B, Hormazabal V (1998) Antibacterial properties of extracts from selected planktonic freshwater cyanobacteria – a comparative study of bacterial bioassays. J Appl Microbiol 84:1117–1124

Ouyang Y, Andersson CR, Kondo T, Golden SS, Johnson CH (1998) Resonating circadian clocks enhance fitness in cyanobacteria. Proc Natl Acad Sci USA 95:8660–8664

Paerl HW (1996) Microscale physiological and ecological studies of aquatic cyanobacteria: macroscale implications. Microsc Res Tech 33:47–72
Paerl HW, Priscu JC (1998) Microbial phototrophic, heterotrophic, and diazotrophic activities associated with aggregates in the permanent ice cover of Lake Bonney, Antarctica. Microb Ecol 36:221–230
Palenik B, Haselkorn R (1992) Multiple evolutionary origins of prochlorophytes, the chlorophyll *b*-containing prokaryotes. Nature 355:265–267
Palenik B, Brahamsha B, Larimer FW, Land M, Hauser L, Chain P, Lamerdin J, Regala W, Allen EE, McCarren J, Paulsen I, Dufresne A, Partensky F, Webb EA, Waterbury J (2003) The genome of a motile marine *Synechococcus*. Nature 424:1037–1042
Palinska KA, Liesack W, Rhiel E, Krumbein WE (1996) Phenotype variability of identical genotypes: the need for a combined approach in cyanobacterial taxonomy demonstrated on *Merismopedia*-like isolates. Arch Microbiol 166:224–233
Pardee AB, Jacob F, Monod J (1959) The genetic control and cytoplasmic expression of "inducibility" in the synthesis of β-galactosidase by *E. coli*. J Mol Biol 1:165–178
Priscu JC, Fritsen CH, Adams EE, Giovannoni SJ, Paerl HW, McKay CP, Doran PT, Gordon DA, Lanoil BD, Pinckney JL (1998) Perennial Antarctic lake ice: an oasis for life in a polar desert. Science 280:2095–2098
Raven JA, Allen JF (2003) Genomics and chloroplast evolution: what did cyanobacteria do for plants? Genome Biol 4:209
Raymond J, Segre D (2006) The effect of oxygen on biochemical networks and the evolution of complex life. Science 311:1764–1767
Richter S, Hess WR, Krause M, Messer W (1998) Unique organization of the *dnaA* region from *Prochlorococcus marinus* CCMP1375, a marine cyanobacterium. Mol Gen Genet 257:534–541
Rikkinen J, Oksanen I, Lohtander K (2002) Lichen guilds share related cyanobacterial symbionts. Science 297:357
Rippka R, Waterbury JB, Cohen-Bazire G (1974) A cyanobacterium which lacks thylakoids. Arch Microbiol 100:419–436
Rippka R, Deruelles J, Waterbury JB, Herdman M, Stanier RY (1979) Generic assignments, strain histories, and properties of pure cultures of cyanobacteria. J Gen Microbiol 111:1–61
Rocap G, Distel DL, Waterbury JB, Chisholm SW (2002) Resolution of *Prochlorococcus* and *Synechococcus* ecotypes by using 16S–23S ribosomal DNA internal transcribed spacer sequences. Appl Environ Microbiol 68:1180–1191
Rocap G, Larimer FW, Lamerdin J, Malfatti S, Chain P, Ahlgren NA, Arellano A, Coleman M, Hauser L, Hess WR, Johnson ZI, Land M, Lindell D, Post AF, Regala W, Shah M, Shaw SL, Steglich C, Sullivan MB, Ting CS, Tolonen A, Webb EA, Zinser ER, Chisholm SW (2003) Genome divergence in two *Prochlorococcus* ecotypes reflects oceanic niche differentiation. Nature 424:1042–1047
Rothrock MJ Jr, Garcia-Pichel F (2005) Microbial diversity of benthic mats along a tidal desiccation gradient. Environ Microbiol 7:593–601
Sano T, Beattie KA, Codd GA, Kaya K (1998) Two (Z)-dehydrobutyrine-containing microcystins from a hepatotoxic bloom of *Oscillatoria agardhii* from Soulseat Loch, Scotland. J Nat Prod 61:851–853
Schopf JW (1993) Microfossils of the Early Archean Apex chert: new evidence of the antiquity of life. Science 260:640–646
Schopf JW, Packer BM (1987) Early Archean (3.3-billion to 3.5-billion-year-old) microfossils from Warrawoona Group, Australia. Science 237:70–73
Scott C, Lyons TW, Bekker A, Shen Y, Poulton SW, Chu X, Anbar AD (2008) Tracing the stepwise oxygenation of the Proterozoic ocean. Nature 452:456–459
Shalapyonok A, Olson RJ, Shalapyonok LS (1998) Ultradian growth in *Prochlorococcus* spp. Appl Environ Microbiol 64:1066–1069
Sheath RG, Vis ML, Hambrook JA, Cole KM (1996) Tundra stream macro-algae of North America: composition, distribution and physiological adaptations. Hydrobiologica 336:67–82
Shi L, Carmichael WW, Miller I (1995) Immuno-gold localization of hepatotoxins in cyanobacterial cells. Arch Microbiol 163:7–15

Shi T, Falkowski PG (2008) Genome evolution in cyanobacteria: the stable core and the variable shell. Proc Natl Acad Sci USA 105:2510–2515

Stanier RY (1980) The journey, not the arrival matters. Annu Rev Microbiol 34:1–48

Stewart I, Webb PM, Schluter PJ, Shaw GR (2006) Recreational and occupational field exposure to freshwater cyanobacteria – a review of anecdotal and case reports, epidemiological studies and the challenges for epidemiologic assessment. Environ Health 5:6

Stock JB, Surette MG, Levit M, Park P (1995) Two-component signal transduction systems: structure-function relationships and mechanisms of catalysis. In: Hoch JA, Silhavy TJ (ed) Two-component signal transduction. ASM, Washington, D.C., pp 25–51

Suzuki JY, Bauer CE (1995) A prokaryotic origin for light-dependent chlorophyll biosynthesis of plants. Proc Natl Acad Sci USA 92:3749–3753

Tamaru Y, Takani Y, Yoshida T, Sakamoto T (2005) Crucial role of extracellular polysaccharides in desiccation and freezing tolerance in the terrestrial cyanobacterium *Nostoc commune*. Appl Environ Microbiol 71:7327–7333

Tandeau de Marsac N, Houmard J (1987) Advances in cyanobacterial molecular genetics. In: Fay P, Baalen CV (eds) The cyanobacteria. Elsevier Science, Amsterdam, pp 251–302

Taton A, Grubisic S, Brambilla E, De Wit R, Wilmotte A (2003) Cyanobacterial diversity in natural and artificial microbial mats of Lake Fryxell (McMurdo Dry Valleys, Antarctica): a morphological and molecular approach. Appl Environ Microbiol 69:5157–5169

Thomas MA (2001) When three's not a crowd. Nature 412:375

Ting CS, Rocap G, King J, Chisholm SW (2002) Cyanobacterial photosynthesis in the oceans: the origins and significance of divergent light-harvesting strategies. Trends Microbiol 10:134–142

Tomitani A, Okada K, Miyashita H, Matthijs HC, Ohno T, Tanaka A (1999) Chlorophyll *b* and phycobilins in the common ancestor of cyanobacteria and chloroplasts. Nature 400:159–162

Tomitani A, Knoll AH, Cavanaugh CM, Ohno T (2006) The evolutionary diversification of cyanobacteria: Molecular-phylogenetic and paleontological perspectives. Proc Natl Acad Sci USA 103:5442–5447

Urmeneta J, Navarrete A, Huete J, Guerrero R (2003) Isolation and characterization of cyanobacteria from microbial mats of the Ebro delta, Spain. Curr Microbiol 46:199–204

van Gremberghe I, Van Wichelen J, Van der Gucht K, Vanormelingen P, D'Hondt S, Boutte C, Wilmotte A, Vyverman W (2008) Covariation between zooplankton community composition and cyanobacterial community dynamics in Lake Blaarmeersen (Belgium). FEMS Microbiol Ecol 63:222–237

Vaulot D, Marie D, Olson RJ, Chisholm SW (1995) Growth of *Prochlorococcus*, a photosynthetic prokaryote, in the equatorial pacific ocean. Science 268:1480–1482

Walker JJ, Pace NR (2007) Endolithic microbial ecosystems. Annu Rev Microbiol 61:331–347

Walsby AE (1994) Gas vesicles. Microbiol Rev 58:94–144

Wang G (2006) Diversity and biotechnological potential of the sponge-associated microbial consortia. J Ind Microbiol Biotechnol 33:545–551

Welker M, Khan S, Haque MM, Islam S, Khan NH, Chorus I, Fastner J (2005) Microcystins (cyanobacterial toxins) in surface waters of rural Bangladesh: pilot study. J Water Health 3:325–337

Wharton RA, Vinyard WC, Parker BC, Simmons GM, Seaburg KG (1981) Algae in cryoconite holes on Canada Glacier in southern Victoria Land, Antarctica. Phycologia 20:208–211

Whitton BA, Potts M (2000) Introduction to cyanobacteria. In: Whitton BA, Potts M (eds) The ecology of cyanobacteria. Klewer Academic, Amsterdam, pp 1–11

Williams SB (2007) A circadian timing mechanism in the cyanobacteria. Adv Microb Physiol 52:229–296

Williams SB, Stewart V (1999) Functional similarities among two-component sensors and methyl-accepting chemotaxis proteins suggest a role for linker region amphipathic helices in transmembrane signal transduction. Mol Microbiol 33:1093–1102

Williams SB, Vakonakis I, Golden SS, LiWang AC (2002) Structure and function from the circadian clock protein KaiA of *Synechococcus elongatus*: a potential clock input mechanism. Proc Natl Acad Sci USA 99:15357–15362

Woelfle MA, Ouyang Y, Phanvijhitsiri K, Johnson CH (2004) The adaptive value of circadian clocks: an experimental assessment in cyanobacteria. Curr Biol 14:1481–1486

Wong FC, Meeks JC (2002) Establishment of a functional symbiosis between the cyanobacterium *Nostoc punctiforme* and the bryophyte *Anthoceros punctatus* requires genes involved in nitrogen control and initiation of heterocyst differentiation. Microbiology 148:315–323

Yamaoka T, Satoh K, Katoh S (1978) Photosynthetic activities of a thermophilic blue-green alga. Plant Cell Physiol 19:943–954

Yoon HS, Golden JW (2001) PatS and products of nitrogen fixation control heterocyst pattern. J Bacteriol 183:2605–2613

Yoshizawa S, Matsushima R, Watanabe MF, Harada K, Ichihara A, Carmichael WW, Fujiki H (1990) Inhibition of protein phosphatases by microcystins and nodularin associated with hepatotoxicity. J Cancer Res Clin Oncol 116:609–614

Zak E, Norling B, Maitra R, Huang F, Andersson B, Pakrasi HB (2001) The initial steps of biogenesis of cyanobacterial photosystems occur in plasma membranes. Proc Natl Acad Sci USA 98:13443–13448

Zehr JP, Waterbury JB, Turner PJ, Montoya JP, Omoregie E, Steward GF, Hansen A, Karl DM (2001) Unicellular cyanobacteria fix N_2 in the subtropical North Pacific Ocean. Nature 412:635–638

Zouni A, Witt HT, Kern J, Fromme P, Krauss N, Saenger W, Orth P (2001) Crystal structure of photosystem II from *Synechococcus elongatus* at 3.8 A resolution. Nature 409:739–743

Chapter 3
Circadian Rhythm of *Cyanothece* RF-1 (*Synechococcus* RF-1)

Tan-Chi Huang and Rong-Fong Lin

Abstract *Cyanothece* RF-1 (*Synechococcus* RF-1) is a unicellular N_2-fixing cyanobacterium isolated from a rice field in Taiwan. The activity of nitrogen fixation in RF-1 revealed circadian rhythms when the cultures were placed in continuous light after diurnal regimen. RF-1 is the first prokaryotic organism shown to exhibit circadian rhythms regulated by a "biological clock." In addition to nitrogen fixation, the uptake rate of several amino acids and the activity of photosynthesis in RF-1 also exhibit circadian rhythmic patterns in free-running conditions. Finally, several rhythms of various proteins are described in this chapter.

3.1 Introduction

Cyanothece RF-1 (*Synechococcus* RF-1) is a unicellular N_2-fixing cyanobacterium isolated from a rice field in Taiwan. The activity of nitrogen fixation in RF-1 revealed circadian rhythms when cultures entrained under diurnal light/dark (LD) regimen were transferred to continuous light. RF-1 is the first prokaryotic organism shown to exhibit circadian rhythms regulated by a circadian "biological clock." The rhythm of nitrogenase activity is controlled at the transcriptional level. In a diurnal LD regimen, the expression of the *nif* and *nif*-associated genes is cyclic and occurs mainly during the dark periods. In addition to nitrogen fixation, the uptake rate of several amino acids and the activity of photosynthesis in RF-1 also exhibit circadian rhythmic patterns in free-running conditions. Membrane protein COP23 (*c*ircadian *o*scillating *p*rotein of *23* kDa) in RF-1 is regulated by circadian protein synthesis and circadian protein degradation. Extracellular Ca^{2+}, light (especially blue light) and new protein synthesis are involved in the regulation of circadian

T.-C. Huang
Institute of Plant and Microbial Biology, Academia Sinica, Taipei, Taiwan, Republic of China (retired in February 2004)

R.-F. Lin(✉)
Institute of Medical BioTechnology, Central Taiwan University of Science and Technology, TaiChung, Taiwan, Republic of China, e-mail: rflin@ctust.edu.tw

J.L. Ditty et al. (eds.), *Bacterial Circadian Programs*.

degradation of COP23. The overt rhythms in RF-1 can also be entrained by a small temperature changes within the growth-permissive range, but the phase is somewhat different from LD entrainment. When the RF-1 cells were entrained simultaneously by LD and temperature fluctuation, the effect of LD predominated over the temperature fluctuation.

3.2 First Discovery of an Endogenous Rhythm in a Prokaryote

A long-term project studying the N_2-fixing cyanobacteria in rice paddy fields was initiated in my laboratory (Dr. Huang) in 1982. In the beginning works, the N_2-fixing cyanobacteria were isolated, purified, and characterized for their nitrogenase activity. The nitrogenase activity was assayed by the acetylene-reduction method as described by Dilworth (1966). As expected, most of the isolates belonged to the heterocyst-forming filamentous type; however, we were very excited to discover that some of the N_2-fixing isolates were unicellular. Since only a few non-heterocystous filamentous and unicellular types are known to fix nitrogen aerobically, these unicellular isolates were further characterized (Huang and Chow 1988). Among them, strains RF-1 and RF-2 were sheathless, while strains RF-6 and RF-7 were ensheathed (RF indicates that the organisms were isolated from a *r*ice *f*ield). Both types fixed nitrogen continuously under constant light (LL). They fixed nitrogen mainly during the dark period when the LL culture was transferred to diurnal LD conditions. The specific N_2-fixing activity and growth rate of the sheathless type were much higher than that of the other types (Huang and Chow 1988).

Nitrogenase is an oxygen-labile enzyme, but oxygen is produced by the cyanobacterium during photosynthesis. Therefore, understanding how effective fixation of N_2 and CO_2 was accomplished under aerobic conditions in the unicellular cyanobacteria is of both theoretical interest and practical importance. Because some isolates of the ensheathed type, such as *Gloeocapsa* sp. or *Gloeothece* sp., had already been well studied as N_2-fixing organisms (Gallon 1980), we focused upon the study of the sheathless RF-1 in my laboratory. As shown in Fig. 3.1, when RF-1 cells were cultured in alternating cycles of light and darkness of various photoperiods, N_2-fixing activity was detected mainly during the dark periods (Chou et al. 1989). As a researcher working on nitrogen fixation, I was glad to find temporal separation between nitrogen fixation and photosynthesis in RF-1 because it is an excellent mechanism to allow two incompatible reactions within one cell. At that time, I had no background in circadian rhythms and was simply interested in the biochemical properties of nitrogenase. By coincidence, an unexpected event happened in my laboratory in 1986: Professor N. Grobbelaar of the University of Pretoria (South Africa) decided to visit our Institute (Institute of Botany, Academia Sinica) for 3 months. It was his first trip to Taiwan and we had no contact until he arrived at our Institute. Dr. Grobbelaar is a well known plant physiologist. His hobby is to collect and plant cycads, which are very popular garden plants in South Africa. It is known that some N_2-fixing cyanobacteria establish symbiotic

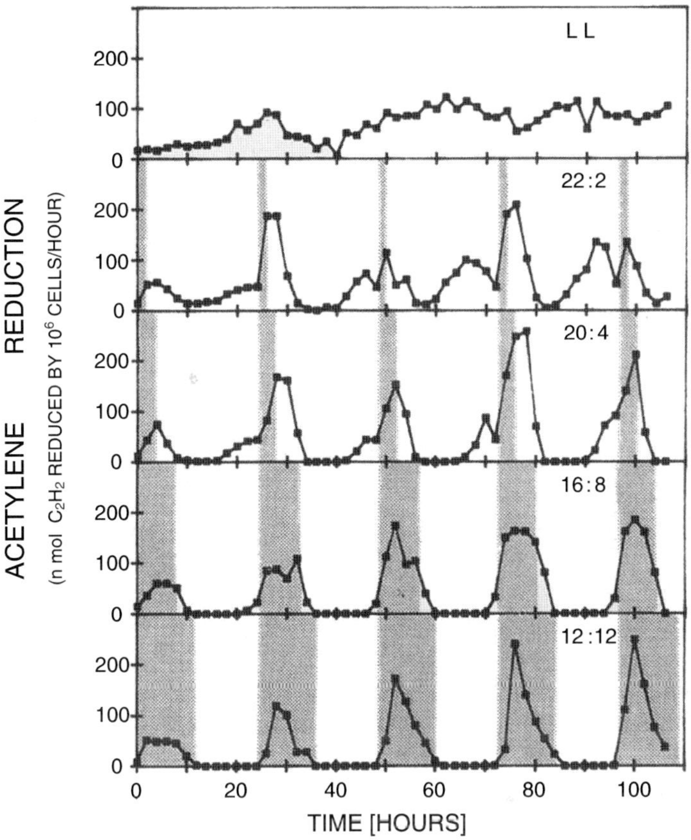

Fig. 3.1 Nitrogen-fixing patterns of RF-1 in LL and different diurnal LD regimens. Cultures adapted to LL condition were transferred to LD22:2, 20:4, 16:8, and 12:12 regimens, respectively (*shaded* portions indicate the dark intervals). The nitrogenase activity of the culture maintained under LL and those newly transferred to different LD conditions were assayed at 2-h intervals for a period of 5 days. Figure from Chou et al. (1989); reproduced by permission

relationships with cycads by inhabiting the coralloid roots of cycads. Dr. Grobbelaar was very interested in this topic and had studied some research on N_2-fixing cyanobacteria isolated from cycad nodules. He joined my laboratory when he learned that I was working on the N_2-fixing cyanobacteria. At that time, I had just published a paper concerning the diurnal N_2-fixing pattern of RF-1 growing under 12 h light/12 h dark (LD12:12) conditions (Huang and Chow 1986).

As a plant physiologist, Dr. Grobbelaar was quite familiar with biological clocks in plants and animals. Thus, he started to study the physiological properties of RF-1. I did not know his whole idea, but I was certain that he was not only interested in how the nitrogenase was protected from the O_2 evolved during photosynthesis. He started to re-examine the diurnal N_2-fixing pattern of RF-1. He transferred the RF-1 culture from LD cycles to LL conditions and proved that diurnal N_2-fixation is a truly endogenous rhythm (Grobbelaar et al. 1986).

Next, I gradually picked up knowledge of circadian rhythms by talking about with Dr. Grobbelaar and by reading related publications. Later, Dr. Tsung-Hsien Chen, also a plant physiologist working as my colleague at Academia Sinica, joined the project about RF-1. Thus, the study of the circadian rhythm in RF-1 would not have been so smooth or rapid without the cooperation of both Dr. Grobbelaar and Dr. Chen.

3.3 Taxonomic Classification of *Cyanothece* RF-1 (*Synechococcus* RF-1)

Due to the fact that cyanobacteria have been classified by various authorities using both botanical and bacteriological criteria, the taxonomic treatment of these organisms has undergone numerous changes. The sheathless unicellular diazotrophs, including *Synechococcus* BG043511 (Mitsui et al. 1986), *Synechococcus* RF-1 (Huang and Chow 1986), and *Cyanothece* ATCC51142 (Reddy et al. 1993), have the same morphology and similar physiological properties. Ultrastructure (Chou and Huang 1991) and rRNA sequence (Turner et al. 2001) also indicated that RF-1 and ATCC51142 are closely related, but they were assigned to different genera. This is because they were discovered and named at different times. BG043511 and RF-1 were both reported in 1986, at which time the most common bacteriological criteria for cyanobacterial classification were as described by Rippka et al. (1979). According to that classification system, unicellular cyanobacteria that undergo cell division in one plane and lack a sheath were assigned to *Synechococcus*. Later, Waterburg and Rippka (1989) proposed a new classification system for Chroococales. They moved the N_2-fixing unicellular cyanobacteria from *Synechococcus* to a separate genus, *Cyanothece*. The genus *Cyanothece* was first proposed by Komárek (1976) to accommodate some species previously placed in *Synechococcus*. The major distinction is that *Cyanothece* species are present as single cells or in pairs, but never grouped into chains as some *Synechococcus* are. The newly created *Cyanothece* is defined as those unicellular cyanobacteria that fix nitrogen and have a diameter larger than 3 μm. Thus, the RF-1 strain of *Synechococcus* should be classified to *Cyanothece* and named as *Cyanothece* RF-1.

3.4 Setting the Phase of the Circadian Rhythm of Nitrogen Fixation in RF-1

3.4.1 Induction by the Light/Dark Regimen

For an endogenous rhythm to qualify as a circadian rhythm, a fundamental property is that the rhythm must reveal a period of about 24 h in a constant environment, i.e., it must have a "free-running period" around 24 h. Several biochemical reactions,

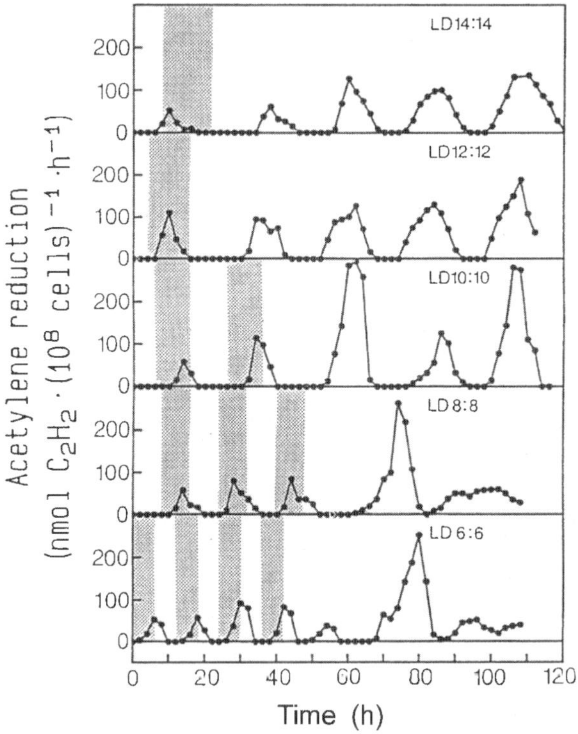

Fig. 3.2 Rhythms of nitrogenase activity of RF-1 in several different LD cycles followed by LL. The cultures were exposed to LD conditions for 1 week and then transferred to LL. Figure from Huang et al. (1990); reproduced by permission

including nitrogen fixation, photosynthesis, dark respiration, and amino acid uptake, were found to be regulated by the circadian clock in RF-1. Because the assay method for measuring N_2-fixation is relatively simple, it was used as a model reaction to study the conditions required for setting the phase of circadian rhythms in RF-1.

Like eukaryotes, the circadian rhythm of the prokaryote RF-1 can be entrained by different environmental factors. Among them, the LD regimen seems to be the most effective one. When the entrained LD cycle is not very different from 24h, RF-1 can adjust the phase of its N_2-fixing oscillation and generate a characteristic circadian rhythm. As shown in Fig. 3.2, RF-1 cultures that were entrained to either a prior LD14:14, LD12:12, or LD10:10 regimen each exhibited an endogenous N_2-fixing rhythm with free-running periods around 24h after transferring to LL. But if the period of the prior LD cycles was considerably different from 24h, as with LD8:8 or LD6:6 in Fig. 3.2, the N_2-fixing activity fluctuated without a 24h period under subsequent LL conditions (Huang et al. 1990). Although diurnal LD12:12 or LD 16:8 regimens are effective for setting the circadian rhythm in RF-1, the latter regimen is often preferred in order to provide more illumination time for faster RF-1 growth. During the LD entraining process, a brief (30min) interruption of the dark or light period did not significantly affect the phase of the nitrogenase activity rhythm.

The establishment of a circadian rhythm is a gradual process that is influenced by the intensity of the induction treatment and by the organism's physiological condition. The rhythm in RF-1 can be induced by one cycle of LD treatment, such as by exposing a culture to 8 h darkness. However, RF-1 requires more than one consecutive LD cycle for maximal induction of the rhythm. Total darkness is not required in the dark portion of the LD cycle for the induction of a circadian rhythm. For example, a circadian rhythm in LL can be induced in RF-1 by entraining to a prior light cycle which alternates bright light with dim light, e.g., three consecutive cycles of a regimen of 12 h at 3,000 lux and 12 h at 1,000 lux is effective (Chen et al. 1993; 1 lux = 1 cd m-2).

The circadian rhythm of nitrogen fixation can also be induced by exposing a culture pre-adapted to continuous red light (680 nm; half-band width of 10 nm) to a single 12-h dark period (Chen et al. 1993). In plants, phytochromes are involved in regulating some circadian rhythms (Lumsden 1991). For example, the circadian expression of the wheat *cab-1* gene, which encodes the major light-harvesting chlorophyll-binding protein of chloroplasts, is disturbed if the plant is exposed to 730 nm far-red light for 30 min during the dark period (Nagy et al. 1988). In RF-1, however, the rhythm was not affected by a 30-min pulse of far-red light applied at various intervals during the dark treatment (Chen et al. 1993).

3.4.2 Induction by a Small Temperature Change Within the Growth-Permissive Range

The growth-permissive temperature for RF-1 cultured in nitrate-free BG-11 medium (BG-11_0) ranges from 20°C to 37°C. In addition to LD regimens, the circadian N_2-fixation rhythm can be induced by temperature changes within the growth-permissive conditions. For instance, an arrhythmic culture adapted to 25°C for a long time was exposed to 35°C for 8 h on three consecutive days (raised-temperature cycles: 16 h 25°C/8 h 35°C), or transferred between 30°C and 20°C (lowered temperature cycles). The rhythm in free-running conditions for both treatments was found to have a period of about 24 h (Huang et al. 1994). The results indicated that the period of the circadian rhythm in RF-1, as in eukaryotes, was insensitive to different constant ambient temperatures (i.e., "temperature compensation"). However, temperature cycles can entrain this circadian rhythm. Further experiments revealed that a 5°C temperature step was effective for the entrainment, for example, exposing a LL culture either to 12 h 30°C/12 h 25°C or 12 h 30°C/12 h 35°C.

Knowing the characteristics of temperature induction for circadian rhythms in RF-1 enabled us to design experiments for examining the influence of organism's physiological condition during entrainment. It is known that microbes can generally cease proliferation under unfavorable growth conditions and switch to the lowest levels of metabolism and energy expenditure, thereby surviving harsh physiological conditions in the "suspended state" designated by some microbiologists. Cells surviving in the suspended state may return to active growth if conditions become

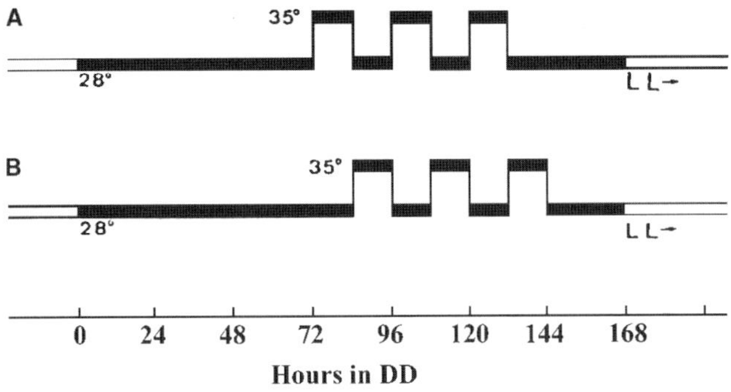

Fig. 3.3 Diagram showing the entrainment protocol for RF-1 cultures by temperature cycles while the cells were in the "suspended state." The cultures were entrained by three cycles of a 12h 28°C/12h 35°C regimen started on day 3 after the cultures were transferred from LL to DD. One was exposed to 35°C from 0800 to 2000 hours (**A**), the other from 2000 to 0800 hours (**B**). Figure from Huang and Pen (1994); reproduced by permission

favorable once again. The cell density of RF-1 cultures remains almost constant after being transferred from LL to constant darkness (DD). No dividing cells were observed in the culture microscopically from the third day of darkness but cell growth resumed after the DD cultures were re-exposed to light. Thus, physiologically, the RF-1 cells were in a suspended state after they had adapted to DD. RF-1 can be maintained in such a suspended state for more than 2 weeks by transferring a culture grown in LL to DD within its growth-permissive temperature. As shown in Fig. 3.3, two RF-1 cultures maintained in 3-day darkness were entrained by a temperature cycle (12h 28°C/12h 35°C) regimen for three consecutive days. One was exposed to 35°C for three cycles from 0800 hour to 2000 hour and the other from 2000 hour to 0800 hour. Both cultures were then exposed to LL for 1 day after completion of entrainment. When the two suspended-state cultures adapted through continuous dark and then entrained by raised-temperature were exposed to LL, both cultures exhibited a circadian rhythm of nitrogenase activity (Fig. 3.4). The cultures whose raised-temperature entrainment were applied from 0800 hour peaked at about 15h after the onset of continuous light, while the first peak of nitrogenase activity in the other culture commenced about another 12h later (27h after the onset of the LL). The 12-h difference of the induction treatment therefore resulted in rhythms about 12h out of phase. In contrast, the other two cultures were similarly entrained but exposed to temperature cycle entrainment with a lowered temperature (12h 28°C/12h 20°C). The results were similar to the above treatment (Huang and Pen 1994). The four suspended-state cultures simultaneously resumed their active growth with the same growth curve after transfer to LL. The approximately 12-h phase difference of the circadian nitrogenase activity rhythm entrained by the same type of temperature regimen was apparently caused by the different timing of high versus low temperature while the cells were in the suspended state. Since no cell growth or cell division was detected in the suspended-state cultures, active cell growth is not essential for the induction of the circadian rhythm in RF-1.

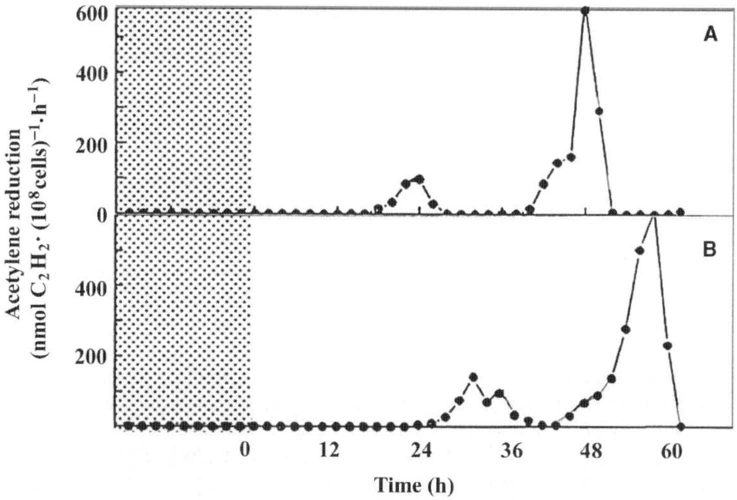

Fig. 3.4 Initiation of the circadian nitrogenase activity rhythm in RF-1 after the cultures were transferred from DD to LL. Two cultures were entrained by the temperature cycle regimens shown in Fig. 3.3. The nitrogenase activity of both cultures was assayed after exposure to LL. **A** Entrainment from 0800 to 2000 hours, as in Fig. 3.3A. **B** Entrainment from 2000 to 0800 hours, as in Fig. 3.3B. Figure from Huang and Pen (1994); reproduced by permission

3.4.3 Priority of LD Entrainment over Temperature in Setting the Circadian Rhythms of RF-1

Illumination and temperature are two major entraining agents for circadian rhythms. The input pathways of these two environmental factors for the entrainment of circadian rhythms in RF-1 are different, as the overt rhythms in the mutant strain CR-1 (one of the circadian-rhythm mutants of RF-1; Huang et al. 1993) could be established by temperature cycles but not by LD cycles (Lin et al. 1999). Therefore, it was of interest to investigate the phases of RF-1 cells under simultaneous entrainment by both LD and temperature regimens. As shown in Fig. 3.5, when the RF-1 cultures growing at 30°C in LL were exposed to either 12 h L/12 h D or 12 h 30°C/12 h 25°C (lowered-temperature cycle), the peaks of circadian nitrogenase activity were around circadian time (CT)18 (Fig. 3.5A) and CT14 (Fig. 3.5B), respectively. When the culture was entrained simultaneously by LD and lowered-temperature cycles (12 h L 30°C/12 h D 25°C; Fig. 3.5C), the peak was around CT18, similar to the phase entrained by the LD regimen alone. When raised-temperature cycles such as 30°C/35°C were used instead of lowered-temperature cycles in the above experiments, the peak of nitrogenase activity was around CT4 (Fig. 3.6B), about 10 h behind (or 14 h ahead) compared with that entrained by LD cycles. However, the phase of nitrogenase activity entrained simultaneously by 12 h L 30°C/12 h D 35°C had its peak around CT20 (Fig. 3.6C), which is a phase closer to that obtained with the LD cycle (Fig. 3.6A) alone than to that obtained with the temperature cycle alone (Fig. 3.6B).

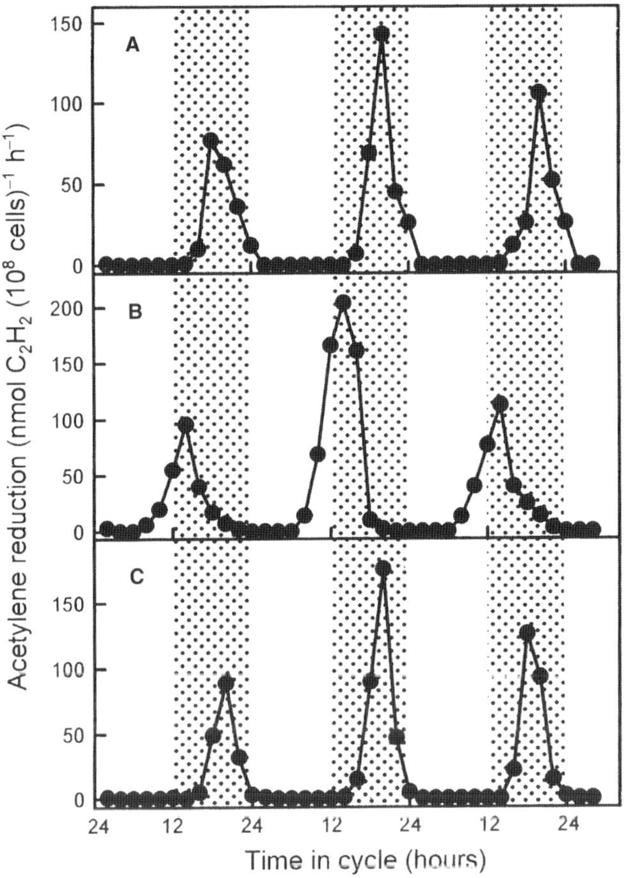

Fig. 3.5 Phase setting of the circadian rhythm of nitrogenase activity in RF-1 entrained by LD, lowered temperature cycles, or LD and lowered temperature cycles simultaneously. The nitrogenase activity was assayed at 2-h intervals after 3 days of entrainment. Cultures growing at 30°C under LL were entrained by 12 h L/12 h D (**A**), 12 h 30°C/12 h 25°C (**B**), or 12 h L 30°C/12 h D 25°C (**C**). The *shaded areas* represent the dark periods (**A**), the 25°C periods (**B**), or the combination of both (**C**). Figure from Lin et al. (1999); reproduced by permission

Since temperature fluctuations within a single day could be more than 5°C, the nitrogenase activity rhythm in the cultures entrained by temperature steps >5°C was also investigated. When the temperature step was increased by 10°C (12 h L 30°C/12 h D 20°C), the result was consistent with that observed with 5°C differences (12 h L 30°C/12 h D 25°C).

3.5 Circadian Rhythm of Leucine Uptake in RF-1

In an attempt to observe amino acid incorporation in RF-1, the uptake rate for 20 different [^{14}C]-labeled amino acids was measured with a liquid scintillation counter in the middle of a light and a dark period for LD12:12 entrained cultures. It was found

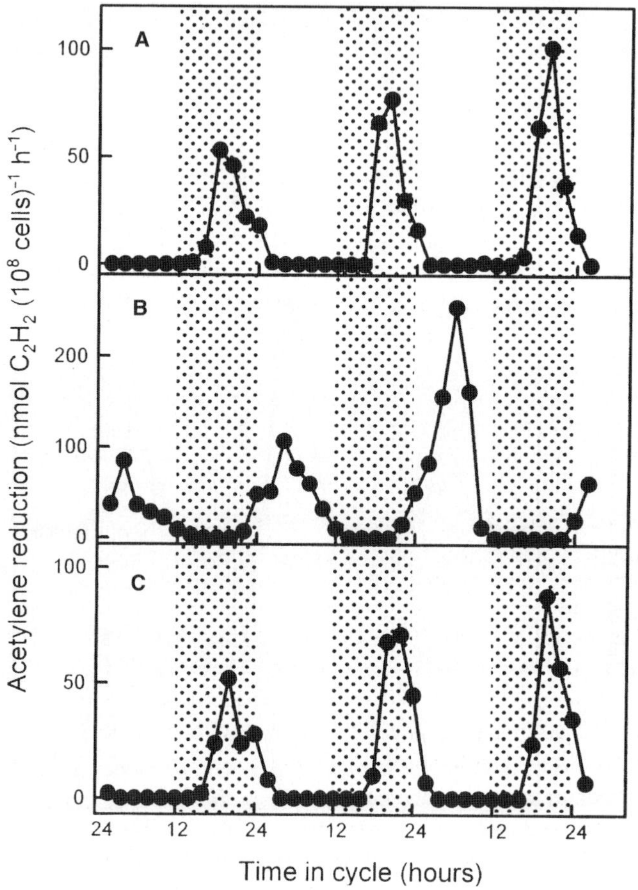

Fig. 3.6 Phase setting of the circadian rhythm of nitrogenase activity in RF-1 entrained by LD, raised temperature cycles, or LD and raised temperature cycles simultaneously. The nitrogenase activity was assayed at 2-h intervals after 3 days of entrainment. Cultures growing at 30°C under LL were entrained by 12h L/12h D (**A**), 12h 30°C/12h 35°C (**B**), or 12h L 30°C/12h D 35°C (**C**). The *shaded areas* represent the dark periods (**A**), the 35°C periods (**B**), or the combination of both (**C**). Figure from Lin et al. (1999); reproduced by permission

that the uptake rate of leucine during the light period was about eight times higher than that during the dark period. Leucine revealed the highest uptake difference among the 20 natural amino acids. Thus, the uptake rate of leucine in RF-1 was used to study whether it is controlled by circadian clock. The non-metabolizable leucine analog, 2-amino isobutyric acid (AIB) was also investigated to show whether the uptake rate of leucine could be attributed to a change in leucine metabolism (Chen et al. 1991). When RF-1 cells were cultured in $BG11_0$ under LD12:12 cycles, the uptake rates of leucine and AIB fluctuated periodically and were several times higher during the light period than the dark period. If the cultures were subsequently exposed to LL, the periodic variation in leucine and AIB uptake persisted without a noticeable change. The average period of the rhythm under free-running conditions was about

24h. The peaks of leucine uptake are about 12h out of phase if compared with that of nitrogenase activity. The rhythm of leucine or AIB uptake rate was not affected by the presence of 25mM $NaNO_3$, which represses nitrogenase activity.

The circadian rhythm of leucine uptake could also be induced by temperature changes within the growth-permissive range; both a raised- or a lowered-temperature cycle had this effect (Huang et al. 1994). In addition to leucine, the uptake rate of L-valine, L-isoleucine, L-proline, L-phenylalanine, L-tyrosine, L-methionine, and L-tryptophan may also be controlled by the oscillator (Chen et al. 1991). To find out whether the rhythmic uptake of amino acids is a general property of unicellular cyanobacteria, the uptake rate of L-[^{14}C]-leucine in PCC 7942 and PCC 6803 was also determined. As in RF-1, the rate of L-leucine uptake in these two cyanobacterial species was higher during the light period than the dark period of LD cycles. But unlike RF-1, a persistent rhythm in L-leucine uptake could not be observed in these two organisms when they were transferred from LD to LL (Chen et al. 1991).

3.6 Observation of the Circadian Photosynthetic Rhythm and Dark Respiration in RF-1

The general activities of photosynthesis and dark respiration were investigated using a Clark oxygen electrode to measure oxygen production in the light and uptake in the dark. The assay procedures are relatively complicated and are therefore not suitable for studying the circadian rhythm in samples that will be assayed every hour or two for several days. Yen et al. (2004) proposed a method using a dissolved-oxygen (DO) meter to continuously and automatically record the fluctuation of DO level in cyanobacterial cultures. The DO meter probe consisted of a Clark-type polarographic sensor covered with a permeable membrane. When the probe was inserted into the culture, oxygen diffused through the membrane at a rate proportional to the concentration of oxygen in the culture, causing a current flow that could be measured. A magnetic stirrer with proper stirring speed was employed to ensure a uniform distribution of oxygen concentration throughout the culture. The culture was exposed to the atmosphere so the oxygen concentration in the culture was equilibrated between atmospheric diffusion and production or consumption of oxygen by cyanobacteria. Because the oxygen concentration of the atmosphere is nearly constant, the DO variation of the culture therefore reflected photosynthetic activity during the light cycle and dark respiration activity during the dark cycle. The samples were measured every 20min and the DO values were automatically recorded by a computer.

Figure 3.7 illustrates the variation in the DO values of RF-1 cells cultured in $BG11_0$. During LD, a significant dip was observed during the dark interval, revealing that oxygen was drastically consumed in the culture. Such a dip implied a considerable variation in the dark respiration rate. At the onset of the light period, photosynthesis actively produced a large amount of oxygen and resulted in a sharp DO peak. About 2h later, the oxygen-producing rate of photosynthesis decreased, achieving an equilibrium in the second half of the light interval between photosynthetic oxygen production and diffusion of oxygen from the atmosphere. When cultures were

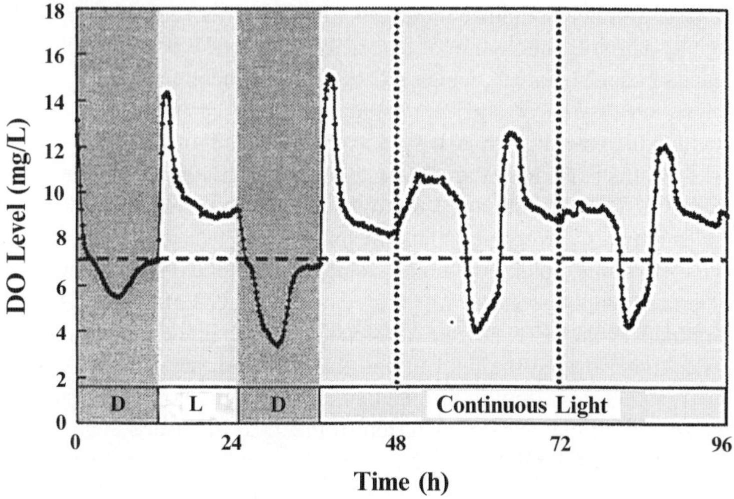

Fig. 3.7 Circadian photosynthetic activity and dark respiration activity of RF-1 in BG-11_0 medium. The *horizontal dashed line* indicates the background dissolved oxygen (*DO*) level. *L* Liter. Figure from Yen et al. (2004); reproduced by permission

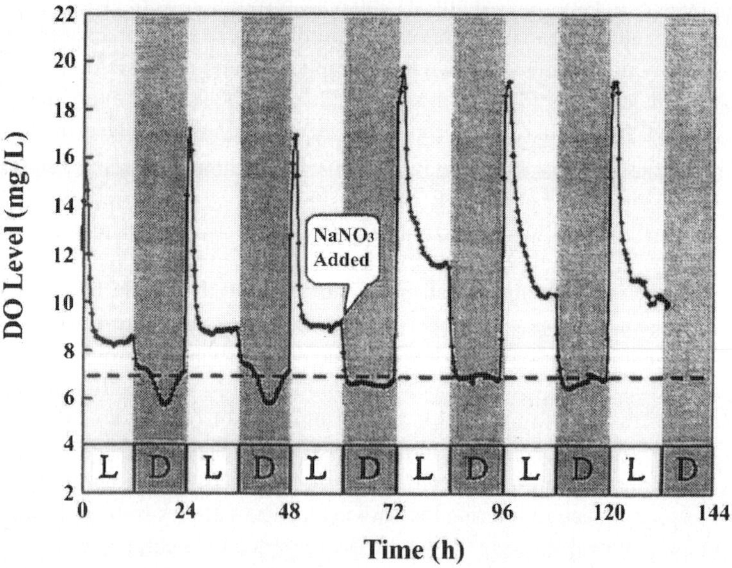

Fig. 3.8 Effects of nitrate on the dark respiration activity of RF-1 cells grown in BG-11_0 medium. The final concentration of $NaNO_3$ was 0.02%. Figure from Yen et al. (2004); reproduced by permission

switched to LL, a rhythmic variation with a period of about 24 h persisted for at least 3 days, indicating the presence of a circadian rhythm of photosynthetic and dark respiration rates. However, when $NaNO_3$, an inhibitor of nitrogenase, was added to the culture at the beginning of a dark interval, the DO immediately dipped in that dark period and in the following dark periods disappeared, resulting in a constant DO level which indicated a constant dark respiration (Fig. 3.8). The result revealed a close relationship between nitrogenase activity and dark respiration rate in RF-1. A similar result had also been reported previously (Grobbalaar et al. 1991). These results indicate that the increase in the dark respiration rate is not directly driven by the biological clock because nitrate did not prevent the establishment of the clock-controlled leucine-uptake rhythm (Chen et al. 1991) or the circadian photosynthesis rhythm. Instead, increased respiration activity is coupled to nitrogen fixation. Since nitrogenase is oxygen-labile, the increase in the respiration rate has been suggested to be essential for protection of nitrogenase activity in RF-1 (Grobbelaar et al. 1987).

3.7 Circadian Rhythms in RF-1 at a Biochemical Level

3.7.1 Examining the Polypeptides Synthesized with a Circadian Oscillating Rate

The circadian rhythm in RF-1 can be detected by examining the synthetic rate of polypeptides. Since the observation of nitrogenase at the gene expression level is discussed in Sect. 3.8, we discuss here the polypeptides not involved in N_2-fixation. Thus, RF-1 cells were grown in nitrate-containing medium (BG-11) to repress the *nif* (N_2-fixing) and *nif*-associated genes (see Sect. 3.7.2).

The protein synthesis rate was labeled with [^{35}S]-methionine and examined by autoradiography (Huang et al. 1994). As shown in Fig. 3.9, the synthesis of several polypeptides expressed a circadian oscillating rate under free-running condition when entrained by diurnal LD regimen. Ten of the polypeptides with relative molecular masses of 65, 61, 58, 38, 36, 33, 24, 23, 20, and 18 kDa were identified as having rhythmic patterns. Among them, the synthesis of polypeptides with molecular masses of 61, 38, and 36 kDa oscillated in phase with that of the 65 kDa polypeptide. The synthetic phase of the 58-, 24-, 23-, and 20-kDa polypeptides was advanced by about 4 h relative to that of the 65-kDa polypeptide; nevertheless, the synthetic phase of the 33-kDa polypeptide was delayed by about 4 h. The synthesis of the 18-kDa polypeptide was about 12 h out of phase with respect to that of the 65-kDa polypeptide.

When RF-1 cultures were entrained by environmental factors other than LD regimen, such as raised (Fig. 3.10) or lowered (Fig. 3.11) temperature cycle regimens, the synthesis rhythm of the polypeptides under the new entrainment condition were identical to that entrained by a diurnal LD regimen. The pattern of phase relationships among those polypeptides with circadian synthesis rates was also similar in the individual cases. We therefore suggest that the circadian rhythm induced by LD or temperature regimen is initiated by the same oscillator, and possibly there is only one circadian oscillator operating in RF-1. The results in Figs. 3.9–3.11 also indicate

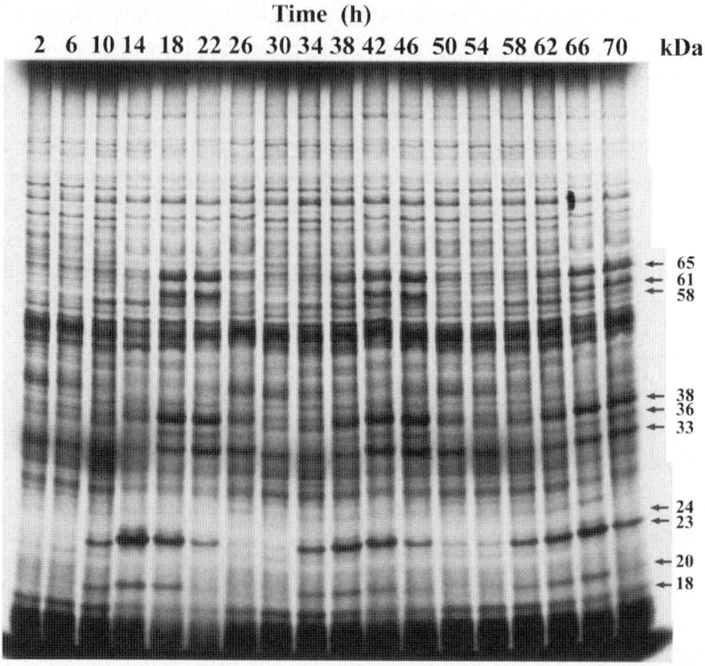

Fig. 3.9 Autoradiography of [^{35}S]-methionine-labeled polypeptides synthesized by RF-1 in free-running conditions after the cultures were entrained by LD regimens (LD16:8). Samples were collected at 4-h intervals over 70 h after the onset of free-running conditions. The apparent molecular masses of some polypeptides showing circadian rhythm patterns are indicated by *arrows*. Figure from Huang et al. (1994); reproduced by permission

that there may be fixed phase relationships among the above clock-controlled polypeptides, which can be diagramed by a rotated 24-h cycle (Fig. 3.12).

3.7.2 Regulation of the Circadian Synthesis of COP23

The control mechanism of the biological clock appears to be relatively complex. Consequently, it may be investigated from different approaches. One approach involves the elucidation of the components of the central oscillator, followed by unraveling the coupling pathways between the oscillator and the observed rhythms. An alternative approach is to identify the clock-controlled genes and then to determine how these genes are regulated. Among the circadian oscillating polypeptides identified in RF-1, the polypeptide COP23 of 23 kDa was chosen for further study because it exhibited a circadian oscillation of abundance with a robust amplitude.

COP23 is possibly located on the cell membrane (Chen et al. 1996a, b), because it can be extracted by boiling RF-1 cells in a buffer of 62.5 mM Tris (pH 6.8),

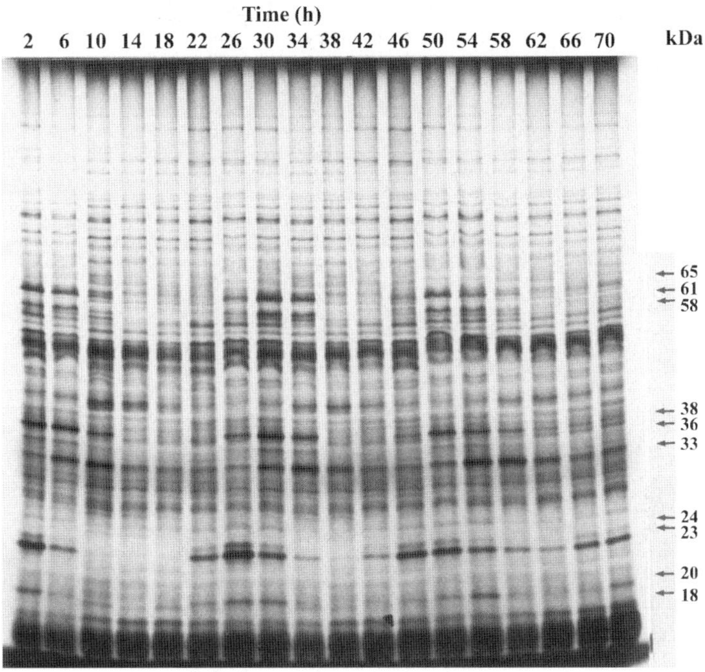

Fig. 3.10 Autoradiography of [^{35}S]-methionine-labeled polypeptides synthesized by RF-1 in free-running conditions after the cultures were entrained by a raised-temperature cycle regimen (16 h 25°C/8 h 35°C). Samples were collected at 4-h intervals over 70 h after the onset of free-running conditions. The apparent molecular masses of some polypeptides showing circadian rhythm patterns are indicated by *arrows*. Figure from Huang et al. (1994), reproduced by permission

whether or not that buffer contains 2% SDS. By boiling RF-1 cells, the content of COP23 dominates in the crude extract and can be separated as a single band with SDS-PAGE. The N-terminal amino acid sequence is Asp-Asp-Lys-Tyr-Pro-Gln-Tyr-Asn-Met-Ile-The-Glu-Gly-Phe-Pro. Southern hybridization using a degenerate probe of a synthetic oligonucleotide with 45 nucleotides deduced from the N-terminal sequence identified the coding region with 699 bp of the COP23 gene and part of the upstream sequence (see GenBank U29340).

When RF-1 cells entrained by LD12:12 regimens were taken at 4 h intervals after the LD-entrained cultures were transferred to LL, the abundance of the COP23 protein exhibited circadian fluctuations under free-running conditions (Fig. 3.13A). To assay the rhythm in the synthesis rate of COP23, RF-1 cells were labeled with [^{35}S]-methionine for 30 min. Proteins were extracted by heating and separated with SDS-PAGE. Figure 3.13B shows that the synthesis rate of COP23 also exhibited a circadian rhythm. The results also indicate that the synthesis of COP23 initiated slightly earlier than its accumulation. In this study, expression of the COP23 gene was also examined; and the mRNA level of COP23 detected by Northern hybridization was found to exhibit a circadian rhythm pattern (Fig. 3.13C). The peak of COP23 mRNA expression occurred at a time corresponding

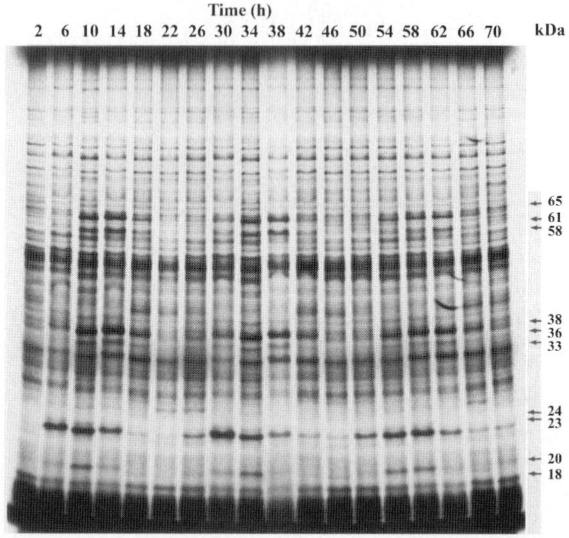

Fig. 3.11 Autoradiography of [^{35}S]-methionine-labeled polypeptides synthesized by RF-1 in free-running conditions after the cultures were entrained by lowered-temperature cycle regimen (16 h 30°C/8 h 20°C). Samples were collected at 4-h intervals over 70 h after the onset of free-running conditions. The apparent molecular masses of some polypeptides showing circadian rhythm patterns are indicated by *arrows*. Figure from Huang et al. (1994); reproduced by permission

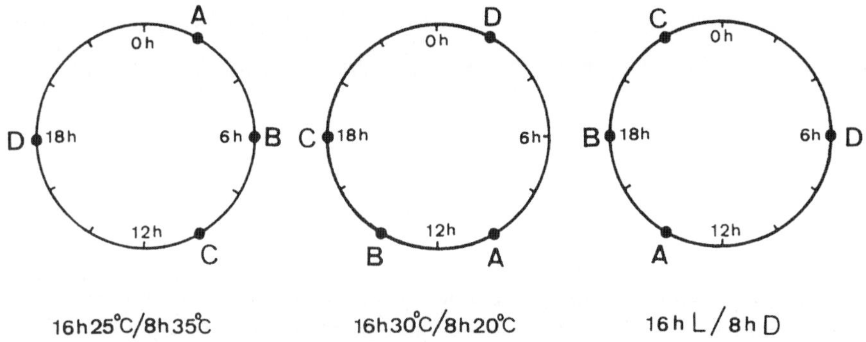

Fig. 3.12 The apparently fixed phase-relationships for "clock"-controlled polypeptide synthesis in RF-1, based on the data from Figs. 3.9–3.11, can be represented by a 24-h cyclic diagram. The *dots* on the circumference represent the timepoints of the peak of polypeptide synthesis rate after the cultures were transferred to free-running conditions. Zero hour (*0 h*) indicates the transition timepoint after entrainment. **A** 20-, 23-, 24-, and 58-kDa polypeptides; **B** 36-, 38-, 61-, and 65-kDa polypeptides; **C** 33-kDa polypeptide; **D** 18-kDa polypeptide. Figure from Huang et al. (1994); reproduced by permission

to that of the circadian synthesis rate of the COP23 protein. The results revealed that the circadian protein-synthesis rhythm of COP23 is controlled at the transcriptional level.

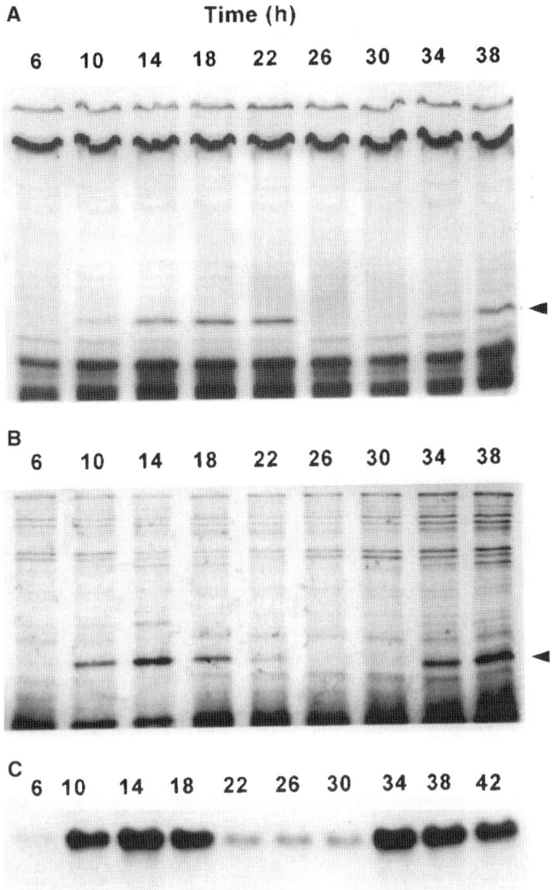

Fig. 3.13 Circadian rhythms of protein abundance, protein synthetic rate, and the gene expression of COP23 in RF-1 after entrainment of LD regimen. The RF-1 cultures were collected at 4-h intervals under free-running conditions after LD12:12 entrainment. **A** Total proteins were extracted in boiling water and analyzed by SDS-PAGE; *arrowhead* indicates COP23. **B** Total proteins in RF-1 cells were labeled in vivo with [^{35}S]-methionine, then proteins were extracted in boiling water and detected by autoradiography after SDS-PAGE; *arrowhead* indicates COP23. **C** Northern hybridization of RF-1 mRNA with digoxygenin-labeled COP23 probe

3.7.3 Factors Affecting the Circadian Degradation of COP23

3.7.3.1 New Protein Synthesis is Needed Before the Initiation of Rapid COP23 Degradation

Fig. 3.13A shows that COP23 accumulated significantly at 14 h and vanished rapidly after 22 h. In order to clarify the regulation of COP23 degradation, the stability of the COP23 protein in vivo was investigated by the [^{35}S]-methionine pulse-abeling technique. Because the cyclic peak of COP23 synthesis occurred at 10–18 h after the onset of free-running conditions for a culture entrained by LD (Fig. 3.13B), the LD-entrained cultures were pulse-labeled with [^{35}S]-methionine at 13 h and 17 h after exposure to LL. The COP23 band of the two experiments declined rapidly at almost the same circadian phase (Fig. 3.14A, B), even though

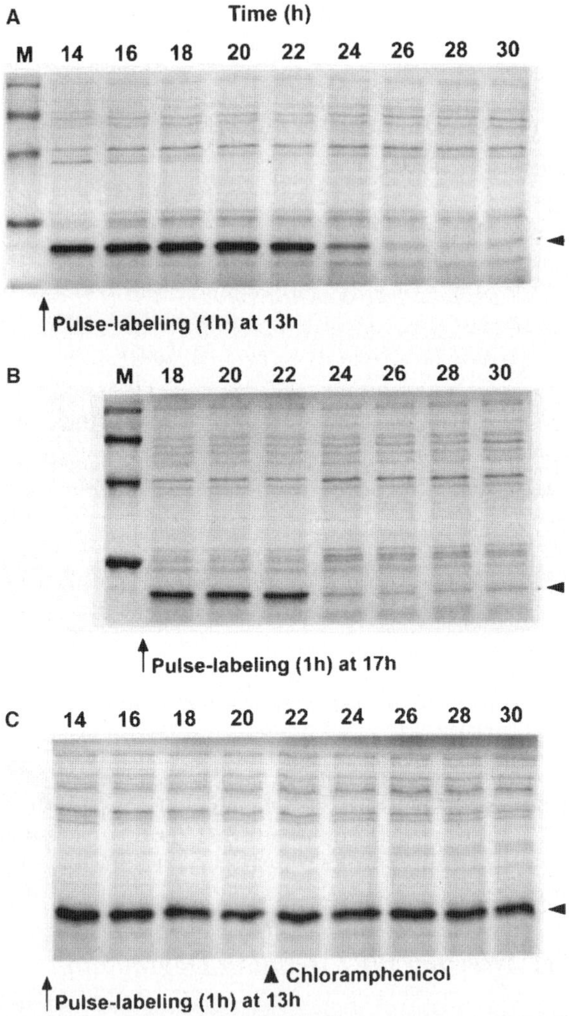

Fig. 3.14 Stability of the COP23 protein in RF-1. The LD-entrained cultures were pulse-labeled for 1 h with [^{35}S]-methionine at 13 h (**A**) or 17 h (**B**) after the cultures were transferred to LL. The proteins of RF-1 were extracted at 2-h intervals and separated by SDS-PAGE. Labeled proteins on the electrophoresis gels were detected by autoradiography. In corollary experiment, RF-1 cultures were pulse-labeled for 1 h at 13 h in LL conditions and chloramphenicol (100 μg ml^{-1}) was then added at 21 h (**C**). Figure from Chen et al. (1996a); reproduced by permission

they were labeled at different phases. This result indicated that the degradation of COP23 did not occur at a constant rate at the two circadian phases; instead, rapid degradation was initiated at the circadian phase corresponding to about 22h of LL. However, when chloramphenicol (100 μg ml^{-1}) was added to the RF-1 culture immediately after pulse labeling (at 21h), COP23 was more stable and did not exhibit the rapid decline (Fig. 3.14C). Thus, new protein synthesis must be involved in promoting proteolysis of COP23, possibly by de novo synthesis of a new protease or a COP23-modifying enzyme.

3.7.3.2 Extracellular Ca^{2+} is Required for COP23 Degradation

When the LD-entrained cultures were transferred to LL, COP23 started to accumulate at 14h and decreased rapidly after 22h (Fig. 3.13A). However, if EGTA (2mM) was added to the cultures before the start of the rapid degradation, the decline of COP23 was inhibited, thereby preventing the circadian fluctuation of COP23 abundance (Lin et al. 2003). If Ca^{2+} was supplemented before the onset of COP23 degradation, the rapid degradation resumed at the same phase as the cultures without EGTA treatment.

Although addition of EGTA to the RF-1 cultures disturbed the circadian rhythm of COP23 abundance, it did not significantly affect the circadian synthesis of COP23 (Lin et al. 2003). These results indicated that the synthesis of COP23 is not feedback-inhibited by COP23 itself but is controlled by an "oscillator," and this oscillator runs even when extracellular Ca^{2+} is chelated by EGTA. Nevertheless, these results do not imply that calcium is not required for clock function. Since EGTA cannot penetrate into cytosol, intracellular Ca^{2+} is probably just influenced slowly and gradually. As a consequence, the concentration of cytosolic Ca^{2+} may still remain high enough for the "oscillator" to function normally over the duration of the EGTA-addition experiment.

3.7.3.3 Light is Essential for COP23 Degradation

The protein COP23 has been shown to accumulate in the dark phase and decrease during the light phase in the LD-entrained cultures (Fig. 3.15A, left panel). However, COP23 was not degraded when the LD-entrained cultures were transferred to darkness during the light phase instead of maintaining the cultures in light (Fig. 3.15A, middle panel). Degradation of COP23 resumed when the cultures transferred to dark were re-exposed to light (Fig. 3.15A, right panel). These results indicated that light was essential for COP23 degradation.

Comparing the effects of different light spectra, including white light from fluorescent lamps, red light (660nm), and blue light (425nm), revealed that blue light is more effective than red light for COP23 degradation (Fig. 3.15B). This result was obtained even when the blue light intensity was lowered to 20 μmol m^{-2} s^{-1}; however, no significant effect of red light occurred even with a light intensity as high as 500 μmol m^{-2} s^{-1}. The white light with spectra mainly covering the blue light range was also effective, but its effect on COP23 degradation was less than that of blue light.

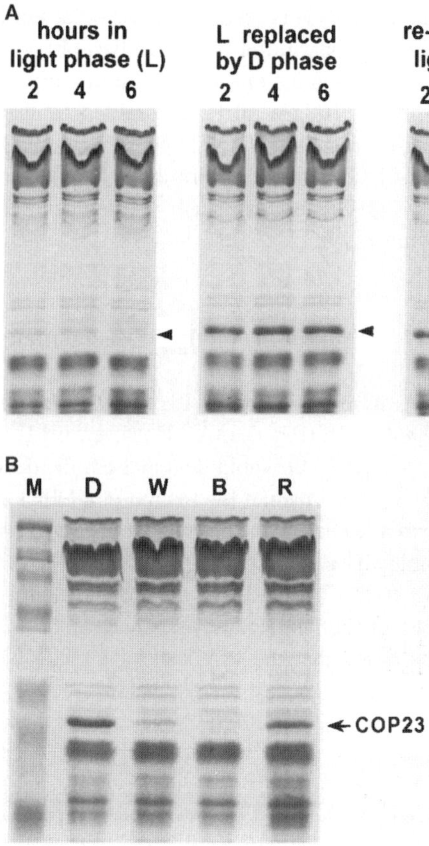

Fig. 3.15 COP23 degradation is light-dependent. **A** Light effects on COP23 degradation. Content of COP23 (indicated by *arrowhead*) in the diurnal LD-entrained RF-1 cultures were analyzed: at 2, 4, and 6 h of the light interval of LD (*left panel*), at the same timepoints as the left panel but with the light phase replaced by darkness (*middle panel*), and when cultures after light to dark transfer were re-exposed to light at the 3-h timepoint, indicated by the vertical arrow (*right panel*). **B** Spectra of light that influence COP23 degradation. Content of COP23 was analyzed: at 2 h after the light interval of LD-entrained cultures was replaced by darkness (*D*); when illuminated either by white light (50 μmol · m^{-2} · s^{-1}) from a fluorescent lamp (*W*), by blue light (50 μmol · m^{-2} · s^{-1}; *B*), or by red light (50 μmol · m^{-2} · s^{-1}; *R*). Figure from Lin et al. (2003); reproduced by permission

3.8 Circadian Rhythm of Nitrogenase Activity in RF-1 at the Molecular Level

Nitrogenase is a complex protein consisting of three different kinds of polypeptides encoded by the *nifH*, *nifD*, and *nifK* genes, respectively. Studying the time course of nitrogenase synthesis using chloramphenicol and nitrate revealed that renitiation of nitrogenase synthesis was required for nitrogen fixation during the dark period in a diurnal LD regimen (Huang and Chow 1990). In order to identify the source of rhythmic regulation of nitrogenase genes, the nucleotide sequence of the *nifHDK* operon in RF-1 was determined (Chen et al. 1996a, b). Northern hybridization using *nif* genes as a probe showed that RF-1 cells synthesize nitrogenase-specific mRNA constitutively in arrhythmic LL cultures. In contrast, in a diurnal LD12:12 regimen the synthesis of the nitrogenase mRNA is cyclic and occurs exclusively during the dark periods. A circadian rhythmic expression of *nif* genes, which corresponds in phase to the nitrogenase activity, persists after the LD-entrained culture is transferred to continuous light (Huang and Chow 1990). These results indicate that the rhythm of nitrogenase activity in RF-1 is controlled at the transcriptional level.

In addition to the structural genes of nitrogenase, other *nif* and *nif*-associated genes, including the *nifB* operon (*nifB-fdxN-nifS-nifU*), *nifP*, *nifE-nifN*, *nifX-orf*, *nifW-hes*A-*hes*B, and the "*fdx*"-containing operon were also cloned and sequenced. As described by Huang et al. (1999), all *nif* and *nif*-associated genes in RF-1 are arranged in a continuous cluster spanning approximately 18 kb and containing seven operons. The genes *hesA*, *hesB,* and "*fdx*" were assigned as *nif*-associated genes because they were expressed only under N_2-fixing conditions. Like the *nifHDK* operon, all *nif* and *nif*-associated genes were expressed in a rhythmic pattern with peaks during the dark phase when the culture was grown in a LD regimen. A circadian rhythm persisted after the culture was transferred to LL (Huang et al. 1999).

It is known that the *nif* genes in RF-1 are repressed in the presence of nitrate. Therefore, a culture of RF-1 growing in nitrate-containing medium does not express nitrogenase activity. When a culture growing in nitrate-containing medium is exposed to a diurnal LD regimen, the rhythm of nitrogenase activity manifests itself after the culture is transferred to nitrate-free medium and incubated in constant illumination (Huang and Chou 1991). These results indicate that the circadian N_2-fixing rhythm of RF-1 can be induced while the *nif* genes are repressed. The results support the idea that the circadian rhythms of N_2-fixation, leucine uptake, photosynthesis, and others represent only the "hands" of the oscillator. Each "hand" may be controlled by the oscillator directly, step-wise, or in combination. The setting of the oscillator is independent of the expression of its "hands." However, the results obtained with the expression of the *nif* genes indicate that, if any "hand" is allowed to be expressed after the oscillator has been set, the phase of the "hand" then follows the timing of the pre-set oscillator.

In conclusion, *Cyanothece* RF-1 is a unicellular N_2-fixing cyanobacterium that is the first prokaryotic organism proven to exhibit circadian rhythmicity. In particular, nitrogen fixation in RF-1 reveals circadian rhythms when the cultures are placed in continuous light. In addition to nitrogen fixation, a number of other output rhythms have been characterized, such as the uptake rate of several amino acids, the activity of photosynthesis, the abundance of several proteins, and the expression of the *nif* gene.

References

Chen HM, Huang TC, Chien CY (1996a) Nucleotide sequence of the *nifHDK* operon in the aerobic nitrogen-fixing unicellular *Synechococcus* RF-1. Bot Bull Acad Sin 37:99–105

Chen HM, Chien CY, Huang TC (1996b) Regulation and molecular structure of a circadian oscillating protein located in the cell membrane of the prokaryote *Synechococcus* RF-1. Planta 199:520–527

Chen TH, Pen SY, Huang TC (1993) Induction of nitrogen-fixing circadian rhythm in *Synechococcus* RF-1 by light signals. Plant Sci 92:55–59

Chen TS, Chen TL, Hung LM, Huang TC (1991) Circadian rhythm in amino acid uptake by *Synechococcus* RF-1. Plant Physiol 97:55–59

Chou HM, Huang TC (1991) Ultrastructure of the aerobic, nitrogen-fixing unicellular cyanobacterium *Synechococcus* sp. RF-1. Algolog Stud 64:53–59

Chou HM, Chow TJ, Tu J, Wang HR, Chou HC, Huang TC (1989) Rhythmic nitrogenase activity of *Synechococcus* sp. RF-1 established under various light-dark cycles. Bot Bull Acad Sin 30:291–296

Dilworth JJ (1966) Acetylene reduction by nitrogen-fixing preparations from *Clostridium pasteurianum*. Biochim Biophys Acta 127:285–294

Gallon JR (1980) Nitrogen fixation by photoautotrophs. In: Stewart WDP, Gallon JR (eds) Nitrogen fixation. Academic, London, pp 197–238

Grobbelaar N, Huang TC, Lin HY, Chow TJ (1986) Dinitrogen-fixing endogenous rhythm in *Synechococcus* RF-1. FEMS Microbiol Lett 37:173–177

Grobbelaar N, Lin HY, Huang TC (1987) Induction of a nitrogenase activity rhythm in *Synechococcus* and the protection of its nitrogenase against photosynthetic oxygen. Curr Microbiol 15:29–33

Grobbelaar N, Lin WT, Huang TC (1991) Relationship between the nitrogenase activity and dark respiration rate of *Synechococcus* RF-1. FEMS Microbiol Lett 83:99–102

Huang TC, Chou WM (1991) Setting of the circadian N_2-fixing rhythm of the prokaryotic *Synechococcus* sp. RF-1 while its *nif* gene is repressed. Plant Physiol 96:324–326

Huang TC, Chow TJ (1986) New type of N_2-fixing unicellular cyanobacteium (blue-green algae). FEMS Microbiol Lett 36:109–110

Huang TC, Chow TJ (1988) Comparative studies of some nitrogen-fixing unicellular cyanobacteria isolated from rice fields. J Gen Microbiol 134:3089–3097

Huang TC, Chow TJ (1990) Characterization of the rhythmic nitrogen-fixing activity of *Synechococcus* sp. RF-1 at the transcription level. Curr Microbiol 20:23–26

Huang TC, Pen SY (1994) Induction of a circadian rhythm in *Synechococcus* RF-1 while the cells are in a "suspended state". Planta 194:436–438

Huang TC, Tu J, Chow T, Chen TH (1990) Circadian rhythm of the prokaryote *Synechococcus* sp. RF-1. Plant Physiol 92:531–533

Huang TC, Wang ST, Grobbelaar N (1993) Circadian rhythm mutants of the prokaryotic *Synechococcus* RF-1. Curr Microbiol 27:249–254

Huang TC, Chen HM, Pen SY, Chen TH (1994) Biological clock in the prokaryotic *Synechococcus* RF-1. Planta 193:131–136

Huang TC, Lin RF, Chu MK, Chen HM (1999) Organization and expression of nitrogen-fixation genes in the aerobic nitrogen-fixing unicellular cyanobacterium *Synechococcus* sp. strain RF-1. Microbiology 145:743–753

Komárek J (1976) Taxanomic review of the genera *Synechocystis* SAUV. 1892, *Synechococcus* NÄG. 1849, and *Cyanothece* gen. nov. (Cyanophyceae). Arch Protistenk 118:119–179

Lin RF, Chou HM, Huang TC (1999) Priority of light/dark entrainment over temperature in setting the circadian rhythms of the prokaryote *Synechococcus* RF-1. Planta 209:202–206

Lin RF, Tsai KD, Huang TC (2003) Factors affecting the circadian degradation of COP23 in *Synechococcus* RF-1. Bot Bull Acad Sin 44:151–158

Lumsden PJ (1991) Circadian rhythms and phytochrome. Annu Rev Plant Physiol Plant Mol Biol 42:351–371

Mitsui A, Kumazawa S, Takahashi A, Ikemoto H, Cao S, Arai T (1986) Strategy by which nitrogen-fixing unicellular cyanobacteria grow photoautotrophically. Nature 323:720–722

Nagy F, Kay SA, Chou NH (1988) A circadian clock regulates transcription of the wheat *cab*-1 gene. Genes Dev 2:376–382

Reddy KJ, Haskell JB, Sherman DM, Sherman LA (1993) Unicellular, aerobic, nitrogen-fixing cyanobacteria of the genus *Cyanothece*. J Bacteriol 175:1284–1292

Rippka R, Deruelles J, Waterbury JB, Herdman M, Stanier RY (1979) Generic assignments, strain histories, and properties of pure culture of cyanobacteria. J Gen Microbiol 111:1–61

Turner S, Huang TC, Chaw SM (2001) Molecular phylogeny of nitrogen-fixing unicellular cyanobacteria. Bot Bull Acad Sin 42:181–186

Waterbury JB, Rippka R (1989) Order Chrococales Wettstein 1924, emend. Rippka et al. 1979. In: Staley JT, Bryant MP, Pfenning N, Holt JG (eds) Bergey's manual of systematic bacteriology, vol 3. Williams & Wilkins, Baltimore, pp 1728–1746

Yen UC, Huang TC, Yen TC (2004) Observation of the circadian photosynthetic rhythm in cyanobacteria with a dissolved-oxygen meter. Plant Sci 166:949–952

Chapter 4
The Decade of Discovery: How *Synechococcus elongatus* Became a Model Circadian System 1990–2000

Carl Hirschie Johnson and Yao Xu

Abstract The coincidence of good fortune, clever ideas, and hard work has transformed the *Synechococcus elongatus* system into one of the best characterized circadian clock systems, even though it is the newest comer to molecular clock analyses. Only 20 years ago, the consensus among circadian clock researchers was that prokaryotes were incapable of circadian rhythmicity. Now the *S. elongatus* system has caught up with eukaryotic clock systems, and in some areas it is surfing the leading wave of circadian clock research. This chapter is the story of how that happened.

4.1 Before Cyanobacteria (B.C.), it was *Chlamydomonas*

Takao Kondo is not only an excellent biologist, he is also very clever at designing apparatuses and writing computer programs for data acquisition and analysis. It was those skills that first stimulated Carl Johnson's desire to collaborate with Takao on the circadian clock in the eukaryotic alga *Chlamydomonas* in the mid-1980s. At that time, Carl was a postdoctoral student in J.W. ("Woody") Hastings' laboratory and was continuing the development of the *Chlamydomonas* circadian system that had begun with the studies of Victor Bruce (1970). Carl had inherited Victor's apparatus for measuring *Chlamydomonas* phototaxis rhythms, but he was not obtaining clean rhythms from these eukaryotic algae. Upon Carl's first visit to Japan in 1984, he visited Hideaki Nakashima at the National Institute of Basic Biology (NIBB) and Hideaki showed Carl some of the *Chlamydomonas* phototaxis data of his NIBB colleague, Takao Kondo (Takao happened to be out of town at the

C.H. Johnson(*)
Department of Biological Sciences, Vanderbilt University, Nashville, Tennessee 37235, USA, e-mail: carl.h.johnson@vanderbilt.edu

Y. Xu
Department of Biological Sciences, Vanderbilt University, Nashville, Tennessee 37235, USA, e-mail: yao.xu@vanderbilt.edu

J.L. Ditty et al. (eds.), *Bacterial Circadian Programs.*

time of Carl's visit). Takao's phototaxis data were better than those Carl had been collecting, so Carl came back from Japan in a very enthusiastic mood about Takao's data, which demonstrated that excellent rhythms could be measured from *Chlamydomonas*. Carl therefore convinced Woody Hastings to invite Takao to Woody's laboratory for 3 months in 1985 to initiate a collaboration on *Chlamydomonas* circadian rhythms (this visit was when Carl and Takao met for the first time). Takao's visit to the USA was followed by a 3-month visit of Carl to the NIBB in 1986 to work with Takao and Hideaki under the auspices of a Jean and Katsuma Dan Fellowship. The primary projects accomplished during that 1986 visit were: (i) to study the action spectroscopy of light-induced phase-resetting of the *Chlamydomonas* clock with Takao, a project that ultimately resulted in three publications (Johnson et al. 1991; Kondo et al. 1991; Johnson and Kondo 1992) and (ii) to block the phase shift by light using translational inhibitors in *Neurospora* with Hideaki, which led to another publication (Johnson and Nakashima 1990).

4.2 Searching for a New "Model System"

Takao came to Carl's laboratory with his family for a 10-month sabbatical in 1990–1991. Carl assumed that Takao planned to continue our study of rhythms in *Chlamydomonas*, so it was a surprise when Takao announced upon his arrival in the USA that he wanted to search for a new model system for studying circadian systems. Apparently Takao had been conferring with Masahiro Ishiura, who had convinced Takao that an organism with more molecular genetic tools than *Chlamydomonas* would be better for intensive circadian investigation. Therefore, Takao came to the USA to explore the possibility that *Escherichia coli* or yeast might have a circadian clock. Carl does not remember all the different assays that Takao tried on *E. coli* and yeast, but one of his ideas was to look for daily rhythms of sensitivity to ultraviolet (UV) light. Takao did not find rhythms of UV sensitivity in *E. coli* or yeast, but his approach stimulated the later discovery in my laboratory of rhythmic UV sensitivity in *Chlamydomonas* (Nikaido and Johnson 2000). To this day, the existence (or non-existence) of circadian/daily rhythms in *E. coli* or yeast remains an open question.

About halfway through Takao's sabbatical in Carl's laboratory, Carl attended the annual meeting of the American Society for Cell Biology. At this meeting he presented a poster, and by a stroke of luck, the neighboring poster was from the laboratory in Taiwan that had discovered circadian rhythms of nitrogen fixation in the cyanobacterium *Synechococcus* RF-1 (see Chap. 3). That poster's presenter was Tsung-Hsien Chen, who was collaborating with Tan-Chi Huang on the study of *Synechococcus* RF-1. As Carl and Tsung-Hsien started to discuss circadian rhythms in algae while they stood by their posters, Carl forgot all about the rest of the meeting in his excitement about the Taiwanese group's research on cyanobacteria, which was the first persuasive demonstration of circadian rhythms in a prokaryote (Grobbelaar et al. 1986; Johnson et al. 1996). Carl returned to his laboratory and tried to convince Takao that cyanobacteria were the new "model system" to investigate. At first Takao seemed reluctant, but after conferring with Masahiro long-distance, he was reassured that there were enough genetic tools available to cyanobacteriologists to make this organism an excellent system to pursue. Carl and Takao contacted Drs. Huang and Chen to initiate a collaboration, at which point *Synechococcus* RF-1 was mailed to Carl's laboratory. The idea at that time was to clone the promoter for the nitrogenase gene that Dr. Huang had shown to be rhythmically expressed (Huang and Chow 1990), fuse it to a luciferase gene, and introduce it into the organism to create a rhythmically luminescent organism. Based on Carl's experience with measuring rhythms of luminescence in Woody Hastings' laboratory from the endogenously luminescent alga *Gonyaulax* (Johnson et al. 1984), a genetically malleable prokaryote that expressed luminescence rhythms that could be measured for many cycles non-invasively sounded like a winner. The flaw in this plan was that techniques for genetic transformation of *Synechococcus* RF-1 had not been worked out, but we hoped that the methods that had been developed for the transformation of other cyanobacterial species would be successful with *Synechococcus* RF-1.

About a month after receiving the sample of *Synechococcus* RF-1, Carl visited a friend in New York City and decided while there to "drop in" on Steve Kay, who was then a postdoctoral fellow with Nam-Hai Chua at Rockefeller University. At that time, Andrew Millar was a graduate student with Dr. Chua, but Andrew was spending most of his time working together with Steve to develop *Arabidopsis* as an excellent genetic system for elucidating plant circadian clocks. While Carl was visiting Steve and Andrew that day, he mentioned the plans to make a luminescence reporter strain of *Synechococcus* RF-1. Remarkably, Steve and Andrew had already received a sample of *Synechococcus* RF-1 and were underway in the process of making a luminescence reporter strain of this cyanobacterium! (Steve and Andrew were also making a luciferase reporter strain of *Arabidopsis*, which was a project that has been spectacularly successful.) This was very depressing news for Carl and Takao. At that time, neither Carl nor Takao had much experience with molecular genetic techniques, so it was hopeless to compete on the identical approach with Steve and Andrew, who were molecular genetic "jocks."

4.3 Homing in on *S. elongatus*

Though discouraged, Carl and Takao did not give up. They decided to drop further work with *Synechococcus* RF-1 and focus instead upon a cyanobacterium for which molecular genetic techniques had already been developed. The problem was, what to assay as a circadian output in an uncharacterized strain? In *Synechococcus* RF-1, nitrogen fixation or nitrogenase activity were known to be rhythmic, but in a new cyanobacterium it was anybody's guess as to what process might be rhythmic. Because Carl's laboratory was doing a lot of 2-D gel electrophoresis to discover circadian-regulated protein expression in *Chlamydomonas* at that time, they chose to look for rhythmic protein expression in a genetically malleable cyanobacterium. Once found, they reasoned that they could clone a rhythmic protein's promoter, make a luminescent reporter construct, and transform it into the organism, but they expected to be far behind the Kay and Millar team that was using *Synechococcus* RF-1. In retrospect, this episode is reminiscent of advice to scientists from Dr. Efraim Racker, who wrote a book in 1976 about mitochondrial electron transport that included the wise statement that "troubles are good for you" scientifically (as long as you respond to them constructively!; Racker 1976).

In this case, the reason that these troubles were good for Carl and Takao is that they led to contacting Susan Golden at Texas A&M University. To determine the optimal conditions for extracting proteins from cyanobacteria for 2-D gel electrophoresis, Carl called several cyanobacteriologists who encouraged him to call Susan. Susan was working on the regulation of gene expression in response to changes of light intensity in the genetically tractable cyanobacterium *S. elongatus* PCC 7942, and she was a "card-carrying" molecular geneticist. When Carl explained to Susan on the telephone what he and Takao had in mind, Susan casually mentioned that a postdoctoral student in her laboratory, Michael Schaefer, had done a 48-h time course on the abundance of mRNA from the *psbAI* gene, and *psbAI* expression had appeared to show a daily rhythm! (*psbAI* encodes the predominant form of the D1 protein of photosystem II.) In addition, Susan said that Carl Strayer, a technician in her laboratory (who would later clone TOC1 from *Arabidopsis* in Steve Kay's laboratory) had just produced a luminescence reporter strain in which the bacterial luciferase gene set (*luxAB*) was fused to the *psbAI* promoter and transformed into *S. elongatus*. This was a windfall, and it established a Johnson/Kondo/Golden team that was well on its way. (Except that we did not yet know whether *S. elongatus* had a circadian clock; the *psbAI* mRNA data that Michael Schaefer had collected was unpersuasive because of lapses in collection time points when he had slept and because of noise in the traces that was later explainable in terms of post-transcriptional regulation of *psbAI* mRNA that obscures the exquisite rhythm of *psbAI* promoter activity.)

Susan was able to send the P*psbAI*::*luxAB* reporter strain of *S. elongatus* (strain name = AMC149) to Carl and Takao just before Takao's return to Japan at the end of his sabbatical in Carl's laboratory. Takao was departing from the USA via a brief visit to Woody Hastings' laboratory. Woody had a custom apparatus that was designed and built by Dr. Walter Taylor (and later refined by Drs. Hellmuth Broda and Till Roenneberg) for the specific purpose of long-term, continuous, non-invasive measurements of circadian luminescence from *Gonyaulax*. In honor of Walter, this apparatus is affectionately called the "Taylortron" (Fig. 4.1). Takao had an opportunity to collect two days of data from AMC149 in the Taylortron before returning to Japan. Carl's best recollection of those data is shown in Fig. 4.2a. The barest trace of an oscillation could be discerned in those data, but they were far from clearly rhythmic.

Fig. 4.1 Photograph of a "Taylortron" for automated measurement of luminescence rhythms (this particular Taylortron is the one in Carl's laboratory, circa 1992; the original Taylortron is in J.W. Hastings' laboratory at Harvard University). There are 30 positions for monitoring rhythms of luminescence from samples in 20-ml scintillation vials. A computer-controlled cart with a photomultiplier tube moves from position to position and measures luminescence from the bottom of the vials. In this photograph, two vials have been placed on top of the Taylortron, but during measurement cycles, they are placed down inside the Taylortron at the round openings. The cart with the photomultiplier tube is to the right and beyond the vials. A fluorescent lamp illuminates the samples from below. The first Taylortron was created by Dr. Walter Taylor in J.W. Hastings' laboratory and later refined by Drs. Hellmuth Broda and Till Roenneberg. Takao built his own Taylortron in Japan to make the first measurements of robust rhythmicity from cyanobacteria (Kondo et al. 1993; see Fig. 4.2b). In Carl's laboratory, this apparatus has been used to monitor luminescence rhythms from cyanobacteria, but with the addition of very sensitive photon-counting circuitry, it was also used to study circadian rhythms of Ca^{2+} fluxes in plants using the luminescent Ca^{2+} indicator aequorin (Johnson et al. 1995)

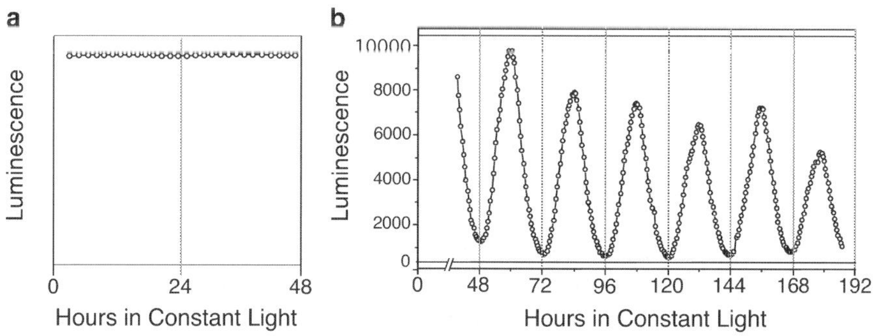

Fig. 4.2 **(a)** Carl's recollection of the first observations of the luminescence "rhythm" of the reporter strain of *S. elongatus* (specifically, the AMC149 reporter strain). These data were collected by Takao using the Taylortron in Woody Hastings' laboratory. **(b)** Luminescence rhythms of AMC149 after Takao's refinement of the methodology using a Japanese version of the Taylortron

4.4 Victory!

Takao was not discouraged by the data obtained in Woody's laboratory and after his return to Japan, Takao constructed a clever dual-channel luminometer that automatically closed a lid for a luminescence measurement and reopened the lid for white light irradiation to keep the photosynthetic cyanobacteria happy. Based on suggestions from Woody about handling the decanal substrate for bacterial luciferase, Takao dissolved decanal in vegetable oil and placed it in a microcentrifuge tube that was inside a closed vial with cyanobacteria in liquid culture. Because decanal is a vapor, it was able to saturate the cyanobacterial culture with decanal (Fig. 4.3). With those innovations, Takao was able to measure rhythms that appeared to be entrainable to light/dark cycles. Encouraged by those results, Takao constructed a Japanese version of the Taylortron and was finally able to measure the beautiful rhythm of *psbAI* promoter activity as assayed with the *luxAB* luminescence reporter (Fig. 4.2b). In retrospect, the combination of the *psbAI* promoter fused to *luxAB* and expressed in *S. elongatus* was a happy coincidence due to Susan's laboratory making the AMC149 strain for a completely different investigation of light-inducible gene expression. Subsequent experiments using other species of cyanobacteria have found rhythms (Aoki et al. 1995), but the reporters in those strains are not bright. And even in *S. elongatus*, many promoters do not show as robust rhythms of luminescence as does P*psbAI* (Liu et al. 1995). The combination of the P*psbAI*::*luxAB* reporter and the *S. elongatus* strain remains one of the most robustly rhythmic combinations in cyanobacteria, even after 15 years of intensive research.

Fig. 4.3 A close-up of two 20-ml vials with a liquid culture of cyanobacteria. A micro-centrifuge tube containing a decanal/oil mixture is placed within each vial to provide a volatile decanal atmosphere inside the vial

Takao's demonstration of robust rhythmicity in AMC149 elicited a flurry of activity in our various labs. Takao was continuing to collect luminescence data to show the salient properties of circadian rhythms: persistence, phase-shifting, and temperature compensation. Susan's laboratory conclusively demonstrated that *psbAI* mRNA levels were rhythmic by the commonly accepted northern blot assay. Carl's laboratory was working with Walter Taylor in Woody Hastings' laboratory to determine that the rhythm of luminescence was at least partly due to changes in luciferase levels (and therefore a reporter of the *psbAI* promoter's activity) and not to rhythmic changes in luciferase's substrates (decanal, O_2, $FMNH_2$). We had a last-minute fright when Takao reported that his cultures of AMC149 were contaminated with fungi – suddenly it seemed possible that all the daily patterns we had observed could be due to the secretion of a rhythmic factor from the eukaryotic fungi that merely stimulated the prokaryotic AMC149 to glow cyclically! If that had been true, all our hopes of circadian rhythms in prokaryotic cyanobacteria were jeopardized. Fortunately, after the fungal contamination was cleaned up, the AMC149 cells still proudly displayed their rhythmic luminescence.

With the story complete, we rushed to assemble a manuscript. We were in a hurry because it seemed possible to publish the first use of a genetically encodable luciferase as a reporter of circadian rhythms (in a prokaryote, no less!), because Kay and Millar had not yet published the use of luciferase as a circadian reporter in plants. We first submitted the manuscript to *Nature*, and then to *Science* – in both cases, the editors did not choose to review the manuscript. In the meantime, Kay and Millar published their first studies using genetically encoded luciferase as a circadian reporter in the plant *Arabidopsis* (Millar et al. 1992a, b). (Although we did not publish the first report of a genetically encoded luciferase used as a circadian reporter, the way it worked out was more fair because Kay and Millar deserved to have the first publication on this topic due to the fact that they had been working on this technology in plants for a longer time.) With the appearance of the Kay and Millar papers, publication of our study became less urgent. After the disappointing experiences with *Science* and *Nature*, we sent the manuscript to *Proceedings of the National Academy of Sciences of the USA* (*PNAS*), where it was enthusiastically accepted and published in 1993 (Kondo et al. 1993). The Kay and Millar team never published a paper about rhythms in *Synechococcus* RF-1, so we assume they abandoned that project after the appearance of our 1993 paper on *S. elongatus*' luminescence rhythms.

4.5 As Time Glows By: Rhythms in Single Colonies and the Triumph of Technology

The 1993 *PNAS* paper established that liquid cultures of *S. elongatus* exhibit circadian rhythms of luminescence. In one respect, this was a seminal report because it used luciferase reporter technology and demonstrated in a single paper all three salient properties of circadian rhythms in a prokaryote (persistence, phase-shifting, and temperature compensation persistence; see Chap. 1). On a more fundamental level, however, our investigation had not made the key discovery of circadian rhythms in prokaryotes – the *Synechococcus* RF-1 studies of Huang and collaborators had that distinction (see Chap. 3). While it might not have been obvious at the time, however, the 1993 *PNAS* paper established a new model system for circadian studies that could go further than most other model systems. To take full advantage of a prokaryotic system for analyzing circadian rhythms, a method to facilitate the identification of mutants was necessary. Therefore, with the first publication finished, Takao and Masahiro Ishiura turned to the task of developing an optimal mutant identification procedure for *S. elongatus* rhythms.

Masahiro encouraged Takao to utilize the standard microbial strategy of mutagenesis followed by screening of colonies growing on Petri dishes. But how to screen? The two keys to a solution were: (i) the luminescence reporter and (ii) Takao's aforementioned talent for designing apparatuses and writing computer programs for data acquisition and analysis. Fortunately, single colonies of AMC149 are luminescent on agar plates (Fig. 4.4). Takao and Masahiro were able to demonstrate proof of principle by using a cooled charge-coupled device (CCD) camera to visualize the rhythms of luminescence of individual AMC149 colonies on agar plates and to demonstrate that those rhythms could be tracked with excellent precision for at least 4 days (later, for many days; Kondo and Ishiura 1994). Not content to observe rhythms from just one plate, however, Takao brought his unique talents to bear and designed and built the turntable-screening apparatus. This apparatus placed twelve 100-mm Petri dishes on a Macintosh computer-operated turntable (Fig. 4.5). The samples were sequentially rotated on the turntable beneath a sensitive CCD camera for 3-min exposures. Takao wrote an elegant program for data acquisition and another program for data analysis. The data acquisition program could automatically locate each colony and keep track of its luminescence rhythm. Up to 1,000 colonies/plate could be monitored and with 12 plates monitored in a single experiment, up to 12,000 colonies could be screened in an assay. Of course,

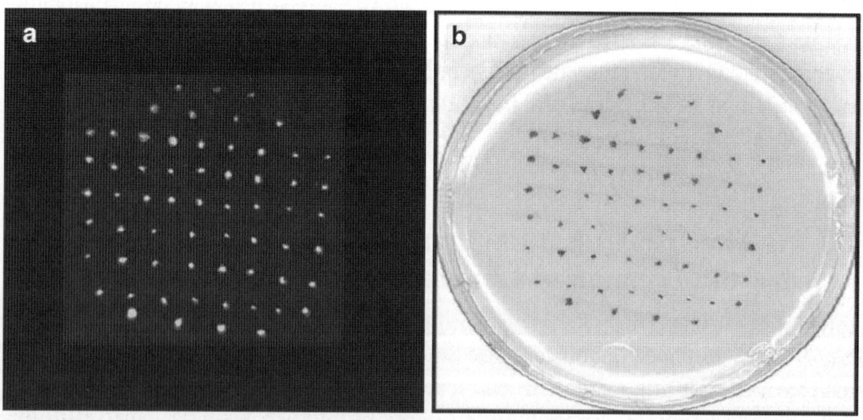

Fig. 4.4 AMC149 colonies are luminescent on agar plates: (**a**) luminescence, (**b**) brightfield

Fig. 4.5 Photograph of a "Kondotron" (this is the Kondotron in Carl's laboratory; the original Kondotron is in Takao's laboratory at Nagoya University). The turntable has 12 positions for 100-mm Petri dishes. The CCD camera is to the right (and just barely out of view), on top of a felt-encased baffle system to exclude incidental light during the imaging of bioluminescent cyanobacterial colonies on the Petri dishes. Three circular fluorescent light fixtures are suspended above the turntable to provide light for photosynthesis to the cyanobacterial plates that are not being imaged. The turntable is rotated by a computer-controlled stepper motor that is underneath the Kondotron and therefore out of view

the circadian assay still required 4–5 days to complete, but the turntable apparatus allowed up to 12,000 potentially mutant colonies to be screened in less than 1 week for unusual phenotypes of period, phase, or amplitude. Our laboratory respectfully calls this turntable-screening apparatus the "Kondotron" in honor of its inventor.

The development of the Kondotron was a breakthrough. It was the first high-throughput screening apparatus for circadian rhythms based on luminescence and enabled Masahiro's molecular genetic expertise to be directed towards developing methods for mutagenesis and complementation of *S. elongatus* that were specifically designed for circadian goals. Masahiro was already an expert of genetic techniques for bacterial and animal cells, and he visited Susan's laboratory for 2 weeks in 1993 to round out his knowledge of genetic techniques that were tailored to *S. elongatus*. With the Kondotron, Takao and Masahiro undertook an extensive mutagenesis project. For these early mutagenesis experiments, they chose to mutagenize chemically with ethyl methanesulfonate (EMS) to create point mutations. This approach led to the isolation of many interesting mutants exhibiting short periods, long periods, and arhythmia (Kondo et al. 1994). The largest range of variation for circadian period mutants for any organism was reported in that 1994 paper: from 16h to 60h (Carl's laboratory also contributed to the isolation of

EMS-provoked circadian mutants, but 98% of this early work came from the Kondo and Ishiura team). Ultimately, the mutagenesis screens identified several *hundred* period mutants, which eventually led to the identification of the central clock gene cluster as described below. The fact that there were so many period mutants led to the hope that lots of central clock genes were involved and that we would be able divide those genes up among the four groups to lessen collaborative overlaps that might lead to competitions (alas, that was not to be).

Identifying mutants is an interesting endeavor in its own right (and the mutants proved to be useful for many projects, including the adaptive significance studies; see Chap. 12; Ouyang et al. 1998; Woelfle et al. 2004), but the main goal of the mutant hunt was ultimately to identify clock genes. In this arena, Masahiro's expertise paid off. He applied genetic complementation to identify the regions of wild-type DNA that would "rescue" the mutants by restoring a wild-type phenotype. Masahiro made a number of complementation libraries and rescued mutants by direct genetic transformation and complementation. This was a low efficiency method, but the Kondotron created enough screening power to allow this approach to be successful. At about the same time, Susan's laboratory was developing a toolbox of other mutagenesis and complementation methods including transposon mutagenesis and random insertion of over-expression promoters (Andersson et al. 2000; Clerico et al. 2007). One particularly clever method was complementation of *S. elongatus* mutants by conjugation with *E. coli* harboring wild-type cyanobacterial DNA on plasmids that could be mobilized (what Carl likes to call "microbial sodomy"). The conjugation method for complementation was of limited usefulness with Kondotron screening for technical reasons (conjugation with *E. coli* led to "messy" plates that were optically poor for Kondotron visualization of *S. elongatus* colonies), but proved to be very helpful in other applications, e.g., in identifying via an insertional mutagenesis screen a circadian role for the sigma factor RpoD2 (Tsinoremas et al. 1996). Susan's laboratory continued to study the involvement of sigma factors in the global gene expression of cyanobacterial genes (Nair et al. 2002).

4.6 Carl's Sabbatical: from a Small Town in Japan to a Small Town in Texas

In 1994–1995, Carl became an "ambassador" of the collaborative team by undertaking a 5-month sabbatical with Takao at the National Institute of Basic Biology (Okazaki, Japan; June–October 1994), followed immediately by a 5-month sabbatical with Susan at Texas A&M University (College Station, Tex., USA; November 1994–March 1995). This back-to-back sabbatical helped to coordinate research efforts between the Japanese and American groups. It also led to the construction of an American Kondotron; stimulated by Takao's generous offer to share the Kondotron technology with Carl, a good portion of this sabbatical was dedicated to Carl learning how to construct and operate a Kondotron, which was assembled in Texas while in Susan's laboratory. On a more personal note (because: (i) Carl grew up in Texas, (ii) Carl's wife is Japanese, and (iii) their son had been born in January 1994), this international sabbatical gave the grandparents, aunts, and uncles in Texas and Tokyo the opportunity to get to know the Japanese/American hybrid progeny.

In 1994, both Takao and Masahiro were Research Associates (roughly the Japanese equivalent of an Assistant Professor) who were associated with different laboratories and were collaborating on circadian clock projects at the NIBB in the small but charming town of Okazaki (birthplace of a famous Shogun, Ieyasu Tokugawa). Two graduate students, Shinsuke Kutsuna and Setsuyuki Aoki, were working with Takao and Masahiro, and these four scientists constituted the circadian biologists at NIBB in 1994. At that time, the four of them were energetically pursuing the screens for clock period mutants and complementation towards identifying the genes responsible for the biological clock in *S. elongatus*. While Carl was in Okazaki in 1994, he had three goals: (i) to learn how to make an American Kondotron, (ii) to study the circadian photobiology of *S. elongatus*, and (iii) to finish the "random library" project that resulted in our surprising discovery of global gene expression by the cyanobacterial clock (Liu et al. 1995).

To accomplish the first goal, Takao trained Carl in the use of the Kondotron software and also tutored Carl in the construction of interface boxes that allowed the Macintosh computer to "talk" to the cooled CCD camera and the stepper motor that controls the turntable upon which the Petri-dish cultures are placed (Fig. 4.5). Takao did not have a schematic of the interface boxes he had constructed previously, and he wanted to keep his two Kondotrons operating continuously. Therefore, in a brief interval between two experiments, we quickly photographed the printed circuit boards of the interface boxes of one of the Kondotrons and then used those photographs as a schematic to solder together the components for the new interface boxes – an unorthodox but effective method!

Carl's second goal was to study the circadian photobiology of *S. elongatus* using action spectroscopy by taking advantage of the NIBB's Okazaki Large Spectrograph, one of the very few large spectrographs in the world that can be used for generating visible light spectra for eliciting wavelength-dependent biological responses. This is the same facility that Carl and Takao had used to study the circadian photobiology of *Chlamydomonas* (Johnson et al. 1991; Kondo et al. 1991; Johnson and Kondo 1992). The luminescence rhythm of *S. elongatus* can only be monitored in a background of illumination (because photosynthesis is needed to endogenously generate one of the substrates for bacterial luciferase, namely reduced $FMNH_2$), and the phase-resetting response of cyanobacterial cells is not very sensitive to light pulses when there is a background of constant illumination. Therefore, at first it was not clear how to measure the spectral response of light-induced phase shifting when light pulses gave very little phase shift. However, a dark pulse in constant white light (LL) does give a strong phase-resetting response. Therefore, Carl and Takao hit upon the idea of using light of different spectra to *reverse* the phase shift elicited by a dark pulse. In this case, there is no phase shift for the white LL control and a big phase shift with the dark pulse; the assay is to determine which spectra of light given instead of the dark pulse will prevent phase-shifting (active spectra) versus which spectra of light will give a phase shift like that of the dark pulse (inactive spectra). Using a 5–6h dark/light treatment protocol, Carl and Takao found that blue and red light were most effective at reversing a dark-induced phase shift. In fact, the action spectrum looked roughly like that expected for photosynthesis (similar to the action spectrum we published for *Chlamydomonas* in LL; Johnson et al. 1991); however, unlike the case for *Chlamydomonas*, inhibitors of photosynthetic electron transport do not seem to affect the phase resetting in *S. elongatus*. The action spectrum data for *S. elongatus* remain unpublished because Takao's laboratory later refined the dark-pulse phase shifting protocol and used it for subsequent action spectra measurements. Therefore, Takao felt that the newer data (also presently unpublished!) superseded the action spectrum data obtained during Carl's sabbatical in 1994.

The third goal of Carl's sabbatical in Japan was to finish the data analyses and write the manuscript for the "random library" experiments that discovered global circadian gene expression in *S. elongatus* (Liu et al. 1995). This project was our best example of a truly collaborative project. Susan had the original inspiration that bacterial luciferase (*luxAB*) could be adapted to a promoter trap protocol to discover how many promoter/enhancer regions in the *S. elongatus* genome were regulated by the circadian clock system. Susan designed a strategy to use a vector that would randomly insert *luxAB* throughout the *S. elongatus* genome by single recombination. This concept exploited the very useful homologous recombination that is characteristic of *S. elongatus*. The single homologous recombination allowed the duplication of the region around the insertion so that the original insertion site was reconstructed, preventing the disruption of the insertional site (although this method does not prevent polar effects). Yi Liu, a graduate student in Carl's laboratory (and later of *Neurospora* clock fame), constructed the vector and in combination with Nikos Tsinoremas, one of Susan's postdoctoral students, made a random library of *S. elongatus* DNA that was inserted into the vector (this sounds easy, but it was not). Then, the "random library" of genomic pieces attached to *luxAB* in the single-recombination vector was shipped to Takao and Masahiro for transformation and screening by the Kondotron. Therefore, each of the laboratories contributed significantly and crucially to the accomplishment of this project. The delightful (and unexpected) result was that whenever *luxAB* inserted close enough to a promoter/enhancer to be expressed (and thereby confer luminescence), the luminescence pattern *always* displayed a circadian modulation (Liu et al. 1995). Even heterologous promoters that were expressed in *S. elongatus* (such as the promoter of *conII* from *E. coli*) were expressed rhythmically. This observation and other work led to the hypothesis that circadian gene expression in *S. elongatus* is at least partially due to clock-regulated changes in chromosomal topology that rhythmically orchestrate global gene patterns (Mori and Johnson 2001; Min et al. 2004; Smith and Williams 2006; Woelfle et al. 2007; but also see Takai et al. 2006). Takao and Masahiro's screening of the "random library" clones was finishing as Carl's sabbatical began and therefore Carl assisted the analyses of those data and wrote the manuscript. Yi Liu visited Japan during that time to help

Fig. 4.6 Photograph of the "Cyanobacterial Circadian Quadrumvirate" taken at Takao's mountain home in Japan in the summer of 1994: Masahiro Ishiura, Susan Golden, Carl Johnson, Takao Kondo

with the analyses and writing (and almost could not return to the USA due to visa problems!). From the perspective of integrated collaborative efforts, the "random library" project was our team's finest hour.

During Carl's sabbatical in Okazaki, Susan also visited Okazaki for a strategy pow-wow of the "Cyano Circadian Quadrumvirate," and a photograph was taken of the foursome at that meeting in 1994 (Fig. 4.6). At the beginning of November 1994, Carl and his family packed their bags and continued the sabbatical odyssey in College Station.The home of Susan's laboratory, College Station (combined with the adjacent Bryan, Tex.) is a relatively small town that is overwhelmed by the huge Texas A&M University. During Christmas holidays when the students are away from the campus, the normally busy streets of College Station are practically deserted.

While Carl was in College Station in 1994–1995, his primary goals were to assemble an American Kondotron and to use it to perform novel screens for clock mutations. At that time, Susan's laboratory had two excellent postdoctoral students who were contributing to the clock projects, Dr. Nikos Tsinoremas (who had been involved in the "random library" project) and Dr. Carol Andersson. While Carl was in Okazaki, a machine shop in Nashville (Tennessee; Carl's home and the location of Vanderbilt University) had been constructing the turntable portion of the Kondotron (slightly modified and improved from Takao's original design). Carl transported the turntable and cooled-CCD camera from Nashville to College Station and began to assemble the Kondotron with the interface boxes he had constructed with Takao. Although there were minor glitches and unexpected problems that needed to be solved, within a few weeks the American Kondotron was operating and ready for screening. Its first application was the screening of an "overexpression" library that Carol had made to provide an alternative approach for discovering clock genes in *S. elongatus*. The concept behind this screen was that clock genes whose function was essential for viability might be undiscoverable by chemical mutagenesis because the mutations obtained would be lethal. However, if those genes were overexpressed, they might have a circadian phenotype without causing lethality. Carol and Susan chose to use the *E. coli* promoter *conII* that is expressed in *S. elongatus* – by random insertion of the *conII* promoter throughout the genome, overexpression of genes would be possible and the site of overexpression would be tagged for identification. Ultimately, this approach led to the discovery of a factor (PsfR) that influenced *psbAI* expression levels, but it has not yet discovered central clock genes (Thomas et al. 2004).

4.7 Getting to the Marrow of the Clock: the *kaiABC* Clock Gene Cluster

As mentioned above, Takao and Masahiro started in 1993 EMS mutagenesis screens using the Kondotron (later, several Kondotrons) on *S. elongatus* transformed with the P*psbAI*::*luxAB* reporter (and later, other reporters; Kondo et al. 1994). Ultimately, several hundred mutants (some contributed by Carl's laboratory for EMS mutagenesis, others from Susan's laboratory by transposon mutagenesis and other methods) were identified that exhibited short periods, long periods, and arhythmia (Kondo et al. 1994). Takao and Masahiro were confident that the many mutants indicated many different loci that contribute to the clock, and expected to divide up the various genes to the participating laboratories for further analyses. To identify which genes had been mutated to result in the various aberrant clock phenotypes, Masahiro made a number of complementation libraries and rescued mutants by direct genetic transformation and complementation. A few of the first mutants to be successfully complemented were several long-period mutants: p30, p38, p48, p60 (exhibiting periods of 30, 38, 48, 60 h, respectively). At this stage, Carl and Susan were "lobbying" for a more active role in the clock mutant project, and were concerned that decisions should be made about dividing up the project if all mutants pointed to the same locus. After some initial resistance, Takao and Masahiro agreed to provide Susan's laboratory with the four plasmids that successfully complemented the p30, p38, p48, and p60 mutants to determine whether any of the sequences were similar among these complementing plasmids (Takao and Masahiro favored the alternative that these four mutations were in four separate genes, so they were pessimistic that there would be overlapping sequences).

As soon as Susan received the four plasmids from Takao and Masahiro, Carol did a quick experiment over a weekend to determine if there was any overlap in hybridization patterns among the complementing plasmids with chromosomal DNA. Carol ran a Southern blot with the chromosomal DNA that had been cut with a variety of restriction enzymes and then probed the blot with each plasmid individually. This experiment was done while Carl was still on sabbatical in Susan's laboratory (1995), so he was there when Carol immediately got the astonishing result that all four plasmids hybridized to exactly the same bands of restriction-cut chromosomal DNA! This was the first indication that central clock mutations in *S. elongatus* map to what is now recognized as the three-gene *kaiABC* cluster. This was both a very exciting scientific result but also foreboding because it meant there would not be a plethora of genes that could be divided up among the four laboratories for further analyses. After Takao consulted with Masahiro about Carol's data, they asked Susan and Carl to stop further analysis of the plasmids (i.e., the *kaiABC* cluster). Takao and Masahiro continued the characterization of the *kaiABC* cluster in Japan without further experimental input from Susan or Carl (both of whom did participate in the writing and editing of the resulting publication of the *kaiABC* cluster; Ishiura et al. 1998).

The three *kai* genes are immediately next to each other and are controlled by two promoters, one for *kaiA* and another that drives *kaiB* and *kaiC* expression as a dicistronic message. Takao and Masahiro named the three gene cluster "kai" for the Japanese word meaning "cycle or rotation number." Takao and Masahiro were finally successful in complementing a number of mutants, 19 of which were reported in the 1998 paper: three mapped to *kaiA*, two mapped to *kaiB*, and 14 mapped to the largest gene, *kaiC* (Ishiura et al. 1998; see Chap. 5). Surprisingly, each of the EMS mutations described in the 1998 paper for *kaiABC* were missense mutations – no nonsense mutations were identified and there is still no explanation for the absence of nonsense mutations because deletion mutant strains in which each *kai* gene singly or the entire *kaiABC* cluster together is deleted are completely viable (albeit arhythmia).

As more graduate and postdoctoral students joined the laboratories of the Quadrumvirate, it was necessary to identify projects for the new personnel. Because there was only one central clock gene cluster, it was difficult to split up further analyses of the *kaiABC* cluster into four equal parts. While there were many experiments to do in the period between the discovery of the *kaiABC* cluster (1995) and its publication in 1998, Takao and Masahiro tended to claim priority for their own laboratory personnel to undertake the standard experimental approaches that had been productive in the investigations of eukaryotic clock systems, such as studies of clock gene expression (mRNA and protein), clock protein interactions as assayed by yeast hybrid methods, clock protein phosphorylation studies, and so on. Therefore, the personnel of Carl's and Susan's laboratories often felt overly restricted in the kinds of experiments that they could perform, and this led to some friction. In some cases, both Japanese and American laboratories followed similar lines of enquiry that led to overlapping but separate publications (e.g., the relationship between the circadian system and the cell division cycle in *S. elongatus*; Mori et al. 1996; Kondo et al. 1997), while in other cases, converging lines of investigation led to publications in which authorship was shared on a single publication between Japanese and American laboratories (e.g., the discovery of the KaiC-interacting kinase, SasA (Iwasaki et al. 2000)).

4.8 "Troubles Were Good For Us": Stimulating the Creative Juices

Although the restrictions upon the laboratory personnel of the Quadrumvirate in the interest of preserving the four-way collaboration were sometimes frustrating, they often had an unexpected benefit. In this case, avoiding overlaps in the experiments done by the four collaborating laboratories led to the necessity to try imaginative ideas and/or to break new ground as far as circadian research was concerned. This stimulation of creativity evoked Efraim Racker's maxim once again that "troubles are good for you" (Racker 1976).

One example of a line of enquiry that benefited from parallel but different approaches and led to a serendipitous discovery was the search for homologs to the *kai* genes in eukaryotes. This was an important question, especially considering the reasonable prediction based on the "endosymbiotic hypothesis" for the evolution of eukaryotes that there might be an evolutionary conservation of clock genes between cyanobacteria and eukaryotic plants and algae. Masahiro and Takao chose to search for *kai* homologs in the genetic model plant *Arabidopsis*, without success. Because of Carl's prior work with *Chlamydomonas*, his laboratory members tested whether there might be *kai* homologs in either the nuclear or chloroplast genomes of this green alga. This test again came up empty-handed (in fact, we now know that the genome of *Chlamydomonas* has few genes that are homologous to plant clock genes, suggesting the possibility that the circadian clock mechanisms in *Chlamydomonas* vs plants vs cyanobacteria were derived independently; Mittag et al. 2005). However, Yao Xu, an experienced plant molecular biologist who joined Carl's laboratory in 1995, was not satisfied to test only *Arabidopsis* and *Chlamydomonas* for *kai* homologs. Therefore, Yao screened a genomic library from tobacco for DNA sequences that hybridized to a *kaiABC* probe. From the entire tobacco genomic library, Yao found only two positively hybridizing clones. These two clones had overlapping sequences and Yao determined that the *kaiABC*-hybridizing sequence was a gene that he named *ZGT* (from the Chinese for "clock- and light-controlled," namely *Z*hong-*G*uang-*T*iaokong; Xu and Johnson 2001). At first, we thought that *ZGT* might be a homolog of *kaiC* because there was approximately 25% similarity between the *ZGT* and *kaiC* sequences. Further work discouraged that interpretation. Nevertheless, *ZGT* became interesting in its own right as a light- and clock-regulated gene that acts as a coupling agent between the central circadian oscillator and output rhythms in plants (Xu and Johnson 2001). If the Kondo/Ishiura labs *versus* the Johnson laboratory had not been pursuing parallel but different approaches to the search for *kai* homologs, *ZGT* may not have been discovered.

4.9 Development of a New Protein Interaction Method: no FRET, do BRET!

Perhaps a clearer example of overlapping goals leading to creative approaches was the development of bioluminescence resonance energy transfer (BRET) as a method to monitor protein interactions. Hideo in Takao's laboratory was using yeast two-hybrid methods to study Kai protein interactions. Carl and his laboratory was also interested in this question, but Hideo and Takao had already "staked out the territory" of using the yeast two-hybrid method. So in order to address the scientific question without jeopardizing the collaboration, Carl's laboratory had to develop a different approach. At about that same time (November 1996), Carl met Dr. Ammasi Periasamy at the University of Virginia and learned about his experiments with fluorescent resonance energy transfer (FRET) as a technique to monitor protein–protein interactions. Ammasi and other investigators were using green fluorescent protein (GFP) variants as genetically encodable fluorophores for FRET experiments. FRET has many advantages for imaging and quantifying protein interactions in vivo and in vitro (Periasamy and Day 2005). However, a liability of FRET is that it requires fluorescence excitation of the sample; the FRET excitation light would be likely to reset the phase of the circadian rhythm in cyanobacteria, causing undesirable perturbations. As Carl was flying back to Nashville from the visit to the University of Virginia, he remembered from his bioluminescence background as a postdoctoral student with Woody Hastings that, in nature, GFP participates in resonance energy transfer with a luciferase (in the case of GFP, with the Ca^{2+}-dependent luciferase aequorin in the jellyfish *Aequorea*). Therefore, Carl had the aerial inspiration to replace the donor fluorophore of the FRET technique with a luciferase. In the presence of a substrate, bioluminescence emanating from the luciferase could potentially excite an acceptor fluorophore if the luciferase and fluorophore were close enough. In this scenario, FRET would become "BRET" and would not require a perturbing fluorescence excitation. But would it work?

We did not want to use aequorin itself as the donor luciferase because we were concerned that the natural aequorin/GFP partnership might involve some interfering protein interaction that would complicate our measurements of interactions between candidate proteins. Therefore, we chose *Renilla* luciferase (RLUC) as the donor luciferase for BRET: its emission spectrum was similar to that of aequorin, but because it comes from a different species of coelenterate (*R. reniformis*), it would be unlikely to interact directly with *Aequorea*'s GFP. RLUC was also advantageous because it has a single polypeptide (MW ~35 kDa); a multiple subunit luciferase like the bacterial luciferase (LuxA/LuxB) that we used as a reporter in cyanobacteria would be difficult to use as a fusion protein for BRET studies. As proof of principle, we first tried to make a positive control where we forced interaction between RLUC and an acceptor fluorophore by genetic fusion. We hoped to find a very significant difference in the spectrum of RLUC versus the RLUC•fluorophore fusion protein. Our initial BRET construct was a fusion of RLUC to GFP, and we expressed the resulting fusion protein in *E. coli* (Fig. 4.7).

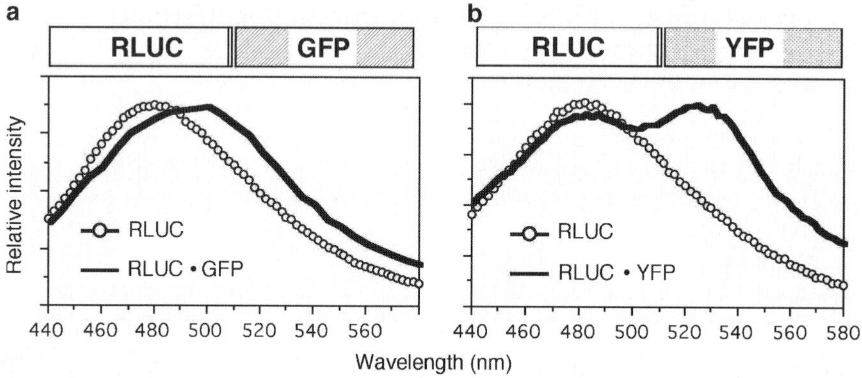

Fig. 4.7 Luminescence spectral profiles of *Renilla* luciferase (RLUC) and fusion proteins of RLUC fused to GFP (RLUC • GFP; **a)** and RLUC fused to YFP (RLUC • YFP; **b)**

Unfortunately, only a slightly detectable spectral shift was added to the RLUC emission spectrum (Fig. 4.7a). This result indicated that there was not enough spectral separation between the luminescence emission of RLUC as compared to the fluorescence of GFP to detect a significant BRET signal.

The fluorescence emission peak of GFP is at 504 nm, but our biophysicist collaborator, David Piston, had just received a new red-shifted mutant of GFP that was not yet commercially available with a fluorescence emission of 527 nm. This mutant was yellow fluorescent protein (YFP). David suggested using the then-new YFP as BRET's acceptor fluorophore in the hope that there would be enough spectral separation between RLUC's emission and YFP's emission to detect a clear BRET signal. Moreover, David reasoned that: (i) the emission spectrum of RLUC was broad enough to provide good excitation of YFP and (ii) the calculated spectral overlap between RLUC and YFP would yield a critical Förster radius (R_0) of ~50 Å, which is a useful distance for molecular interactions. Thus, Yao made an RLUC • YFP construct, and the fusion protein was expressed in *E. coli*. As we hoped, a significant BRET signal was observed between RLUC and YFP both in vivo and in vitro as a second peak of luminescence emission at ~527 nm (Fig. 4.7b; Xu et al. 1999). This result suggested that a significant proportion of the RLUC energy was transferred to YFP by resonance energy transfer and emitted at the characteristic wavelength of YFP (527 nm). We were happy to conclude that RLUC and YFP could be effective BRET partners.

Once we had determined that RLUC and YFP were good BRET partners, we were ready to use BRET as an assay of candidate protein interactions in vivo. Based on analogous experiments using FRET, RLUC was genetically fused to one candidate protein and YFP was fused to another candidate protein. If RLUC and YFP are brought close enough by a putative interaction between the candidate proteins, the bioluminescence energy generated by RLUC can be resonance-transferred to YFP, which then emits yellow light. However, if there is no interaction between the two candidate proteins, RLUC and YFP remain too far apart for significant BRET and only the blue-emitting spectrum of RLUC is detected. Thus, BRET/FRET can be used as a "molecular ruler" to measure distances between proteins that might be interacting (within 50–60 Å) by quantifying the emission ratio at 527 nm/480 nm (480 nm is the peak RLUC emission). We chose the cyanobacterial Kai proteins to test whether BRET could be used as a protein–protein assay in *E. coli*. Yao tested a total of 64 combinations of KaiA, KaiB, or KaiC fusions with RLUC or EYFP, including each combination of N- versus C-terminal fusions. Among these cases, all of the KaiB↔KaiB combinations reproducibly showed a strong BRET signal, and KaiB interactions were also observed in vitro (in extracts of *E. coli* cells; Xu et al. 1999).

Then, we tried to use BRET to test Kai protein interactions over the circadian cycle in *S. elongatus* cells in vivo. However, when RLUC was expressed in *S. elongatus* cells, the emission spectrum of RLUC alone was so broad that any BRET signal that might have been present was obscured by the broad RLUC emission. This broadening of the RLUC emission spectrum in *S. elongatus* cells (which was not observed from *E. coli* cells in vivo) is probably due to the presence of interfering pigments in the cyanobacterial cells. Consequently, BRET was not useful for our original goal of studying clock protein interactions in *S. elongatus* cells in vivo. Nevertheless, BRET has now been successfully applied to many other cell types and scientific problems and is one of the methods of choice for high-throughput screening of G protein-coupled receptor interactions (Soutto et al. 2005; Pfleger and Eidne 2006; Xu et al. 2007). Improvement in the sensitivity of CCD cameras has also recently enabled imaging of BRET signals in single cells, plant seedlings, etc. (Xu et al. 2007). Therefore, the pressure to avoid competition among the cyanobacterial collaborators led to the development of a new way to detect protein interactions that has been of general usefulness. When given lemons, make lemonade!

4.10 Other Examples of Breaking New Ground and Trajectories Beyond 2000

One of the important stories to emerge from the study of cyanobacterial clocks was a rigorous test of adaptive significance. Does this circadian clock actually enhance fitness? This is not a rhetorical question – most of the evidence that supports the adaptiveness of clocks in general is not rigorous and falls into the category of "adaptive storytelling" (Johnson 2005). *S. elongatus* is one of the few systems in which the adaptive significance of circadian programs has been rigorously tested. The background to the concept of this test was that a population biologist named Douglas Taylor interviewed for an Assistant Professorship in Carl's department in the spring of 1994, and during the interview process, Doug and Carl started talking about whether the adaptiveness of circadian clocks had been clearly demonstrated. Doug had been a postdoctoral fellow in the laboratory of Richard Lenski, who is famous for studies on experimental evolution in *E. coli*. Doug enlightened Carl as to the advantages of competing microbial strains against each other in different environmental conditions as a rigorous experimental test of fitness (Doug is now Professor and Chair at the University of Virginia). Hence, the idea was born to test the fitness enhancement conferred by the circadian system by competing cyanobacterial strains with differing circadian characteristics against each other under different light/dark regimes (Ouyang et al. 1998; Woelfle et al. 2004). The experiments were largely done in Carl's laboratory using batch cultures, but Carol Andersson in Susan's laboratory contributed the key observation that log-phase continuous cultures display the same competitive characteristics. These studies are described in Chap. 12, but it is worthy of note that our competition/selection experiments with cyanobacteria were the first example of this approach in the circadian clock field and inspired subsequent investigations in plants (Dodd et al. 2005).

This chapter has focused upon the development of the *S. elongatus* system over the decade 1990–2000, but many of the trajectories begun in that decade were not complete by 2000. Therefore, a brief discussion of further developments is included here. In Susan's laboratory, the need to find exciting projects with minimal overlaps to the projects ongoing in the laboratories of Takao and Masahiro led in other directions. In particular, Susan's laboratory has been instrumental in elucidating genes whose proteins appear to be involved in the entrainment pathway of *S. elongatus*, namely *cikA* and *ldpA* (reviewed by Mackey and Golden 2007; see Chap. 8). Starting with the paper in 1996 on sigma factors (Tsinoremas et al. 1996), Susan's laboratory has continued to provide insights into a role for sigma factors in regulating circadian gene expression in *S. elongatus* (Nair et al. 2002). In the context of understanding the phenomenon of global gene expression (Liu et al. 1995), both Susan's and Carl's laboratories became interested in the possibility that the global rhythm of gene expression might be accompanied by (and possibly regulated by) comprehensive changes in chromosomal topology, supercoiling, and/or compaction (Min et al. 2004; Smith and Williams 2006; Woelfle et al. 2007). In general, of the four laboratories in the Quadrumvirate, Susan's laboratory has led the way to new technologies, especially those related to genetics. For example, Susan's laboratory developed a number of new methodologies for working with *S. elongatus*, such as transposon mutagenesis, transformation by conjugation with *E. coli*, and the development of a self-luminescent strain that no longer requires the decanal vapor method shown in Fig. 4.3 (Andersson et al. 2000). This self-luminescent strain was of particular value because it enabled Susan's laboratory to pioneer the use of the TopCount scintillation counter for high-throughput monitoring of cyanobacterial luminescence rhythms (the TopCount is less amenable to the use of decanal vapor than the Taylortron or the Kondotron). These technological advances culminated in an innovative sequencing/knockout project in which the entire genome of *S. elongatus* was sequenced in parallel with a systematic knockout of each gene to assess effects on circadian rhythms (Holtman et al. 2005).

Other research trajectories established in the 1990–2000 decade by the Quadrumvirate have extended into the 2000s. Takao and Hideo have largely led the way with the biochemical analyses of the cyanobacterial clockwork in elucidating Kai protein interactions and KaiC phosphorylation (Iwasaki et al. 1999, 2002), concluding with the demonstration that a transcriptional/translational feedback loop

(TTFL) is *not* necessary to the core clock mechanism in *S. elongatus* (Tomita et al. 2005) and the spectacular demonstration that purified KaiA, KaiB, and KaiC could reconstitute a temperature-compensated circadian oscillation in vitro (Nakajima et al. 2005; see Chap. 5). Realizing that the ultimate explanation for the mechanism of circadian oscillators will require characterizing the structures of the molecular components of circadian clocks in addition to their interactions to finally elucidate their functions, the laboratories of Carl, Susan, and Masahiro initiated the first successful foray into the structural biology of circadian clock proteins. Susan and Masahiro (and their collaborators) focused upon KaiA structure and function (Williams et al. 2002; Vakonakis et al. 2004; Uzumaki et al. 2004), Masahiro elucidated KaiB's tetrameric structure (Iwase et al. 2005), and Carl and his collaborators solved the structure of KaiC, including the first structural determination of its phosphorylation sites (Pattanayek et al. 2004; Xu et al. 2004), while Masahiro's laboratory has contributed additional key information about KaiC's structure and subunit interactions (Hayashi et al. 2004, 2006). Along with structural studies from other laboratories, these studies have established cyanobacteria as the first circadian system for which the 3-D structures of all core clock proteins have been derived (see Chaps. 6, 7).

It has been a great experience and a privilege to be key players in the development of a system that was initiated by a chance conversation with Tsung-Hsien Chen at a poster session and has catapulted cyanobacteria to the status of being a major model system for the study of circadian rhythms. The experience has had its frustrations and heartaches, coupled with delightfully unexpected twists and turns. As in a statement attributed (perhaps incorrectly) to Albert Einstein: "If we knew what we were doing, it wouldn't be called research, would it?"

Acknowledgements This "history" is based on our recollections, which are probably unconsciously biased. It is probably relatively accurate as regards the Johnson laboratory, but as it touches upon the activities of other laboratories, this history is undoubtedly colored. Therefore, dear reader, please take this history as it is intended – as a panorama of reminiscences from our perspective. We thank our collaborators and mentors (Martin Egli, Phoebe Stewart, Hassane Mchaourab, David Piston, Michael Cox, Ross Inman, Rekha Pattanayek, Dewight Williams, Hideo Iwasaki, Mark Byrne, Woody Hastings, Michael Menaker, Colin Pittendrigh) and our present/former laboratory members (Tetsuya Mori, Yi Liu, Yan Ouyang, Mark Woelfle, Vladimir Podust, Ximing Qin) for stimulating, encouraging, assisting, cajoling, and berating us. We especially thank the other members of the "Quadrumvirate" (Takao Kondo, Susan Golden, Masahiro Ishiura) for an exciting collaboration that was made possible by good science and friendship. Moreover, Susan Golden gave helpful suggestions and corrections on a draft of this chapter. Finally, we are grateful for the research support from the National Science Foundation, the Human Frontiers of Science Program, and the National Institute of General Medical Science that has made this project possible.

References

Andersson CR, Tsinoremas NF, Shelton J, Lebedeva NV, Yarrow J, Min H, Golden SS (2000) Application of bioluminescence to the study of circadian rhythms in cyanobacteria. Methods Enzymol 305:527–542

Aoki S, Kondo T, Ishiura M (1995) Circadian expression of the *dnaK* gene in the cyanobacterium *Synechocystis* sp. strain PCC 6803. J Bacteriol 177:5606–5611

Bruce VG (1970) The biological clock in *Chlamydomonas reinhardtii*. J Protozool 17:328–334

Clerico EM, Ditty JL, Golden SS (2007) Specialized techniques for site-directed mutagenesis in cyanobacteria. In: Rosato E (ed) Methods in molecular biology, vol 362. Humana, Totowa, pp 155–171

Dodd AN, Salathia N, Hall A, Kévei E, Tóth R, Nagy F, Hibberd JM, Millar AJ, Webb AA (2005) Plant circadian clocks increase photosynthesis, growth, survival, and competitive advantage. Science 309:630–633

Grobbelaar N, Huang T-C, Lin HY, Chow TJ (1986) Dinitrogen-fixing endogenous rhythm in *Synechococcus* RF-1. FEMS Microbiol Lett 37:173–177

Hayashi F, Itoh N, Uzumaki T, Iwase R, Tsuchiya Y, Yamakawa H, Morishita M, Onai K, Itoh S, Ishiura M (2004) Roles of two ATPase-motif-containing domains in cyanobacterial circadian clock protein KaiC. J Biol Chem 50:52331–52337

Hayashi F, Iwase R, Uzumaki T, Ishiura M (2006) Hexamerization by the N-terminal domain and intersubunit phosphorylation by the C-terminal domain of cyanobacterial circadian clock protein KaiC. Biochem Biophys Res Comm 318:864–872

Holtman CK, Chen Y, Sandoval P, Gonzales A, Nalty MS, Thomas TL, Youderian P, Golden SS (2005) High-throughput functional analysis of the *Synechococcus elongatus* PCC 7942 genome. DNA Res 12:103–115

Huang T-C, Chow T-J (1990) Characterization of the rhythmic nitrogenase activity of *Synechococcus* sp. RF-1 at the transcription level. Curr Microbiol 20:23–26

Ishiura M, Kutsuna S, Aoki S, Iwasaki H, Andersson CR, Tanabe A, Golden SS, Johnson CH, Kondo T (1998) Expression of a gene cluster *kaiABC* as a circadian feedback process in cyanobacteria. Science 281:1519–1523

Iwasaki H, Taniguchi Y, Ishiura M, Kondo T (1999) Physical interactions among circadian clock proteins KaiA, KaiB and KaiC in cyanobacteria. EMBO J 18:1137–1145

Iwasaki H, Williams SB, Kitayama Y, Ishiura M, Golden SS, Kondo T (2000) A *kaiC*-interacting sensory histidine kinase, SasA, necessary to sustain robust circadian oscillation in cyanobacteria. Cell 101:223–233

Iwasaki H, Nishiwaki T, Kitayama Y, Nakajima M, Kondo T (2002) KaiA-stimulated KaiC phosphorylation in circadian timing loops in cyanobacteria. Proc Natl Acad Sci USA 99: 15788–15793

Iwase R, Imada K, Hayashi F, Uzumaki T, Morishita M, Onai K, Furukawa Y, Namba K, Ishiura M (2005) Functionally important substructures of circadian clock protein KaiB in a unique tetramer complex. J Biol Chem 280:43141–43149

Johnson CH (2005) Testing the adaptive value of circadian systems. Methods Enzymol 393:818–837

Johnson CH, Kondo T (1992) Light pulses induce "singular" behavior and shorten the period of the circadian phototaxis rhythm in the CW15 strain of *Chlamydomonas*. J Biol Rhythms 7:313–327

Johnson CH, Nakashima H (1990) Cycloheximide inhibits light-induced phase-shifting of the circadian clock in *Neurospora*. J Biol Rhythms 5:159–167

Johnson CH, Roeber JF, Hastings JW (1984) Circadian changes of enzyme concentration account for rhythm of enzyme activity in *Gonyaulax*. Science 223:1428–1430

Johnson CH, Kondo T, Hastings JW (1991) Action spectrum for resetting the circadian phototaxis rhythm in the CW15 strain of *Chlamydomonas*. II. Illuminated cells. Plant Physiol 97:1122–1129

Johnson CH, Knight MR, Kondo T, Masson P, Sedbrook J, Haley A, Trewavas A (1995) Circadian oscillations of cytosolic and chloroplastidic free calcium in plants. Science 269:1863–1865

Johnson CH, Golden SS, Ishiura M, Kondo T (1996) Circadian clocks in prokaryotes. Mol Microbiol 21:5–11

Kondo T, Ishiura M (1994) Circadian rhythms of cyanobacteria: monitoring the biological clocks of individual colonies by bioluminescence. J Bacteriol 176:1881–1885

Kondo T, Johnson CH, Hastings JW (1991) Action spectrum for resetting the circadian phototaxis rhythm in the CW15 strain of *Chlamydomonas*. I. Cells in darkness. Plant Physiol 95:197–205

Kondo T, Strayer CA, Kulkarni RD, Taylor W, Ishiura M, Golden SS, Johnson CH (1993) Circadian rhythms in prokaryotes: luciferase as a reporter of circadian gene expression in cyanobacteria. Proc Natl Acad Sci USA 90:5672–5676

Kondo T, Tsinoremas NF, Golden SS, Johnson CH, Kutsuna S, Ishiura M (1994) Circadian clock mutants of cyanobacteria. Science 266:1233–1236

Kondo T, Mori T, Lebedeva NV, Aoki S, Ishiura M, Golden SS (1997) Circadian rhythms in rapidly dividing cyanobacteria. Science 275:224–227

Mackey SR, Golden SS (2007) Winding up the cyanobacterial circadian clock. Trends Microbiol 15:381–388

Millar AJ, Short SR, Hiratsuka K, Chua N-H, Kay SA (1992a) Firefly luciferase as a reporter of regulated gene expression in plants. Plant Mol Biol Rep 10:324–337

Millar AJ, Short SR, Chua N-H, Kay SA (1992b) A novel circadian phenotype based on firefly luciferase expression in transgenic plants. Plant Cell 4:1075–1084

Min H, Liu Y, Johnson CH, Golden SS (2004) Phase determination of circadian gene expression in *Synechococcus elongatus* PCC 7942. J Biol Rhythms19:103–112

Mittag M, Kiaulehn S, Johnson CH (2005) The circadian clock in *Chlamydomonas reinhardtii*: What is it for? What is it similar to? Plant Physiol 137:399–409

Mori T, Binder B, Johnson CH (1996) Circadian gating of cell division in cyanobacteria growing with average doubling times of less than 24 hours. Proc Natl Acad Sci USA 93:10183–10188

Mori T, Johnson CH (2001) Circadian programming in cyanobacteria. Semin Cell Dev Biol 12:271–278

Nair U, Ditty JL, Min H, Golden SS (2002) Roles for sigma factors in global circadian regulation of the cyanobacterial genome. J Bacteriol 184:3530–3538

Nakajima M, Imai K, Ito H, Nishiwaki T, Murayama Y, Iwasaki H, Oyama T, Kondo T (2005) Reconstitution of circadian oscillation of cyanobacterial KaiC phosphorylation in vitro. Science 308:414–415

Nikaido SS, Johnson CH (2000) Daily and circadian variation in survival from ultraviolet radiation in *Chlamydomonas reinhardtii*. Photochem Photobiol 71:758–765

Ouyang Y, Andersson CR, Kondo T, Golden SS, Johnson CH (1998) Resonating circadian clocks enhance fitness in cyanobacteria. Proc Natl Acad Sci USA 95:8660–8664

Pattanayek R, Wang J, Mori T, Xu Y, Johnson CH, Egli M (2004) Visualizing a circadian clock protein: crystal structure of KaiC and functional insights. Mol Cell 15:375–388

Periasamy A, Day RN (eds) (2005) Molecular imaging: FRET microscopy and spectroscopy. Oxford University Press, New York

Pfleger KD, Eidne KA (2006) Illuminating insights into protein–protein interactions using bioluminescence resonance energy transfer (BRET). Nat Methods 3:165–174

Racker E (1976) A new look at mechanisms in bioenergetics. Academic, New York

Smith RM, Williams SB (2006) Circadian rhythms in gene transcription imparted by chromosome compaction in the cyanobacterium *Synechococcus elongatus*. Proc Natl Acad Sci USA 103:8564–8569

Soutto M, Xu Y, Johnson CH (2005) Bioluminescence RET (BRET): techniques and potential. In: Periasamy A, Day RN (eds) Molecular imaging: FRET microscopy and spectroscopy. Oxford University Press, New York, pp 260–271

Takai N, Nakajima M, Oyama T, Kito R, Sugita C, Sugita M, Kondo T, Iwasaki H (2006) A KaiC-associating SasA–RpaA two-component regulatory system as a major circadian timing mediator in cyanobacteria. Proc Natl Acad Sci USA 103:12109–12114

Thomas C, Andersson CR, Canales SR, Golden SS (2004) PsfR, a factor that stimulates *psbAI* expression in the cyanobacterium *Synechococcus elongatus* PCC 7942. Microbiology 150:1031–1040

Tomita J, Nakajima M, Kondo T, Iwasaki H (2005) No transcription–translation feedback in circadian rhythm of KaiC phosphorylation. Science 307:251–254

Tsinoremas NF, Ishiura M, Kondo T, Andersson CR, Tanaka K, Takahashi H, Johnson CH, Golden SS (1996) A sigma factor that modifies the circadian expression of a subset of genes in cyanobacteria. EMBO J 15:2488–2495

Uzumaki T, Fujita M, Nakatsu T, Hayashi F, Shibata H, Itoh N, Kato H, Ishiura M (2004) Crystal structure of the C-terminal clock-oscillator domain of the cyanobacterial KaiA protein. Nat Struct Mol Biol 11:623–631

Vakonakis I, LiWang AC (2004) Structure of the C-terminal domain of the clock protein KaiA in complex with a KaiC-derived peptide: implications for KaiC regulation. Proc Natl Acad Sci USA 101:10925–10930

Williams SB, Vakonakis I, Golden SS, LiWang AC (2002) Structure and function from the circadian clock protein KaiA of *Synechococcus elongatus*: a potential clock input mechanism. Proc Natl Acad Sci USA 99:15357–15362

Woelfle MA, Ouyang Y, Phanvijhitsiri K, Johnson CH (2004) The adaptive value of circadian clocks: An experimental assessment in cyanobacteria. Current Biol 14:1481–1486

Woelfle MA, Xu Y, Qin X, Johnson CH (2007) Circadian rhythms of superhelical status of DNA in cyanobacteria. Proc Natl Acad Sci USA 104:18819–18824

Xu Y, Johnson CH (2001) A clock- and light-regulated gene that links the circadian oscillator to *LHCB* gene expression. Plant Cell 13:1411–142

Xu Y, Piston D, Johnson CH (1999) A bioluminescence resonance energy transfer (BRET) system: application to interacting circadian clock proteins. Proc Natl Acad Sci USA 96:151–156

Xu Y, Mori T, Pattanayek R, Pattanayek S, Egli M, Johnson CH (2004) Identification of key phosphorylation sites in the circadian clock protein KaiC by crystallographic and mutagenetic analyses. Proc Natl Acad Sci USA 101:13933–13938

Xu X, Soutto M, Xie Q, Servick S, Subramanian C, von Arnim A, Johnson CH (2007) Imaging protein interactions with BRET in plant and mammalian cells and tissues. Proc Natl Acad Sci USA 104:10264–10269

Chapter 5
The Kai Oscillator

Tokitaka Oyama and Takao Kondo

Abstract Reconstitution of a circadian oscillator in a test tube marked an epoch in the field of chronobiology. Combining the three clock proteins (KaiA, KaiB, KaiC) with ATP is sufficient to generate a robust circadian rhythm. ATP hydrolysis by KaiC, phosphorylation/dephosphorylation of KaiC, and interactions among the Kai proteins can be observed using this oscillatory system. The ATPase activity of KaiC is the foundation of the system, functioning to determine the period length and underlying its temperature independence. An ordered cycle of phosphorylation and dephosphorylation reactions in the KaiC protein, as well as a dynamic protein–protein interaction profile mediated by the changes in the KaiC phosphorylation status, create the robust oscillation. Importantly, the Kai oscillator is likely to work in cyanobacterial cells to synchronize various cellular activities to an approximate daily cycle.

5.1 Introduction

After the three essential clock genes *kaiA*, *kaiB*, and *kaiC* were identified in *Synechococcus elongatus* PCC 7942 (Ishiura et al. 1998), their functions were intensively analyzed by examining various properties of circadian rhythms, including autonomous oscillation, period lengths of approximately 24 h, robustness to ambient perturbations (e.g., temperature compensation of the period), and entrainability (Williams 2007). In these studies, an excess of KaiC in cyanobacterial cells was found to reduce the expression of the *kaiBC* operon, indicating that

T. Oyama(✉)
Kyoto University, Graduate School of Science, Department of Botany, Kitashirakawa-Oiwake-cho, Sakyo-ku, Kyoto, 606–8502, Japan, e-mail: oyama@cosmos.bot.kyoto-u.ac.jp

T. Kondo
Nagoya University, Graduate School of Science, Division of Biological Science, Furo-cho, Chikusa-ku, Nagoya, Aichi 464–8602, Japan, e-mail: kondo@bio.nagoya-u.ac.jp

J.L. Ditty et al. (eds.), *Bacterial Circadian Programs.*

this process is regulated by a transcription/translation negative feedback loop (Ishiura et al. 1998). The negative feedback could have formed a transcription/translation-based oscillator, although how it maintained a stable periodicity in response to changes in culture conditions that markedly affected cellular transcriptional and translational activities was unclear. The molecular features of each gene product were then examined. KaiC was found to be the central oscillator component with two ATPase domains, as well as autophosphorylation and autodephosphorylation activities (Nishiwaki et al. 2000; Iwasaki et al. 2002; Kitayama et al. 2003; Xu et al. 2003). KaiA acts to increase the phosphorylation of KaiC, whereas KaiB inhibits the function of KaiA. The KaiC phosphorylation level in cyanobacterial cells exhibits a circadian oscillation that depends on KaiA and KaiB. Although these findings suggested that a complex network underlies the feedback regulation of *kaiBC* gene expression, KaiC phosphorylation, and interactions among the Kai proteins, they did not explain the basis of the properties of the circadian rhythm.

A breakthrough came when the KaiC phosphorylation rhythm, including temperature compensation, was observed in prolonged darkness (Tomita et al. 2005).

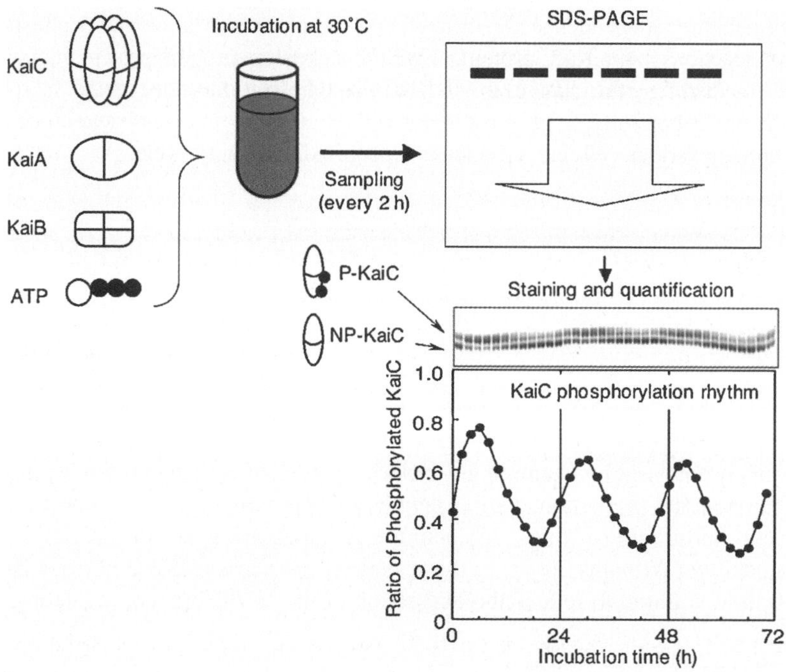

Fig. 5.1 Reconstitution of the KaiC phosphorylation rhythm in vitro. The three Kai proteins (KaiA, KaiB, KaiC) were purified from recombinant *E. coli* cells. The proteins were mixed with ATP and incubated at 30°C. Aliquots were taken every 2 h and subjected to SDS-PAGE. The gel was stained with Coomassie brilliant blue, and the bands of phosphorylated KaiC (P-KaiC) and unphosphorylated KaiC (NP-KaiC) were detected and quantified (Nakajima et al. 2005). The P-KaiC contains the three sorts of phosphorylation states of KaiC (see Fig. 5.2). The ratios of P-KaiC to total KaiC are plotted

The rhythm continued under conditions in which neither transcription of *kaiBC* nor translation of the protein products was permitted. The results suggested that the pacemaker of the cyanobacterial circadian system was not the transcription/translation-based oscillator of *kai* gene expression, but was instead posttranslational machinery that drove the KaiC phosphorylation rhythm. Moreover, recombinant Kai proteins showed features that would likely be required for the phosphorylation rhythm. Namely, KaiA and KaiB can regulate phosphorylation of KaiC positively and negatively, respectively, and both the phosphorylation and dephosphorylation rates showed temperature compensation in vitro (Tomita et al. 2005). Finally, the circadian oscillator was successfully reconstituted in a test tube by incubating the three Kai proteins with ATP (Fig. 5.1; Nakajima et al. 2005). This in vitro Kai oscillation system has enabled direct examination of the physical and chemical properties of the circadian rhythm. This chapter summarizes recent studies about the mechanisms underlying the Kai oscillator.

5.2 Phosphorylation and Dephosphorylation of KaiC

5.2.1 Two Phosphorylation Sites in KaiC and Stepwise Autokinase and Autophosphatase Reactions

Two phosphorylation sites (serine 431, threonine 432) were experimentally identified in the KaiC protein (Nishiwaki et al. 2004; Xu et al. 2004). The four phosphorylation forms of KaiC (S/T, unphosphorylated; pS/T, S431-phosphorylated; S/pT, T432-phosphorylated; pS/pT, S431- and T432-phosphorylated) can be resolved using SDS-PAGE (Fig. 5.2; Nishiwaki et al. 2007). The relative abundance of each form shows a circadian rhythm with the peak levels of the various forms occurring at different times throughout the day. The order of their peaks was found to be S/T, S/pT, pS/pT, and pS/T, before returning to that of S/T. The S/T dominant phase in vitro appears to correspond to the circadian time 4 (CT4), during which the phosphorylation level of KaiC in the cyanobacterial cells is the smallest (Iwasaki et al. 2002; Tomita et al. 2005). Mutational analysis to determine the relationship between the phosphorylation forms and order of the reactions strongly suggested that this process was programmed in the KaiC molecule itself. Substitutions of the amino acids at the KaiC phosphorylation sites to alanine (A) or aspartate (D)/glutamate (E) were made to mimic the dephosphorylated or phosphorylated states, respectively; these variants were then assayed for their phosphorylation and dephosphorylation activities (Nishiwaki et al. 2007). KaiC-T432E, which mimicked the S/pT form, increased the phosphorylation level at the S431 phosphorylation site, whereas KaiC-S431D, which mimicked the pS/T form, reduced the phosphorylation level at T432. These phenomena were observed irrespective of KaiA, the activator of KaiC phosphorylation. Thus, it is likely that the phosphothreonine moiety at amino-acid position 432 of KaiC induces the phosphorylation of S431, and that the phosphoserine moiety at position 431 inhibits the phosphorylation of

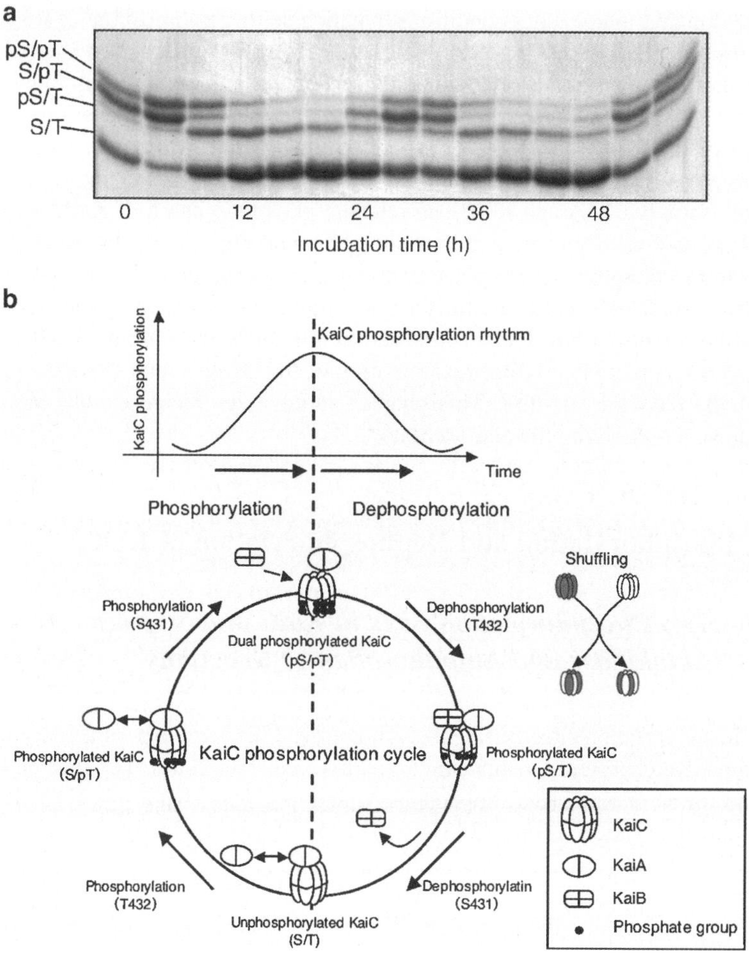

Fig. 5.2 Molecular behaviors observed in the KaiC phosphorylation cycle. (**a**) Four phosphorylation states of KaiC detected using modified SDS-PAGE. The procedures are as described in Fig. 5.1, except the sampling interval was 4 h (Nishiwaki et al. 2007). (**b**) A model of the KaiC phosphorylation cycle. Details are provided in the main text

T432. These results also suggest that doubly phosphorylated KaiC (pS/pT) is first dephosphorylated at the T432 residue.

Other KaiC mutants (KaiC-S431A, KaiC-T432A) were assayed for their phosphorylation rates. While the autophosphorylation activities of both mutant KaiC proteins were enhanced by the addition of KaiA, the phosphorylation rate of KaiC-S431A was eightfold larger than that of KaiC-T432A, which suggests that unphosphorylated KaiC (S/T) is preferentially autophosphorylated at the T432 residue. Taken together these results indicated that, after KaiA promotes the phosphorylation of T432, the four-step phosphorylation cycle proceeds due to mechanisms within the KaiC molecule itself. Similar conclusions regarding the reactions and phosphorylation forms were drawn from kinetic studies of the in vitro KaiC phosphorylation rhythm and a mathematical model (Rust et al. 2007; see Chap. 16).

5.2.2 *Dynamic Formation of Kai Protein Complexes*

Although the KaiC protein per se appears to be able to carry out the ordered phosphorylation and dephosphorylation reactions, a robust phosphorylation rhythm requires interactions among the Kai proteins. Dynamic complex formation has been detected in solution using such methods as immunoprecipitation, gel filtration chromatography, native gel electrophoresis, electron microscopy, and small-angle X-ray scattering (Kageyama et al. 2006; Mori et al. 2007; Nishiwaki et al. 2007; Akiyama et al. 2008). Although estimates of the ratios of the complexes vary among these studies, the general conclusions about the process are consistent: phosphorylated forms of KaiC affect the interactions (Fig. 5.2).

The complex of KaiA, KaiB, and KaiC reaches a maximum level during the middle of the dephosphorylation phase; a small percentage of the KaiC molecules bind to KaiA throughout the cycle, whereas KaiB binds to KaiC preferentially during the dephosphorylation phase. The fact that S431-phosphorylated KaiC was the most prevalent form during the dephosphorylation phase suggested that this form binds to KaiB. This was directly tested in binding assays using KaiC variants bearing mutations at the phosphorylation sites (Kageyama et al. 2006; Nishiwaki et al. 2007). Both KaiC-S431D and KaiC-S431D/T432E showed a strong binding activity to KaiB, whereas KaiC-S431A and KaiC-S431A/T432A did not. Therefore, the phosphoserine at position 431, which was mimicked by aspartate in the mutant proteins, appears to mediate KaiC–KaiB binding. Interestingly, the abilities of the KaiC mutants to bind KaiA increased in the presence of KaiB, suggesting the KaiC–KaiB complex increases the affinity of KaiC for KaiA (Nishiwaki et al. 2007). This difference in the affinities of KaiC and the KaiC–KaiB complex may explain the inactivity of KaiA during the dephosphorylation phase and the reactivation of KaiA at the beginning of the phosphorylation phase.

5.3 ATPase Activity of KaiC as the Basic Timing Mechanism of the Circadian Clock

5.3.1 *Low ATPase Activity of KaiC and its Circadian Oscillation*

The energy source that drives the in vitro Kai oscillator is ATP. The KaiC protein consists of a duplicate pair (CI, CII) of RecA/DnaB type ATPase domains. ATP hydrolysis by KaiC appears to be the only step that supplies energy to this system. Precise quantification of the rate of ATP consumption during the reaction revealed several unique properties of the KaiC ATPase (Terauchi et al. 2007). First, the rate

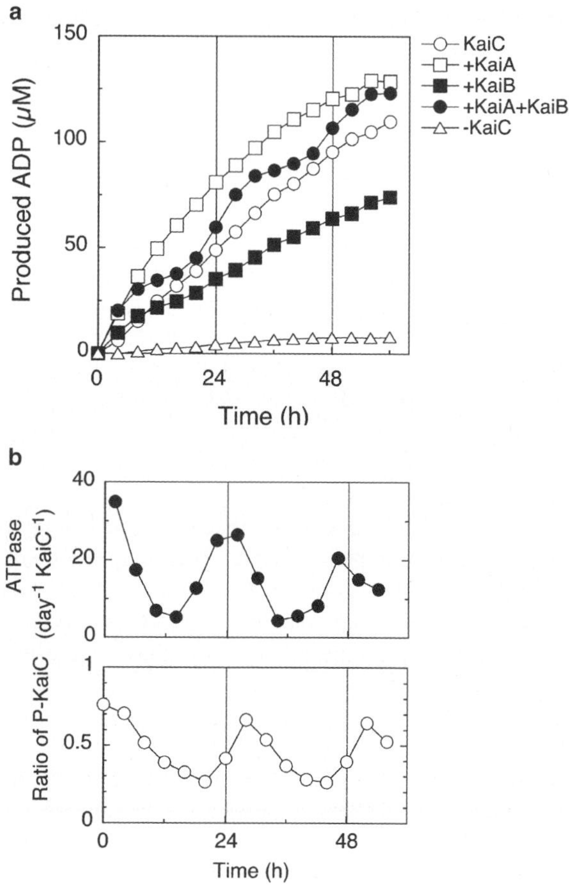

Fig. 5.3 ATPase activity of KaiC. (**a**) Increase in the level of ADP due to the ATPase reaction of KaiC protein. The phosphorylation is included in the reaction in part. KaiC was incubated with or without KaiA and/or KaiB at 30°C and the amount of ADP in the solution was measured by HPLC at 4-h intervals. (**b**) The rhythmic ATPase activity of KaiC determined using the data (+KaiA +KaiB; *upper panel*) shown in (**a**). The KaiC phosphorylation rhythm is also shown (*bottom panel*; Terauchi et al. 2007). The ratios of P-KaiC to total KaiC are plotted

of ATP hydrolysis is extremely low: ~15 molecules/day for each KaiC protein (Fig. 5.3). This rate is much lower than those of other ATPases in the same protein family: RuvB helicase of *Escherichia coli* hydrolyzes 8 × 10^3 molecules/day even under inactive conditions that lack its substrate (Marrione and Cox 1995).

The ATPase activity of KaiC is doubled by adding KaiA to the mixture (~30 molecules/day; Fig. 5.3). Because KaiC autophosphorylation also increases in response to KaiA, the mechanisms regulating these two activities are likely to be related. In contrast, KaiB can reduce the KaiC ATPase activity by ~40% to approximately 9 molecules/day. Because inhibition of the autophosphorylation activity of KaiC by KaiB was observed only when KaiA was included in the reaction mixture, KaiB was thought to function solely as an inhibitor of KaiA (Iwasaki et al. 2002; Kitayama et al. 2003); however, KaiB has been shown to directly affect KaiC function. During the in vitro reaction, the ATPase activity of KaiC shows a robust circadian rhythm, as does KaiC phosphorylation (Fig. 5.3). During the cycle, the ATPase rate alternates between ~30 and ~5 molecules/day. These numbers were similar to those observed with the mixture of KaiC and KaiA, and with the mixture of KaiC and KaiB, respectively. The peak of the ATPase activity was detected ~4 h before the highest phosphorylation level of KaiC and was similar to the observed peak in kinase activity (the phosphate uptake activity), which indicates a close relationship between ATP hydrolysis and the phosphorylation of KaiC.

The crystal structure of the KaiC hexamer demonstrates that the side-chains of the phosphorylation sites (S432, T431) are located on the border of the ATP-binding site in the CII domain (Xu et al. 2004). The coupling of the kinase and ATPase activities implies that the CII domain may be responsible for all of the KaiC ATPase activity. A truncated variant of KaiC containing the CI domain without CII, however, showed ~70% of the ATPase activity of intact KaiC, indicating these two activities can be uncoupled.

5.3.2 *Temperature Compensation of the ATPase Activity of KaiC*

Temperature compensation of the period length is a basic feature of circadian rhythms. The elementary reactions that underlie the KaiC phosphorylation rhythm, i.e., phosphorylation and dephosphorylation, are temperature-insensitive even under non-oscillatory conditions (Tomita et al. 2005). The ATPase activity of KaiC is also temperature-insensitive even under non-oscillatory conditions without KaiA and KaiB (Fig. 5.4). Furthermore, two unphosphorylatable KaiC variants (KaiC-S431A/T432A, which mimics unphosphorylated KaiC; KaiC-S431D/T432E, which mimics doubly phosphorylated KaiC) showed temperature insensitive ATPase activities. These results indicate that temperature compensation can be achieved without the phosphorylation and dephosphorylation reactions. Therefore, this basic feature of the circadian clock appears to be coupled with characteristics of KaiC that enable the temperature-insensitive ATPase activity.

5.3.3 *ATPase Activity as the Determinant of Circadian Period Length*

An approximate 24-h period is another fundamental characteristic of circadian clocks. Identifying the chemical reactions that determine the period length, however, was not an easy task. Under oscillatory conditions, some of the reaction rates within the circadian clock are proportional to the frequency (angular rate) of the clock, but

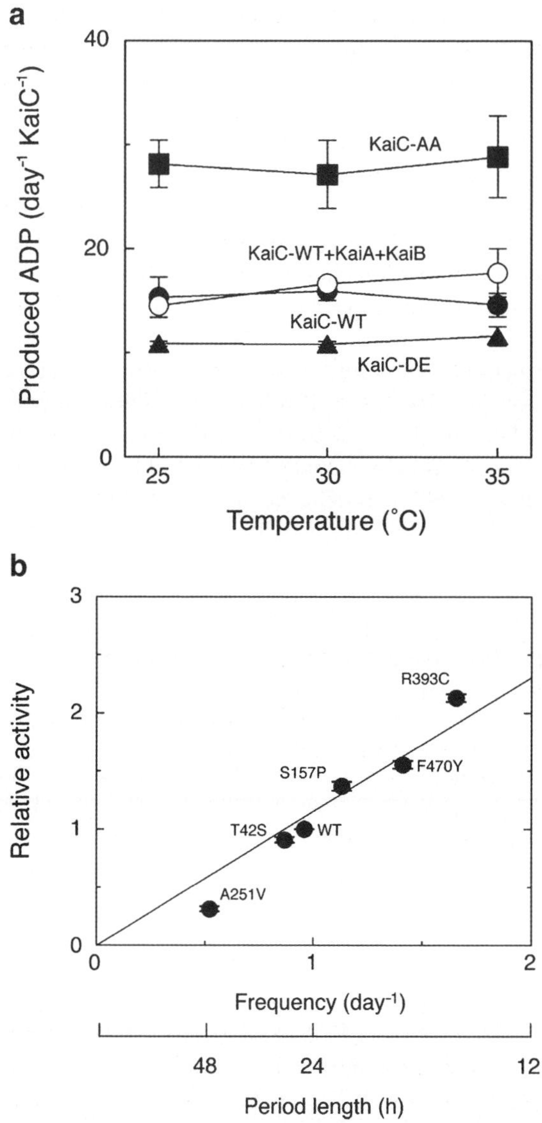

Fig. 5.4 ATPase activities of mutated KaiC proteins. (**a**) Temperature compensation of the ATPase activities of wild-type KaiC (*KaiC-WT*) and variants with mutations at the phosphorylation sites (*KaiC-AA* to mimic unphosphorylated KaiC, *KaiC-DE* to mimic doubly phosphorylated KaiC). Wild-type KaiC was incubated with 1 mM ATP at 25°C, 30°C, or 35°C in the presence (*open circles*) or absence (*filled circles*) of KaiA and KaiB. The ATPase activities of KaiC-AA (*squares*) and KaiC-DE (*triangles*) in the absence of KaiA and KaiB were also examined. (**b**) Correlation between the ATPase activity of KaiC and the circadian period length (*Frequency*). The ATPase activities of wild-type and five KaiC period-mutant proteins (T42S, S157P, A251V, R393C, F479Y) were measured in the absence of KaiA and KaiB. The activities and frequencies are shown as the values relative to that of wild-type KaiC (Terauchi et al. 2007)

determination of which reactions regulate the pace of the clock and which are controlled by the clock required further clarification. Genetic evidence suggested that KaiC is the strongest regulator of period, because mutations in *kaiC* can dramatically change the period length without affecting the amplitude of the cellular circadian rhythm (Ishiura et al. 1998; Nishimura et al. 2002). Five *kaiC* mutants bearing point mutations were identified that showed shorter or longer periods than that of the wild type with robust rhythmicity; the mutant KaiC proteins were purified and assayed for their ATPase activities in the absence of KaiA and KaiB (non-oscillatory conditions). Interestingly, the period length and ATP hydrolysis rate were inversely related. A plot of the hydrolysis rate against the reciprocal of the period (i.e., the frequency) shows a linear correlation between these parameters (Fig. 5.4). This correlation indicates that the circadian pacemaker depends directly on the energy provided by KaiC ATP hydrolysis. In other words, the same amount of energy (hydrolysis of ~15 ATP molecules per KaiC monomer) is required for one circadian cycle irrespective of the period length. This finding represents the first description of a biochemical reaction catalyzed by a single protein that serves as a circadian timekeeper.

5.3.4 Control of KaiC ATPase Activity

Two basic features of the cyanobacterial circadian clock (~24-h period length, temperature compensation) likely result from the ATPase activity of the KaiC protein. Little however is known about how the protein functions mechanistically, although the low ATPase rate may provide some insight. A negative feedback model has been proposed to control the ATPase activity in KaiC (Fig. 5.5; Terauchi et al. 2007). Many ATPases interact with various partners to convert the energy of ATP hydrolysis into mechanical force. Because the average ATPase activity of KaiC is similar in oscillatory and non-oscillatory conditions, the chemical energy from ATP hydrolysis is likely to be transferred to KaiC itself. The mechanical force created by this energy may cause a conformational change in KaiC that reduces its ATPase activity. This type of closed-loop regulation is commonly applied in an electric feedback amplifier to stabilize the output against a drift of the amplifier gain. Larger suppressive effects from the feedback circuit produce higher degrees of stability. Similar to this electric circuit, the ATPase activity (corresponding to the gain of an amplifier) may be highly suppressed by a negative feedback circuit and could be stabilized against drifts created by changes in temperature.

5.3.5 KaiC ATPase-Based Cellular Circadian Rhythms

In this section, we summarize the novel features of the KaiC ATPase as the basic timing mechanism of a circadian clock. The machinery that enables the ATPase activity to be translated into rhythmic phenomena is critical for the cyanobacterial

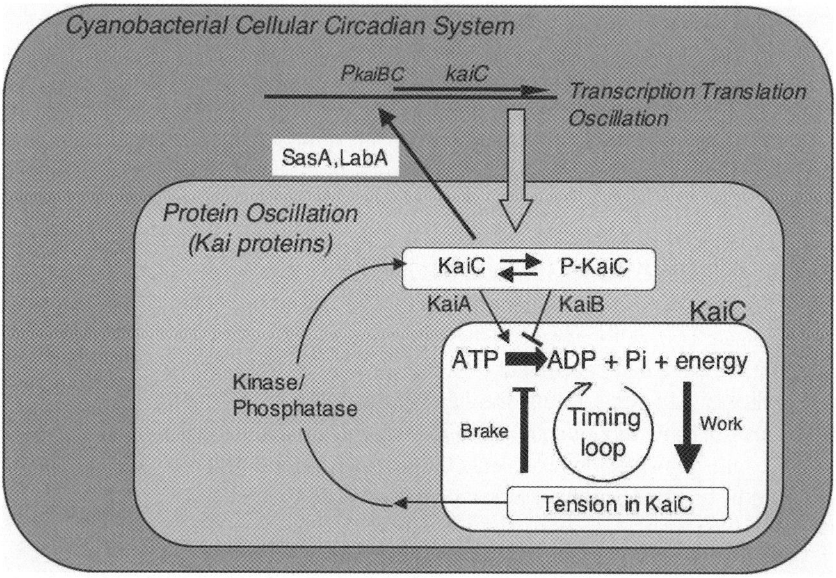

Fig. 5.5 A model of the cyanobacterial circadian system. This schematic diagram shows KaiC ATPase activity, the phosphorylation/dephosphorylation cycle, and transcription/translation-based oscillation. See text for details

circadian system (Fig. 5.5). As mentioned above, the phosphorylation and dephosphorylation of KaiC is coupled with its ATPase activity and the formation of Kai complexes. While the ATPase activity determines the basic rate of the phosphorylation/dephosphorylation cycle, the phosphorylation states of KaiC affect the rate of ATP hydrolysis. Similar to the phosphorylation states, the interactions between KaiC and KaiA/KaiB also affect the rate of ATP hydrolysis. Although it has been presumed that the phosphorylation state of KaiC affects its interaction with KaiA and KaiB (Kageyama et al. 2006; Nishiwaki et al. 2007), the ATPase activity may directly affect binding. The interlaced regulation of these three molecular behaviors serves to produce robust circadian rhythms in vitro and in vivo.

At the cellular level, physiological activities with circadian rhythmicity are coupled with the Kai oscillator (Fig. 5.5). The expression of *kaiBC* is dependent on the dynamic KaiC phosphorylation state (Murayama et al. 2008). SasA and LabA have been identified as regulators of circadian gene expression that are able to transduce timing signals from the Kai oscillator (Iwasaki et al. 2000; Taniguchi et al. 2007; Takai et al. 2009; see Chap. 9). The Kai oscillator is also influenced by the levels of its individual components; which are regulated at the transcription, translation, and degradation levels. Thus, the Kai oscillator is coupled with oscillations in cellular metabolism that quantitatively control the levels of its components. Recent studies suggested that the regulatory circuits containing the Kai proteins may be able to autonomously oscillate without the phosphorylation cycle (Kitayama et al. 2008). Multiple regulatory loops that influence a variety of processes, such as the ATPase activity of KaiC or the cellular metabolic rhythm, are likely to stabilize the circadian system against various perturbations in vivo.

5.4 Synchronization of the KaiC Phosphorylation Rhythm

5.4.1 Synchronization of the Kai Oscillator in vitro

Circadian systems are refractory to quantitative fluctuations in clock component levels that accompany such cellular events as cell growth and division. Studies of the cyanobacterial circadian rhythm at the single cell level have revealed that the circadian clock in individual cells is extremely resilient to such perturbations (Mihalcescu et al. 2004; see Chap. 13). This property is likely generated by the resilience of the Kai oscillator itself (Ito et al. 2007). As shown in Fig. 5.6, the in vitro KaiC phosphorylation rhythm persists for 10 days without damping, indicating that this rhythm is self-sustaining. As previously noted, the phosphorylation cycle of each KaiC molecule is likely to be able to autonomously oscillate with a circadian rhythm. Synchronization among these KaiC molecules drives the self-sustained rhythm of the entire system. To examine the synchronization behavior of the Kai oscillator, six oscillatory samples with different phases were combined (Fig. 5.6). If the KaiC phosphorylation cycle in each sample oscillated independently, the overall phosphorylation ratio should have achieved at a relatively constant value. Immediately after combining the samples, however, the mixture showed a phosphorylation rhythm with an amplitude that was comparable to those observed in the original individual samples, which suggests that the Kai oscillators rapidly synchronized. The phase of the mixture was similar to that of the original sample that was dephosphorylated at the time the samples were mixed. This finding implied that oscillators in the dephosphorylation phase dominate the synchronization process, a result that was clearly demonstrated using fluorescently labeled KaiC proteins to separately trace the phosphorylation transitions of various KaiC proteins in the mixture (Ito et al. 2007).

After mixing anti-phase oscillatory samples at various time-points, KaiC from a sample that was originally in the dephosphorylation phase showed a dephosphorylation rate that was similar to that observed in the unmixed dephosphorylation sample. In contrast, KaiC protein from a sample originally in the phosphorylation phase either underwent dephosphorylation or showed suppression of the phosphorylation reaction. After the phosphorylation ratios of KaiC proteins from the two samples equalized, both began to fluctuate synchronously. Thus, the synchronization processes in the mixtures are a result of the oscillatory samples in the dephosphorylation phase changing the reaction from phosphorylation to dephosphorylation for the other KaiC proteins, after which dephosphorylation persists until the entire pool of KaiC proteins reaches a low level of phosphorylation. These two processes

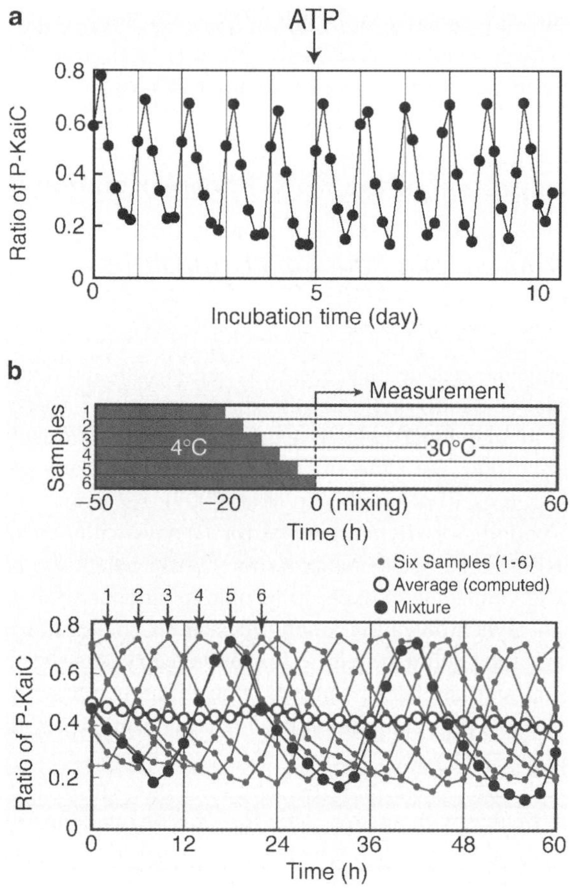

Fig. 5.6 Robust oscillation of KaiC phosphorylation in vitro. (**a**) The KaiC phosphorylation rhythm persisted for 10 days without damping. The solution of the Kai proteins and ATP (1 mM) was incubated at 30°C. After 5 days incubation, the solution was replenished with ATP (*arrow*). An aliquot of the solution was collected every 4 h and subjected to SDS-PAGE. The ratios of P-KaiC to total KaiC are plotted. (**b**) Resilience of the oscillatory system. Six samples of the oscillatory solution with different phases were prepared as follows: the solution was kept at 4°C for more than 30 h, and then the temperature was raised to 30°C at various time-points to yield six samples (*1–6*, *top panel*). At time 0, an equal amount of the samples were mixed. The phosphorylation ratios of the six samples (*gray dots*) and the mixture (*filled circles*) are plotted against the time after mixing. The arithmetic mean of the phosphorylation ratios of the six samples at each time-point is also shown (*open circles*; Ito et al. 2007)

enable heterogeneous samples of KaiC proteins to be synchronized with respect to the reaction cycle. Synchronization appears to be dose-dependent because at least 30% of the sample is required to be in the dephosphorylation phase to entrain the mixture (Ito et al. 2007). Thus, under normal oscillatory conditions, the KaiC proteins that first enter the dephosphorylation phase determine the phase of the oscillatory system by synchronizing the remaining KaiC.

5.4.2 *Synchronization of KaiC Proteins by Monomer Shuffling*

Synchronization likely involves interactions among KaiC proteins with different phosphorylation states. During the oscillations, KaiA and KaiB interact with KaiC; and alteration of these interactions is likely to influence the synchronization of the KaiC proteins (Kageyama et al. 2006; Mori et al. 2007; Nishiwaki et al. 2007; Rust et al. 2007). In addition to this indirect communication among the KaiC proteins, a direct interaction has also been demonstrated. KaiC hexamers have been shown to exchange monomers (Kageyama et al. 2006). This "monomer shuffling" is thought to occur specifically during the early dephosphorylation phase; highly phosphorylated KaiC can exchange monomers with KaiC proteins in other phosphorylation states (Fig. 5.2; Ito et al. 2007). The common phase observed for the synchronization and shuffling processes suggests that monomer shuffling is involved in the synchronization of KaiC.

5.4.3 *Synchronization of the Kai Oscillator in vivo*

The synchronized KaiC phosphorylation rhythm observed in vitro likely contributes to the robustness of the cyanobacterial circadian rhythm. In living cells, although newly synthesized KaiC proteins may be in different phosphorylation states than those of the pre-existing KaiC, which could disturb the circadian system, the period length has appeared to be independent of the protein synthesis rate (Kondo et al. 1997). Newly synthesized KaiC monomers are likely to be less phosphorylated than pre-existing KaiC, suggesting the older proteins at the beginning of the phosphorylation cycle (at the point they begin to be phosphorylated) are at the most advanced phase for the following cycle. Therefore, newly synthesized KaiC would be entrained to the phosphorylation cycle by the pool of pre-existing KaiC.

5.5 Conclusion

The chemical and physical bases of the Kai oscillator have gradually come to light. This oscillator seems to be characterized by the unique features of the KaiC molecule, in particular its ATPase activity. The fact that each KaiC molecule hydrolyzes ~15 ATP molecules/day to drive one circadian cycle suggests that this system is analogous to the mechanical workings of clocks, which are regulated by such devices as a pendulum and balance wheel. It is notable that each swing of these clock components is digitally transmitted to the system as a time unit. Similarly, each hydrolyzed ATP molecule may reflect a time unit for the Kai oscillator. The 15 molecules/day processed by each KaiC protein, however, would appear to provide a less accurate measurement of time than traditional clocks. The processes that synchronize the KaiC molecules are likely to compensate for this potential problem; the average frequency of a large number of oscillators allows an accurate measure of time. In fact, each cyanobacterial cell contains ~10,000 KaiC molecules. The protein structures and dynamic behaviors of KaiC and other Kai proteins should further our understanding of biological clocks.

References

Akiyama S, Nohara A, Ito K, Maeda Y (2008) Assembly and disassembly dynamics of the cyanobacterial periodosome. Mol Cell 29:703–716

Ishiura M, Kutsuna K, Aoki S, Iwasaki H, Andersson CR, Tanabe A, Golden SS, Johnson CH, Kondo T (1998) Expression of a gene cluster *kaiABC* as a circadian feedback process in cyanobacteria. Science 281:1519–1523

Ito H, Kageyama H, Mutsuda M, Nakajima M, Oyama T, Kondo T (2007) Autonomous synchronization of the circadian KaiC phosphorylation rhythm. Nat Struct Mol Biol 14:1084–1088

Iwasaki H, Williams SB, Kitayama Y, Ishiura M, Golden SS, Kondo T (2000) A KaiC-interacting sensory histidine kinase, SasA, necessary to sustain robust circadian oscillation in cyanobacteria. Cell 101:223–233

Iwasaki H, Nishiwaki T, Kitayama Y, Nakajima M, Kondo T (2002) KaiA-stimulated KaiC phosphorylation in circadian timing loops in cyanobacteria. Proc Natl Acad Sci USA 99:15788–15793

Kageyama H, Nishiwaki T, Makajima M, Iwasaki H, Oyama T, Kondo T (2006) Cyanobacterial circadian pacemaker: Kai protein complex dynamics in the KaiC phosphorylation cycle in vitro. Mol Cell 23:161–171

Kitayama Y, Iwasaki H, Nishiwaki T, Kondo T (2003) KaiB functions as an attenuator of KaiC phosphorylation in the cyanobacterial clock system. EMBO J 22:2127–2134

Kitayama Y, Nishiwaki T, Terauchi K, Kondo T (2008) Dual KaiC-based oscillations constitute the circadian system of cyanobacteria. Genes Dev 22:1513–1521

Kondo T, Mori T, Lebedeva NV, Aoki S, Ishiura M, Golden SS (1997) Circadian rhythms in rapidly dividing cyanobacteria. Science 275:224–227

Marrione PE, Cox MM (1995) RuvB protein-mediated ATP hydrolysis: functional asymmetry in the RuvB hexamer. Biochemistry 34:9809–9818

Mihalcescu I, Hsing W, Leibler S (2004) Resilient circadian oscillator revealed in individual cyanobacteria. Nature 430:81–85

Mori T, Williams DR, Byme MO, Qin X, Egli M, Mchaourab HS, Stewart PL, Johnson CH (2007) Elucidating the ticking of an in vitro circadian clockwork. PLoS Biol 5:e93

Murayama Y, Oyama T, Kondo T (2008) Regulation of circadian clock gene expression by phosphorylation states of KaiC in cyanobacteria. J Bacteriol 190:1691–1698

Nakajima M, Imai K, Ito H, Nishiwaki T, Murayama Y, Iwasaki H, Oyama T, Kondo T (2005) Reconstitution of circadian oscillation of cyanobacterial KaiC phosphorylation in vitro. Science 308:414–415

Nishimura H, Nakahira Y, Imai K, Tsuruhara A, Kondo H, Hayashi H, Hirai M, Saito H, Kondo T (2002) Mutations in KaiA, a clock protein, extend the period of circadian rhythm in the cyanobacterium *Synechococcus elongatus* PCC 7942. Microbiology 148:2903–2909

Nishiwaki T, Iwasaki H, Ishiura M, Kondo T (2000) Nucleotide binding and autophosphorylation of the clock protein KaiC as a circadian timing process of cyanobacteria. Proc Natl Acad Sci USA 97:495–499

Nishiwaki T, Satomi Y, Nakajima M, Lee C, Kiyohara R, Kageyama H, Kitayama Y, Temamoto M, Yamaguchi A, Hijikata A, Go M, Iwasaki H, Takao T, Kondo T (2004) Role of KaiC phosphorylation in the circadian clock system of *Synechococcus elongatus* PCC 7942. Proc Natl Acad Sci USA 101:13927–13932

Nishiwaki T, Satomi Y, Kitayama Y, Terauchi K, Kiyohara R, Takao T, Kondo T (2007) A sequential program of dual phosphorylation of KaiC as a basis for circadian rhythm in cyanobacteria. EMBO J 26:4029–4037

Rust MJ, Markson JS, Lane WS, Fisher DS, O'Shea EK (2007) Ordered phosphorylation governs oscillation of a three-protein circadian clock. Science 318:809–812

Takai N, Nakajima M, Oyama T, Kito R, Sugita C, Sugita M, Kondo T, Iwasaki H (2006) A KaiC-associating SasA-RpaA two-component regulatory system as a major circadian timing mediator in cyanobacteria. Proc Natl Acad Sci USA 103:12109–12114

Taniguchi Y, Katayama K, Ito R, Takai N, Kondo T, Oyama T (2007) *labA*: a novel gene required for negative feedback regulation of the cyanobacterial circadian clock protein KaiC. Genes Dev 21:60–70

Terauchi K, Kitayama Y, Nishiwaki T, Miwa K, Murayama Y, Oyama T, Kondo T (2007) ATPase activity of KaiC determines the basic timing for circadian clock of cyanobacteria. Proc Natl Acad Sci USA 104:16399–16381

Tomita J, Nakajima M, Kondo T, Iwasaki H (2005) No transcription-translation feedback in circadian rhythm of KaiC phosphorylation. Science 307:251–254

Williams SB (2007) A circadian timing mechanism in the cyanobacteria. Adv Microb Physiol 52:229–296

Xu Y, Mori T, Johnson CH (2003) Cyanobacterial circadian clockwork: roles of KaiA, KaiB and the *kaiBC* promoter in regulating KaiC. EMBO J 22:2117–2126

Xu Y, Mori T, Pattanayek R, Pattanayek S, Egli M, Johnson CH (2004) Identification of key phosphorylation sites in the circadian clock protein KaiC by crystallographic and mutagenetic analyses. Proc Natl Acad Sci USA 101:13933–13938

Chapter 6
NMR Studies of a Timekeeping System

Ioannis Vakonakis and Andy LiWang

Abstract Cyanobacterial circadian clocks represent perhaps the best studied timekeeping system in terms of the molecular and mechanistic information available; structural biology has contributed significantly in both respects. We present here an overview of progress made using traditional high-resolution nuclear magnetic resonance (NMR) spectroscopy on the structures of these proteins in solution. Combining NMR and a dissection approach yielded high-resolution structures of many clock protein fragments, especially from KaiA, SasA and CikA, and the sole complex available thus far describing the interaction of KaiA with KaiC at high resolution. These structures allowed hypotheses on the mechanism and function of these proteins; we attempt to revisit these here. The development of NMR methodology has created new tools to access increasingly large and dynamic systems. We argue that these new approaches can be used in the study of circadian oscillators.

6.1 Introduction

The cyanobacterial circadian oscillator system or clock (Ishiura et al. 1998) presented structural biologists with unique opportunities and an interesting challenge. The apparent simplicity of three interacting proteins controlling the bacterial oscillator was appealing; and the potential for heterologous expression of these proteins in *Escherichia coli* promised an easier discovery path compared to circadian clock proteins of higher organisms. Yet this system challenged the reductionist approach often followed by structural biologists. The amino acid sequences of the three oscillator proteins, especially KaiA and KaiB, showed little similarity to other known proteins, although the various domain database and protein fold recognition

I. Vakonakis
Department of Biochemistry, University of Oxford, South Parks Road, Oxford, OX1 3QU, United Kingdom, e-mail: ioannis.vakonakis@bioch.ox.ac.uk

A. LiWang(✉)
School of Natural Sciences, University of California at Merced, 4225 N. Hospital Road, Atwater, CA 95301, USA, e-mail: aliwang@ucmerced.edu

J.L. Ditty et al. (eds.), *Bacterial Circadian Programs.*

methods now taken for granted (Kelley et al. 2000; Bateman et al. 2004; Letunic et al. 2004) were little-developed at the time. Thus, given the relative absence of information and eventual problems encountered in preparing stable clock protein samples, it is perhaps not surprising that initial structural information in this system was gained by nuclear magnetic resonance (NMR) spectroscopy. Here, we trace the evolution of NMR research on this circadian system, starting with a short preamble highlighting differences between NMR and other high-resolution structural biology techniques. We try to revisit lessons learned and point to possible avenues of research in the near future. The majority of the material covered addresses KaiA (Ishiura et al. 1998; Nakahira et al. 2004) and its two domains, KaiA135N and KaiA180C. Two other clock-related projects involve the N-terminal domain of SasA (Iwasaki et al. 2000), a clock output protein, and the pseudo-receiver domain of CikA, a protein important for clock resetting (Schmitz et al. 2000).

6.2 NMR: A Short Preamble

NMR is a spectroscopic method that relies on excitation and subsequent relaxation of atomic nuclei in the presence of a strong magnetic field. In biomolecular research it is commonly used on aqueous samples at physiological temperatures, allowing a good approximation of the natural environment of the biomolecules, but often at very high protein concentrations. Because restrictions such as crystallization or grid placement do not apply, NMR samples are often easy to prepare. An important difference between NMR and crystal diffraction and electron microscopy is the nature of the information. Crystallographic and microscopy techniques derive, after appropriate processing, density maps where the shape and form of the biomolecule can often be readily determined. In contrast, NMR experiments provide multiple independent pieces of information, such as chemical environment, inter-atomic distances, amino acid conformation and atomic bond orientation. Biomolecular structures by NMR are represented by ensembles from molecular dynamics simulations where the aforementioned pieces of information act as weak restraints. This renders structure determination a complex process and limits the molecular size to ~30 kDa or smaller; however NMR data can also be used in a structure-independent fashion to answer specific questions quickly and in far larger systems.

A contour plot of NMR spectra from the N-terminal domain of KaiA and an unfolded protein is offered in Fig. 6.1. This type of spectrum, typically acquired in ~20–30 min using modern NMR spectrometers, reports on the local environment of covalently attached hydrogen and nitrogen nuclei. Each peak observed corresponds to a different H–N pair in the protein (usually one per amino acid residue). Different proteins feature different patterns of atomic environments, hence different spectra, allowing the identification of the protein in question much like identifying a person by their fingerprint. Samples of the same protein under different conditions, in complex with ligands or of unfolded/partially folded variants also produce unique

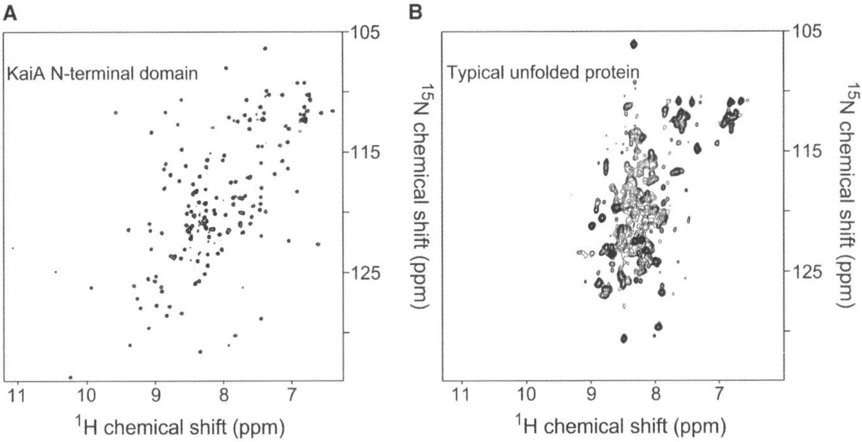

Fig. 6.1 NMR spectra of the folded N-terminal domain of KaiA (**A**) and the thermally denatured state of domain 3Fn3 of human fibronectin (**B**). The structural state of the protein can be readily surmised from the dispersion (or lack thereof) of peaks across the two spectral axes. New cold-probe technology allows the collection of NMR spectra at micromolar, i.e., near physiological, protein concentrations

spectra in each case. As seen in our example, the spectral characteristics of a folded and an unfolded protein are very different and thus indicative of the folding state of the protein.

6.3 KaiA

KaiA acts as the positive element of the cyanobacterial circadian oscillator, overexpression of which increases *kaiBC* expression (Ishiura et al. 1998; Nakahira et al. 2004). A number of residue specific substitutions were found to affect the circadian period (Ishiura et al. 1998; Iwasaki et al. 2002; Nishimura et al. 2002), but KaiA did not show any substantial sequence similarity to other known proteins or sequence motifs. KaiA was shown to interact strongly in vitro and in the yeast nucleus with KaiC (Iwasaki et al. 1999) and, to a much weaker extent, with KaiB, while heterotrimeric KaiABC complexes were demonstrated in vivo (Kageyama et al. 2003). This direct KaiA–KaiC interaction was found to increase the rate of KaiC autophosphorylation (Iwasaki et al. 2002; Williams et al. 2002; Kim et al. 2008), a function important for circadian timekeeping (Nishiwaki et al. 2000; Mori and Johnson 2001; Nishiwaki et al. 2007; Rust et al. 2007). Initial reports suggested that KaiA multimerizes (Iwasaki et al. 1999), while analytical experiments showed that it forms a tight dimer in solution (Vakonakis et al. 2004b).

Despite being expressed in a soluble form in *Escherichia coli*, *Synechococcus elongatus* KaiA proved difficult to purify to homogeneity. Protein stability was low and, despite our efforts, proteolytic degradation over extended periods of time was

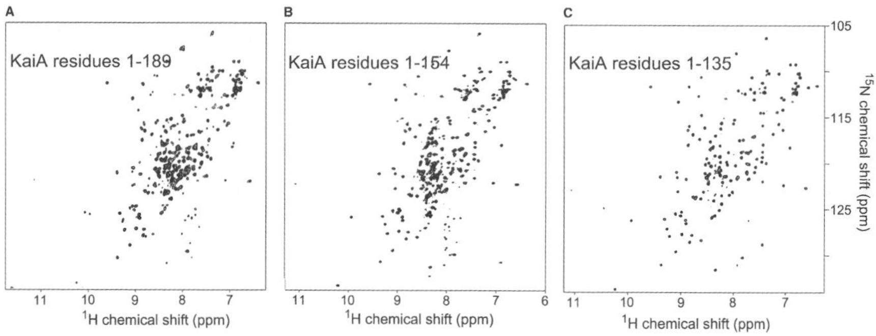

Fig. 6.2 NMR spectra of three different variants from the N-terminus of *S. elongatus* KaiA (**A–C**). Increasing C-terminal truncations from (**A**) to (**C**) result in reduced peak overlap and crowding at the center of the spectrum, indicating reductions in the unfolded portion of the protein. The variant in (**C**) represents the minimum possible size for this domain, as further truncations destabilize the protein fold

common. Initial NMR spectra acquired using full-length KaiA showed little promise for structural studies by NMR, possibly due to the large dimeric particle size. Thus, we resorted to limited proteolysis assays followed by N-terminal sequencing of the resulting fragments to identify stable KaiA subcomponents or domains that might be more amenable to structural studies (Williams et al. 2002). These assays demonstrated that a segment spanning residues 140–180 of *S. elongatus* KaiA is sensitive to proteolytic digestion, separating two relatively stable cores at the protein N- and C-termini. Gene fragments encoding different lengths of KaiA constructs were subcloned and their products tested by NMR (Fig. 6.2). The largest fragment tested, corresponding to KaiA residues 1–189 (Fig. 6.2A), shows many dispersed peaks away from the spectral center, indicative of a well folded protein core. However, the spectral center itself is crowded with peaks indicating that a substantial portion of this construct is not folded. C-terminal truncations of this construct, encoding KaiA residues 1–154 or 1–135 (Fig. 6.2B, C) retained the well dispersed peaks but had fewer peaks in the center of the spectrum, indicating a reduction of the unfolded protein fraction. Further truncated constructs reverted to a fully unfolded form with no peak dispersal in the NMR spectra, similar to that shown in Fig. 6.1B. Thus, we were able to define the minimal well folded core in the *S. elongatus* KaiA N-terminus as residues 1–135 (KaiA135N); a similar process found that the minimal C-terminal core comprises residues 180–284 (KaiA180C; Williams et al. 2002).

Functional analysis of these two domains revealed that both the KaiA–KaiC interaction and the enhancement of KaiC autophosphorylation by KaiA reside with KaiA180C (Williams et al. 2002). No function could be assigned to KaiA135N based on these assays, although mutagenesis experiments suggested it is important for the clock (Ishiura et al. 1998; Iwasaki et al. 2002; Nishimura et al. 2002; Kim et al. 2008). Sequence analysis of cyanobacterial KaiA proteins showed that domains equivalent to KaiA180C are always present and are highly similar (Williams et al. 2002). In contrast, KaiA135N shares less sequence similarity among cyanobacteria and is completely absent in filamentous heterocystous genera, such as *Anabaena* or *Nostoc*, in which KaiA consists of only the C-terminal domain. It has been suggested that evolution of cyanobacterial circadian oscillators was initially driven by the need for temporal separation between photosynthesis and nitrogen fixation in unicellular species (Iwasaki and Kondo 2000; Dvornyk et al. 2003; see Chap. 3). Thus, it seems likely that KaiA135N modulates the clock to better match the different physiological needs of these diverse species. However, we could not form any hypotheses regarding the mode and manner of this modulation prior to determining the three-dimensional structures of KaiA135N and KaiA180C.

6.4 KaiA135N

The relatively small size of KaiA135N (~15 kDa) and the fact that it is monomeric in solution allowed us to determine the structure of this domain using traditional high-resolution NMR methods (Cavanagh et al. 2007). Surprisingly, KaiA135N, the first clock-protein structure to be determined from any organism, was found to be highly similar to receiver domains of bacterial two-component signal-transduction pathways, such as NtrC or CheY (Volz 1993; Fig. 6.3). Although two-component systems are a well studied class of proteins, this similarity could not be predicted ab initio as KaiA135N and NtrC are only ~17% similar in amino acid sequence, below the recognition threshold of alignment algorithms. Signal transduction in two-component systems typically involves Mg^{2+}-dependent phosphoryl-transfer activity (Bourret et al. 1990) that affects the structural state of the α4-helix of the receiver domain (Volkman et al. 2001). However, KaiA135N diverges from traditional receivers as it lacks an aspartate residue required for phosphoryl-transfer (Bourret et al. 1990). In addition, KaiA135N replaces α4 with a long loop (Fig. 6.3) which was determined to be flexible in solution, based on protein dynamics experiments performed by NMR. Thus, we proposed that KaiA135N is a pseudo-receiver domain, a widespread class that can putatively act in signal transduction through direct protein–protein interactions (O'Hara et al. 1999).

We were able to define a potential interaction surface through residue substitutions in KaiA135N known to affect the clock (Nishimura et al. 2002). Examination of the NMR spectra of these substituted proteins showed two general classes. One, when a core structural residue of the protein was replaced, resulted in protein destabilization or unfolding. In these cases, the effect on the clock is believed to originate from the decreased overall stability of the protein, as well as the loss of KaiA135N function. A second, more interesting, class of substitutions does not affect protein stability adversely. Instead, the NMR spectra showed structural changes over a well defined protein surface (Fig. 6.3D) which could serve as a site for protein–protein interactions. These interactions could be involved either in KaiA135N activation, or in signal-transduction between the N- and C-terminal KaiA domains.

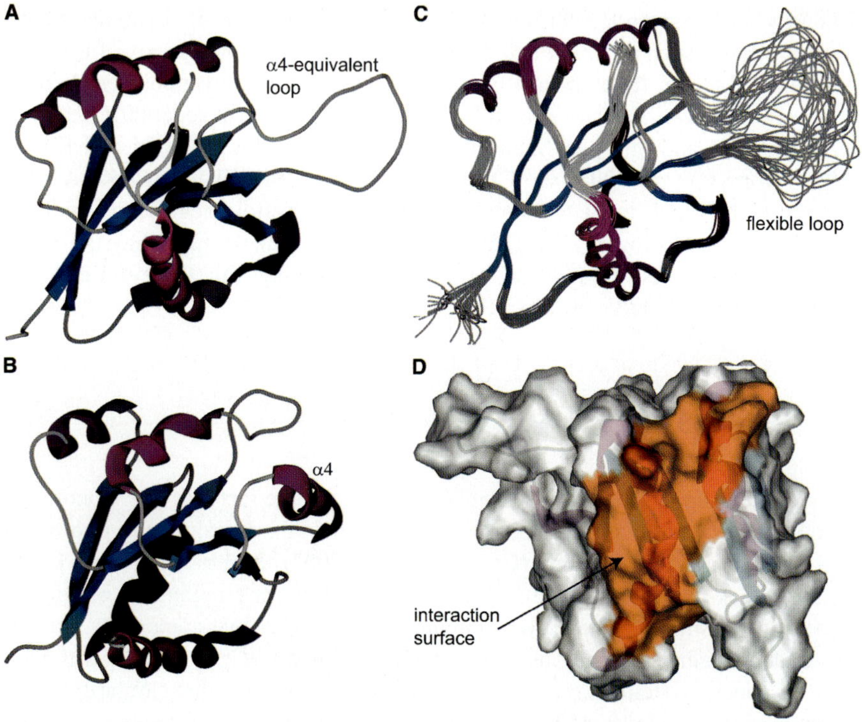

Fig. 6.3 The structure of KaiA135N of *S. elongatus,* KaiA135N^{Se} (**A**) is similar to receiver domains of two-component signal transduction systems (Volz 1993). A representative example, NtrC, is shown in (**B**; Kern et al. 1999). KaiA135N^{Se} differs from typical receivers in some respects, however; the α4-helix is replaced by a long flexible loop, as determined from ^{15}N relaxation experiments. Twenty-five low energy structures are shown superimposed in (**C**). The structural disorder in the loop, as seen in (**C**), arises from the lack of obtainable restraints resulting from loop dynamics. Residues important for activation through phosphoryl transfer are also absent. Instead, protein–protein interactions are likely to be important (O'Hara et al. 1999) and a putative interaction surface is shown *colored* in (**D**)

6.5 KaiA180C

Initially characterized simultaneously along with KaiA135N, KaiA180C proved to be less well behaved in solution, making its structure determination a more difficult proposition. This KaiC-interacting domain was shown to be responsible for KaiA dimerization (Vakonakis et al. 2004b), increasing the molecular size of the system. In addition, the derived *S. elongatus* domain variant had characteristics similar to molten globule proteins, where structural features are obscured by multiple dynamic processes in the protein. These processes manifest in NMR spectra as peak broadening and disappearance; and they typically render the protein unsuitable for high-resolution studies. Fortunately, we were able to show that the highly homologous KaiA C-terminal domain from *Thermosynechococcus elongatus*, KaiA180C^{Te}, exhibits better thermodynamic properties, such as stability and folding cooperativity (Vakonakis et al. 2004b) and provides better NMR spectra. The structure of KaiA180C^{Te} was also determined by X-ray crystallography (Uzumaki et al. 2004). For KaiA180C^{Te} residues K186–S278, the backbone pairwise rmsd between the averaged NMR structure and the single chain deposited for the X-ray crystal structure is 1.07 Å, excluding residues Q246–I256, which connect helices α3 and α4 and are significantly different between the two structures. The interhelical angles are the same for NMR and crystal structures within this subunit of KaiA180C^{Te}.

The solution structure of KaiA180C^{Te} from *T. elongatus* revealed a four-helix bundle monomeric subunit (Fig. 6.4; Vakonakis et al. 2004b), a fold structurally well characterized (Harris et al. 1994). However, as in KaiA135N of *S. elongatus*

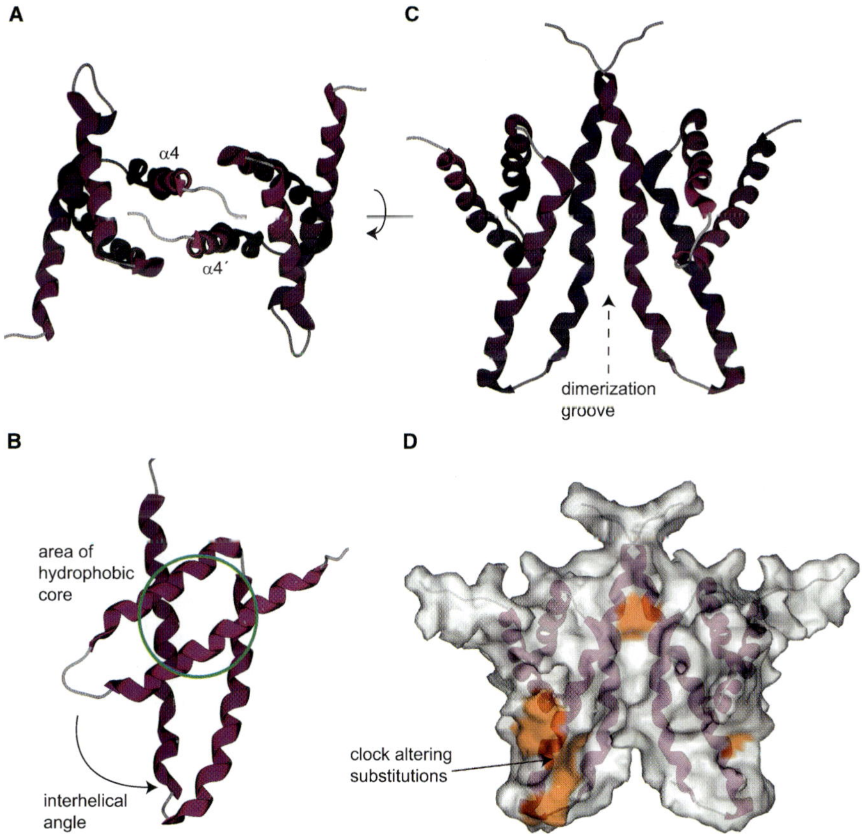

Fig. 6.4 KaiA180C of *T. elongatus* (KaiA180C^{Te}) adopts a four-helix bundle-type structure that forms a dimer along α4 (**A**). The interhelical angle of the bundle is wide, resulting in a small hydrophobic core (**B**). Dimer formation creates a large intersubunit groove (**C**) where many clock period-altering substitutions can be mapped (**D**)

(KaiA135N^{Se}), this fold similarity comes with a twist. The angle defined by the two helical hairpins forming the bundle is very wide (Fig. 6.4B), making KaiA180C a member of the uncommon X-class of bundles (Harris et al. 1994) and the first self-contained example of this class to be described. This high interhelical angle results in a singularly small protein core for the size of this protein. It is likely that the smaller hydrophobic core of the *S. elongatus* KaiA results in its molten globule characteristics, whereas the larger hydrophobic core of the *T. elongatus* KaiA features improved packing of that same core to achieve higher stability.

Two KaiA180C^{Te} subunits interact along the α4-helix (Fig. 6.4A), which partly resembles a coil–coil structure, to form a symmetric dimer with a sizable groove between the two subunits (Fig. 6.4C). Analysis of KaiA180C^{Te} residue specific substitutions known to affect the circadian oscillator showed that a number of nonstructural clock-perturbing residues map at or near this dimer interface groove (Fig. 6.4D). We proposed that this interface would be important for the direct KaiA–KaiC interaction known to occur at this domain (Williams et al. 2002) and proceeded to screen KaiC-derived peptides for binding (Taniguchi et al. 2001). The use of NMR in this system simplified our search, as direct protein–peptide interactions are accompanied by substantial perturbations in the NMR spectra. These perturbations allowed us to quickly identify and optimize the sequence of a binding peptide corresponding to the nonRecA-analogous C-terminus of KaiC (Fig. 6.5A; Vakonakis and LiWang 2004). The structure of the resulting complex shows

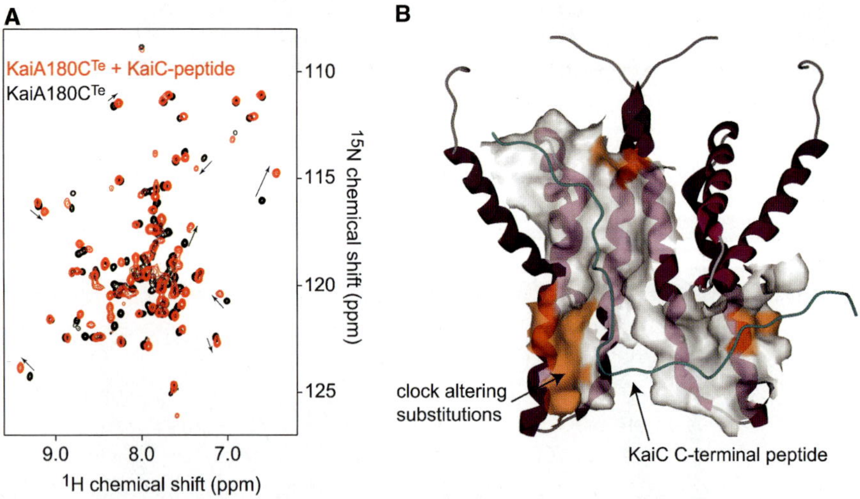

Fig. 6.5 Formation of complexes between KaiA180C^{Te} and a KaiC peptide resulted in substantial perturbations in the NMR spectra, shown in (**A**) as an overlay of spectra in the presence (*red*) or absence (*black*) of this peptide. *Arrows* denote the direction for some of these perturbations. The KaiA180C^{Te}–KaiC peptide interaction surface is shown in (**B**), along with the location of known KaiA clock-altering substitutions in gold

extensive interactions between the KaiC peptide and KaiA, many of which involve residues on KaiA or KaiC known to be important for clock function (Fig. 6.5B; Ishiura et al. 1998; Nishimura et al. 2002). The KaiC peptide was shown to bind along the KaiA180C dimerization groove, form an approximately 90° angle and interact with exposed hydrophobic residues of KaiA180C^{Te}. A single KaiC peptide forms interactions with both KaiA180C^{Te} subunits simultaneously, thus explaining the impact of KaiA dimerization-affecting substitutions on the clock (Vakonakis and LiWang 2004).

6.6 N-SasA

SasA is the sensor histidine kinase component of a two-component signal transduction system (Dutta et al. 1999) that associates with KaiC (Iwasaki et al. 2000; Kageyama et al. 2003) and is responsible for part of the circadian output to clock-controlled genes (Iwasaki et al. 2000; see Chap. 9). The cognate response regulator of SasA is RpaA (Takai et al. 2006). Sensor histidine kinases are typically composed of an N-terminal sensor domain that acts to modulate output of the C-terminal histidine protein kinase (HPK) domain (Dutta et al. 1999; Stock et al. 2000). Direct association between the N-terminal sensor domain (N-SasA; Vakonakis et al. 2004a) and KaiC enhances the autokinase activity of SasA (Smith and Williams 2006) and phosphoryl transfer to RpaA (Takai et al. 2006). N-SasA has significant sequence similarity to KaiB (Kageyama et al. 2003; Dvornyk et al. 2004), which led to the hypothesis that KaiB and N-SasA adopt a similar three-dimensional structure (Dvornyk et al. 2004) and possibly compete for KaiC binding (Iwasaki et al. 2000; Kageyama et al. 2003; Dvornyk et al. 2004).

Contrary to these expectations, solution studies of N-SasA (Vakonakis et al. 2004a) revealed substantially different structural characteristics, compared to KaiB studies and crystal structures (Garces et al. 2004; Hitomi et al. 2005; Fig. 6.6A, B). KaiB is oligomeric in crystals (Garces et al. 2004; Hitomi et al. 2005), in vitro (Iwasaki et al. 1999; Iwase et al. 2005) and in vivo (Xu et al. 1999; Kageyama et al. 2003), while N-SasA is monomeric even at high concentrations (Vakonakis et al. 2004a). N-SasA adopts a canonical thioredoxin-like topology (Vakonakis et al. 2004a; Fig. 6.6A, B), although it lacks the catalytic cysteine residues necessary for redox reactions (Martin 1995). In contrast, the KaiB fold is that of an α–β meander (Garces et al. 2004; Hitomi et al. 2005; Fig. 6.6C). Analysis of N-SasA amino acid sequences from multiple cyanobacteria showed an exposed patch of conserved nonstructural residues around the α2-helix (Fig. 6.6D), an area known to mediate protein–protein interactions in similar thioredoxin-like systems (Doublié et al. 1998; Ma et al. 2003). Thus, we proposed that this area in N-SasA is involved in SasA–KaiC or N-SasA–SasA HPK interactions (Vakonakis et al. 2004a).

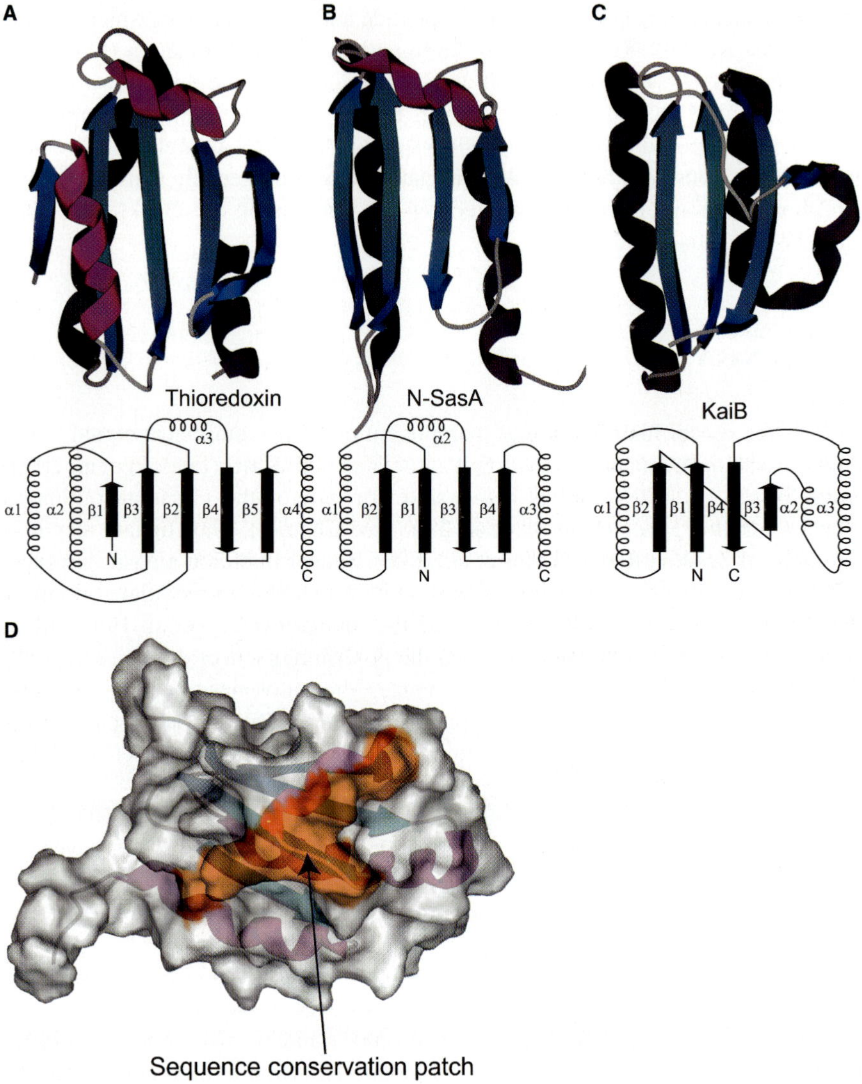

Fig. 6.6 N-SasA is monomeric and adopts a canonical thioredoxin-like topology. Thioredoxin is shown in (**A**) and N-SasA is shown in (**B**). In contrast the KaiB subunit (**C**) features an α–β meander topology (Garces et al. 2004; Hitomi et al. 2005; Iwase et al. 2005; Pattanayek et al. 2008) which then tetramerizes to form the physiological particle. Sequence analysis of N-SasA revealed an exposed patch of conserved residues (**D**) that could constitute a protein interaction interface

6.7 CikA

The phase of the central oscillator of circadian clocks can be reset through environmental cues such as light, necessitating the presence of suitable signal input pathways (Dunlap et al. 2004). Light flux affects the cellular redox state of cyanobacteria through photosynthesis (Mackey and Golden 2007). The circadian input kinase protein (CikA; Schmitz et al. 2000) appears to help synchronize the circadian clock to photosynthetic activity by sensing the redox state of the plastoquinone pool through direct interactions (Ivleva et al. 2006). Quinone-binding occurs through the pseudo-receiver domain of CikA (CikAPsR) and demonstrates that pseudo-receiver domains can function as sensors of small molecules. CikA also has an HPK domain and may be therefore the sensor histidine kinase component of a two-component signal transduction system (Stock et al. 2000). However, the putative cognate response regulator has not yet been identified. CikAPsR attenuates the autokinase activity of the HPK domain of CikA by an order of magnitude (Mutsuda et al. 2003) and is also necessary to localize the protein to the cell pole (Zhang et al. 2006), thus exhibiting multiple functional roles (see Chap. 8).

Insights into quinone-binding were obtained by mapping chemical shift perturbations induced by the quinone analog DBMIB (2,5-dibromo-3-methyl-6-isopropyl-*p*-benzoquinone) onto the structure of CikAPsR (Fig. 6.7A) and were found to form a surface primarily along α1 and β2 (Fig. 6.7B, C; Gao et al. 2007). The NMR structure of CikAPsR has the typical α/β doubly wound topology of receiver (Robinson et al. 2000) and pseudo-receiver domains (Gao et al. 2007). The putative quinone-binding site is on the opposite side of the α4–β5–α5 face that is used by KaiA (Ye et al. 2004) and AmiR (O'Hara et al. 1999) for intersubunit interactions. The structure of CikAPsR was used to develop a model showing how CikAPsR possibly attenuates the autokinase activity of CikA (Gao et al. 2007): CikAPsR

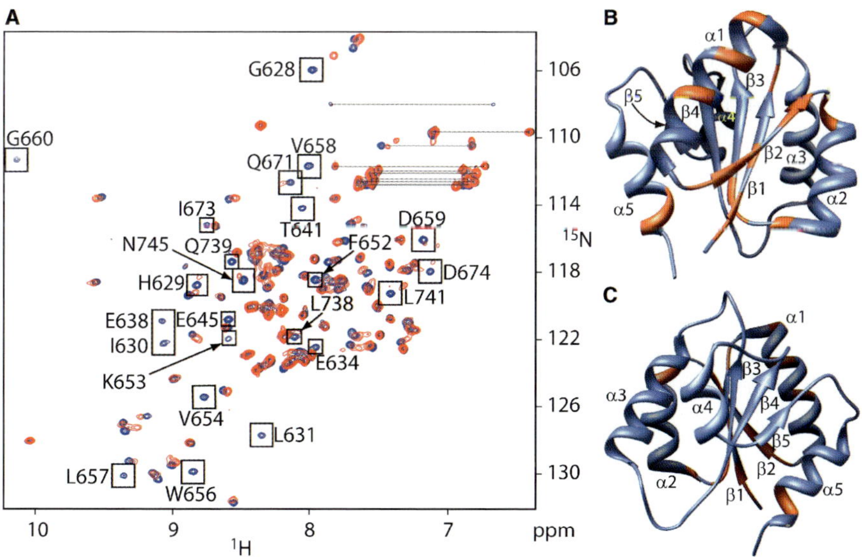

Fig. 6.7 (**A**) DBMIB induced perturbations (*red*) to the NMR spectrum of CikAPsR (apo-state in *blue*). (**B**) Mapping of spectral perturbations by DBMIB onto the structure of CikA in *orange*. (**C**) A view from the side opposite to that shown in (**B**). The spectrum in (**A**) is used with permission from Ivleva et al. (2006)

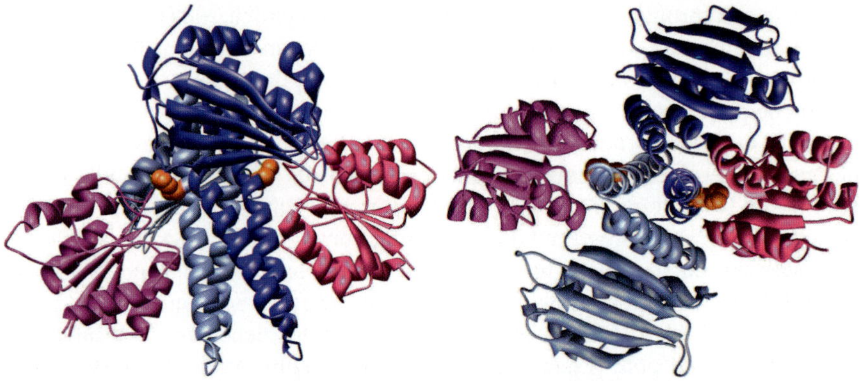

Fig. 6.8 A model of a possible interaction between the PsR and HPK domains of CikA. The PsR domains are shown as differently shaded *pink ribbons*, and the HPK domains are shown as light and dark *blue ribbons*. The side-chain atoms of residue H393 are shown as *orange spheres*. This figure is used with permission from Gao et al. (2007)

docks onto the HPK domain of CikA in a manner similar to how Spo0F binds to Spo0B (Zapf et al. 2000) and thereby physically blocks H393 from phosphorylation (Fig. 6.8).

This model suggests that CikAPsR regulates phosphoryl transfer from the HPK domain to an as yet unidentified response regulator, CikR. Phosphorylation of a conserved aspartate residue in the receiver domain of CikR would activate its effector domain that, in turn, would activate transcription of downstream clock-resetting genes. Alternatively, the phosphorylated form of CikR could directly or indirectly interact with one of the Kai proteins to alter the phase of the KaiC phosphorylation cycle.

6.8 The Road Forward – NMR and Increasingly Large Molecular Complexes

We hope that we have succeeded in illustrating how NMR studies have contributed to our understanding of the structural biology of the cyanobacterial clock. For example, these studies have provided insights into the first clock protein domains – from any organism – in KaiA135N^{Se} and KaiA180C^{Te}. We postulated that KaiA acts as a major clock control molecule (Williams et al. 2002; Vakonakis and LiWang 2004; Vakonakis et al. 2004b) and we provided the first connection between cyanobacterial and plant oscillators through the common extensive use of pseudo-receivers (Strayer et al. 2000). The solution structures and the identification and characterization of the KaiA-interacting fragment of KaiC provided important starting points for later analyses of KaiA and KaiC by X-ray diffraction studies (e.g., Ye et al. 2004; Pattanayek et al. 2006; and as covered elsewhere in this book).

Naturally, not all predictions and hypotheses stood the test of time. Our proposed protein-interaction surface of KaiA135N^{Se} (Williams et al. 2002) was later found to be important in the context of full-length KaiA (Ye et al. 2004), rather than for interactions with upstream proteins. Similarly, the proposed KaiC-binding KaiA groove (Vakonakis et al. 2004b) was offset by 90° degrees (Vakonakis and LiWang 2004). Nevertheless, combining all modern structural methods has improved our understanding of the system tremendously; and, we believe, NMR has played an important role.

How can we move forward to better understand the cyanobacterial oscillators using NMR and other techniques? It is established that the exact state and operation of this mechanism depends on protein–protein interactions among KaiA, KaiB and KaiC. These interactions are determined by the phosphorylation state of KaiC (Nishiwaki et al. 2007; Rust et al. 2007); phosphoform-dependent affinities of KaiC for KaiB and KaiA are likely structurally modulated, as phosphorylation commonly drives large conformational changes (Barford et al. 1991; Russo et al. 1996; Walsh 2006). Thus, a comprehensive understanding of this central oscillator will require determining the dependence of the structure and dynamics of KaiC as a function of its phosphorylation state. KaiC, however, is a homohexamer with a molecular mass of approximately 350 kDa (Kageyama et al. 2003), leading to poor NMR spectra due to fast signal decay. Although a KaiC domain, CII, is known to be monomeric under some conditions (Hayashi et al. 2006), KaiC function and interactions with KaiA or KaiB likely depend on KaiC multimers (Pattanayek et al. 2004; Mori et al. 2007).

Recent advances allow us to overcome the problems of fast signal relaxation and overly complex spectra in large molecules, by combining selective labeling of proteins with a new type of NMR experiment (Ollerenshaw et al. 2003; Tugarinov et al. 2003). The proteins of interest are expressed in ways that allow uniform deuteration of hydrogen sites, except for a single methyl group in valine, leucine and isoleucine residues (Sprangers et al. 2007). This dilution of hydrogen atoms with deuterium atoms reduces spectra complexity and increases signal strength by taking advantage of certain NMR relaxation phenomena (Tugarinov et al. 2003). The remaining methyl group hydrogen atoms are almost uniformly spaced in the protein and serve as reporters of protein dynamics or structural perturbations due to phosphorylation or protein–protein interactions (Fig. 6.9).

Even in the absence of specific assignments for these methyl resonances, recent studies have shown ways to extract important information from the resulting spectra (Velyvis et al. 2007), by detecting changes in the overall protein structure and dynamics under different sample conditions. For example, NMR can be used to determine the distinct spectral signatures for labeled KaiC in its different phosphoforms or, more likely, stable phosphomimics (Nishiwaki et al. 2007) and in the presence and absence of unlabeled KaiB and/or KaiA. Indeed, several methyl groups accessible by this method are located along the subunit interfaces of KaiC and near the sites of phosphorylation, S431 and T432. This would allow us to

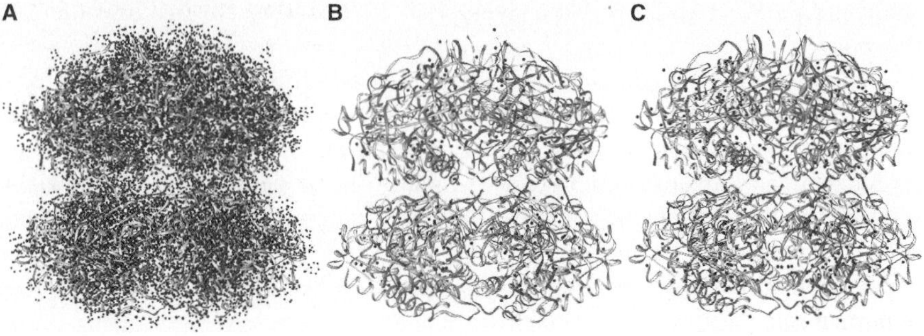

Fig. 6.9 Diluting hydrogen atoms in KaiC for NMR spectroscopy. In **A–C**, the backbone of the KaiC homohexamer is drawn as a ribbon. Black dots show the locations of: 6 × 3500 nonmethyl hydrogens and methyl groups (**A**), 6 × 47 C^{d1} methyl groups of isoleucine residues in the absence of other protons (**B**), and 6 × 33 Cg^1 and 6 × 37 C^{d1} methyl groups of valine and leucine residues, respectively, in the absence of other protons (**C**). For large proteins NMR signals are difficult to observe because extensive proton–proton dipolar interactions lead to very fast relaxation rates, which is true for fully protonated KaiC (**A**). Fast relaxation makes NMR signals broad and weak. Diluting protons with deuterons reduces proton–proton interactions, slowing the relaxation rates of the remaining protons and leading to sharper signals. Leaving only the methyl groups protonated (**B**, **C**) is particularly effective at improving NMR spectra and allows the NMR analyses of proteins far larger than KaiC (Sprangers and Kay 2007). Similar strategies can be employed in studies of KaiA or KaiB

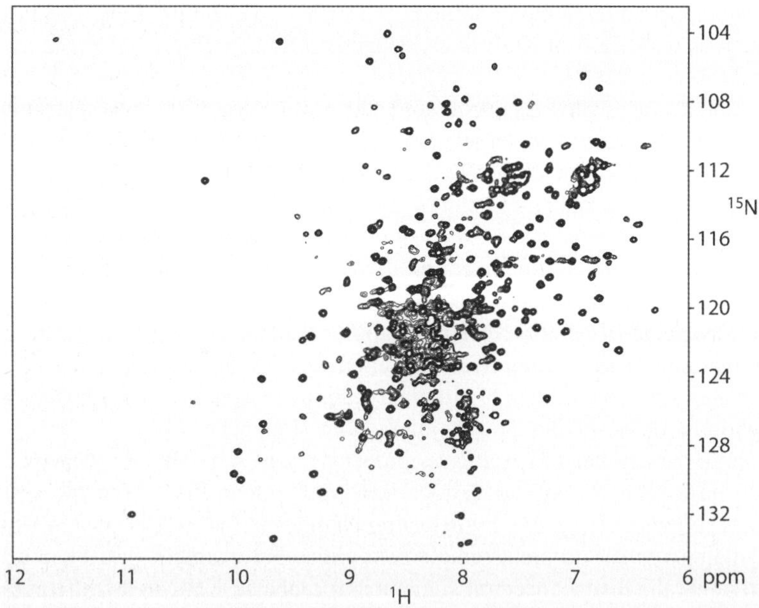

Fig. 6.10 NMR spectrum of the CII domain of *Thermosynechococcus elongatus* KaiC collected at 25°C at 950 MHz (University of Oxford). Spectral assignments from this domain, as well as CI, could be transferred to the whole KaiC, allowing us to map structural perturbations to crystal models. The sample was provided by Yong-Ick Kim

directly observe changes as these proteins interact on a timescale comparable to that of the clock. Additional information can be gathered by spectral assignments that would allow us to map the observed changes onto the available crystal structures. Recently, another large system, the 670-kDa 20S proteasome of Archaea was studied using methods that can probably be applied to the 350-kDa KaiC hexamer (Sprangers and Kay 2007). In this approach, spectral assignments of individual monomeric protein domains, such as CI and CII for KaiC, are determined first and then transferred to the full-length protein. Initial efforts towards that goal have been encouraging, as we have been able to collect good NMR spectra of the CII domain of KaiC in a monomeric state (Fig. 6.10).

As structural studies of the cyanobacterial circadian clock progress from the analysis of single proteins to large complexes, the need for interaction between structural biology approaches will increase. The role of NMR in this context will not likely be that of producing high-resolution structures of complexes de novo. Instead, NMR will reveal details and allow modeling of complexes through mapping onto existing structures binding sites and changes in structures and dynamics resulting from both strong and weak protein–protein interactions.

References

Barford D, Hu S-H, Johnson LN (1991) Structural mechanism for glycogen phosphorylase control by phosphorylation and AMP. J Mol Biol 218:233–260

Bateman A, Coin L, Durbin R, Finn RD, Hollich V, Griffiths-Jones S, Khanna A, Marshall M, Moxon S, Sonnhammer ELL, Studholme DJ, Yeats C, Eddy SR (2004) The Pfam protein families database. Nucleic Acids Res 32:D138–D141

Bourret RB, Hess JF, Simon MI (1990) Conserved aspartate residues and phosphorylation in signal transduction by the chemotaxis protein CheY. Proc Natl Acad Sci USA 87:41–45

Cavanagh J, Fairbrother WJ, Rance M, Skelton NJ (2007) Protein NMR spectroscopy principles and practice. Elsevier Academic, New York

Doublié S, Tabor S, Long AM, Richardson CC, Ellenberger T (1998) Crystal structure of a bacteriophage T7 DNA replication complex at 2.2 Å resolution. Nature 391:251–258

Dunlap JC, Loros JJ, DeCoursey PJ (2004). Chronobiology: biological timekeeping. Sinauer, Sunderland, Mass

Dutta R, Qin L, Inouye M (1999) Histidine kinases: diversity of domain organization. Mol Microbiol 34:633–640

Dvornyk V, Vinogradova O, Nevo E (2003) Origin and evolution of circadian clock genes in prokaryotes. Proc Natl Acad Sci USA 100:2495–2500

Dvornyk V, Deng H-W, Nevo E (2004) Structure and molecular phylogeny of sasA genes in cyanobacteria: insights into evolution of the prokaryotic circadian system. Mol Biol Evol 21:1468–1476

Gao T, Zhang X, Ivleva NB, Golden SS, LiWang A (2007) NMR structure of the pseudo-receiver domain of CikA. Protein Sci 16:465–475

Garces RG, Wu N, Gillon W, Pai EF (2004) *Anabaena* circadian clock proteins KaiA and KaiB reveal a potential common binding site to their partner KaiC. EMBO J 23:1688–1698

Harris NL, Presnell SR, Cohen FE (1994) Four helix bundle diversity in globular proteins. J Mol Biol 236:1356–1368

Hayashi F, Iwase R, Uzumaki T, Ishiura M (2006) Hexamerization by the N-terminal domain and intersubunit phosphorylation by the C-terminal domain of cyanobacterial circadian clock protein KaiC. Biochem Biophys Res Commun 348:864–872

Hitomi K, Oyama T, Han S, Arvai AS, Getzoff ED (2005) Tetrameric architecture of the circadian clock protein KaiB: a novel interface for intermolecular interactions and its impact on the circadian rhythm. J Biol Chem 280:19127–19135

Ishiura M, Kutsuna S, Aoki S, Iwasaki H, Andersson CR, Tanabe A, Golden SS, Johnson CH, Kondo T (1998) Expression of a gene cluster *kaiABC* as a circadian feedback process in cyanobacteria. Science 281:1519–1523

Ivleva NB, Gao T, LiWang A, Golden SS (2006) Quinone sensing by the circadian input kinase of the cyanobacterial circadian clock. Proc Natl Acad Sci USA 103:17468–17473

Iwasaki H, Kondo T (2000) The current state and problems of circadian clock studies in cyanobacteria. Plant Cell Physiol 41:1013–1020

Iwasaki H, Taniguchi Y, Ishiura M, Kondo T (1999) Physical interactions among circadian clock proteins KaiA, KaiB, and KaiC in cyanobacteria. EMBO J 18:1137–1145

Iwasaki H, Williams SB, Kitayama Y, Ishiura M, Golden SS, Kondo T (2000) A KaiC-interacting sensory histidine kinase, SasA, necessary to sustain robust circadian oscillation in cyanobacteria. Cell 101:223–233

Iwasaki H, Nishiwaki T, Kitayama Y, Nakajima M, Kondo T (2002) KaiA-stimulated KaiC phosphorylation in circadian timing loops in cyanobacteria. Proc Natl Acad Sci USA 99:15788–15793

Iwase R, Imada K, Hayashi F, Uzumaki T, Morishita M, Onai K, Furukawa Y, Namba K, Ishiura M (2005) Functionally important substructures of circadian clock protein KaiB in a unique tetramer complex. J Biol Chem 280:43141–43149

Kageyama H, Kondo T, Iwasaki H (2003) Circadian formation of clock protein complexes by KaiA, KaiB, KaiC, and SasA in cyanobacteria. J Biol Chem 278:2388–2395

Kelley LA, MacCallum RM, Sternberg MJE (2000) Enhanced genome annotation using structural profiles in the program 3D-PSSM. J Mol Biol 299:501–522

Kern D, Volkman BF, Luginbuhl P, Nohaile MJ, Kustu S, Wemmer DE (1999) Structure of a transiently phosphorylated switch in bacterial signal transduction. Nature 402:894–898

Kim YI, Dong G, Carruthers CW Jr, Golden SS, LiWang A (2008) The day/night switch in KaiC, a central oscillator component of the circadian clock of cyanobacteria. Proc Natl Acad Sci USA 105:12825–12830

Letunic I, Copley RR, Schmidt S, Ciccarelli FD, Doerks T, Schultz J, Ponting CP, Bork P (2004) SMART 4.0: towards genomic data integration. Nucleic Acids Res 32:D142–D144

Ma Q, Guo C, Barnewitz K, Sheldrick GM, Soling H-D, Uson I, Ferrari DM (2003) Crystal structure and functional analysis of *Drosophila* Wind, a protein-disulfide isomerase-related protein. J Biol Chem 278:44600–44607

Mackey SR, Golden SS (2007) Winding up the cyanobacterial circadian clock. Trends Microbiol 15:381–388

Martin JL (1995) Thioredoxin – a fold for all reasons. Structure 3:245–250

Mori T, Johnson CH (2001) Circadian programming in cyanobacteria. Semin Cell Dev Biol 12:271–278

Mori T, Williams DR, Byrne MO, Qin X, Egli M, McHaourab HS, Stewart PL, Johnson CH (2007) Elucidating the ticking of an in vitro circadian clockwork. PLoS Biol 5:841–853

Mutsuda M, Michel K-P, Zhang X, Montgomery BL, Golden SS (2003) Biochemical properties of CikA, an unusual phytochrome-like histidine protein kinase that resets the circadian clock in *Synechococcus elongatus* PCC 7942. J Biol Chem 278:19102–19110

Nakahira Y, Katayama M, Miyashita H, Kutsuna S, Iwasaki H, Oyama T, Kondo T (2004) Global gene repression by KaiC as a master process of prokaryotic circadian system. Proc Natl Acad Sci USA 101:881–885

Nishimura H, Nakahira Y, Imai K, Tsuruhara A, Kondo H, Hayashi H, Hirai M, Saito H, Kondo T (2002) Mutations in KaiA, a clock protein, extend the period of circadian rhythm in the cyanobacterium *Synechococcus elongatus* PCC 7942. Microbiology 148:2903–2909

Nishiwaki T, Iwasaki H, Ishiura M, Kondo T (2000) Nucleotide binding and autophosphorylation of the clock protein KaiC as a circadian timing process of cyanobacteria. Proc Natl Acad Sci USA 97:495–499

Nishiwaki T, Satomi Y, Kitayama Y, Terauchi K, Kiyohara R, Takao T, Kondo T (2007) A sequential program of dual phosphorylation of KaiC as a basis for circadian rhythm in cyanobacteria. EMBO J 26:4029–4037

O'Hara BP, Norman RA, Wan PTC, Roe SM, Barrett TE, Drew RE, Pearl LH (1999) Crystal structure and induction mechanism of AmiC-AmiR: a ligand-regulated transcription antitermination complex. EMBO J 18:5175–5186

Ollerenshaw JE, Tugarinov V, Kay LE (2003) Methyl TROSY: explanation and experimental verification. Magn Reson Chem 41:843–852

Pattanayek R, Wang J, Mori T, Xu Y, Johnson CH, Egli M (2004) Visualizing a circadian clock protein: crystal structure of KaiC and functional insights. Mol Cell 15:375–388

Pattanayek R, Williams DR, Pattanayek S, Xu Y, Mori T, Johnson CH, Stewart PL, Egli M (2006) Analysis of KaiA–KaiC protein interactions in the cyanobacterial circadian clock using hybrid structural methods. EMBO J 25:2017–2028

Pattanayek R, Williams DR, Pattanayek S, Mori T, Johnson CH, Stewart PL, Egli M (2008) Structural model of the circadian clock KaiB–KaiC complex and mechanism for modulation of KaiC phosphorylation. EMBO J 27:1767–1778

Robinson VL, Buckler DR, Stock AM (2000) A tale of two components: a novel kinase and a regulatory switch. Nat Struct Mol Biol 7:626–633

Russo AA, Jeffrey PD, Pavletich NP (1996) Structural basis of cyclin-dependent kinase activation by phosphorylation. Nat Struct Mol Biol 3:696–700

Rust MJ, Markson JS, Lane WS, Fisher DS, O'Shea EK (2007) Ordered phosphorylation governs oscillation of a three-protein circadian clock. Science 318:809–812

Schmitz O, Katayama M, Williams SB, Kondo T, Golden SS (2000) CikA, a bacteriophytochrome that resets the cyanobacterial circadian clock. Science 289:765–768

Smith RM, Williams SB (2006) Circadian rhythms in gene transcription imparted by chromosome compaction in the cyanobacterium *Synechococcus elongatus*. Proc Natl Acad Sci USA 103:8564–8569

Sprangers R, Kay LE (2007) Quantitative dynamics and binding studies of the 20S proteasome by NMR. Nature 445:618–622

Sprangers R, Velyvis A, Kay LE (2007) Solution NMR of supramolecular complexes: providing new insights into function. Nat Methods 4:697–703

Stock AM, Robinson VL, Goudreau PN (2000) Two-component signal transduction. Annu Rev Biochem 69:183–215

Strayer C, Oyama T, Schultz TF, Raman R, Somers DE, Más P, Panda S, Kreps JA, Kay SA (2000) Cloning of the *Arabidopsis* clock gene *TOC1*, an autoregulatory response regulator homolog. Science 289:768–771

Takai N, Nakajima M, Oyama T, Kito R, Sugita C, Sugita M, Kondo T, Iwasaki H (2006) A KaiC-associating SasA–RpaA two-component regulatory system as a major circadian timing mediator in cyanobacteria. Proc Natl Acad Sci USA 103:12109–12114

Taniguchi Y, Yamaguchi A, Hijikata A, Iwasaki H, Kamagata K, Ishiura M, Go M, Kondo T (2001) Two KaiA-binding domains of cyanobacterial circadian clock protein KaiC. FEBS Lett 496:86–90

Tugarinov V, Hwang PM, Ollerenshaw JE, Kay LE (2003) Cross-correlated relaxation enhanced ^{1}H-^{13}C NMR spectroscopy of methyl groups in very high molecular weight proteins and protein complexes. J Am Chem Soc 125:10420–10428

Uzumaki T, Fujita M, Nakatsu T, Hayashi F, Shibata H, Itoh N, Kato H, Ishiura M (2004) Crystal structure of the C-terminal clock-oscillator domain of the cyanobacterial KaiA protein. Nat Struct Mol Biol 11:623–631

Vakonakis I, LiWang AC (2004) Structure of the C-terminal domain of the clock protein KaiA in complex with a KaiC-derived peptide: implications for KaiC regulation. Proc Natl Acad Sci USA 101:10925–10930

Vakonakis I, Risinger AT, Latham MP, Williams SB, Golden SS, LiWang AC (2001) Sequence-specific 1H, 13C and 15N resonance assignments of the N-terminal, 135-residue domain of KaiA, a clock protein from *Synechococcus elongatus*. J Biomol NMR 21:179–180

Vakonakis I, Klewer DA, Williams SB, Golden SS, LiWang AC (2004a) Structure of the N-terminal domain of the circadian clock-associated histidine kinase SasA. J Mol Biol 342:9–17

Vakonakis I, Sun J, Wu T, Holzenburg A, Golden SS, LiWang AC (2004b) NMR structure of the KaiC-interacting C-terminal domain of KaiA, a circadian clock protein: implications for the KaiA–KaiC interaction. Proc Natl Acad Sci USA 101:1479–1484

Velyvis A, Yang YR, Schachman HK, Kay LE (2007) A solution NMR study showing that active site ligands and nucleotides directly perturb the allosteric equilibrium in aspartate transcarbamoylase. Proc Natl Acad Sci USA 104:8815–8820

Volkman BF, Lipson D, Wemmer DE, Kern D (2001) Two-state allosteric behavior in a single-domain signaling protein. Science 291:2429–2433

Volz K (1993) Structural conservation in the CheY superfamily. Biochemistry 32:11741–11753

Walsh CT (2006). Posttranslational modification of proteins: expanding nature's inventory. Roberts, Englewood

Williams SB, Vakonakis I, Golden SS, LiWang AC (2002) Structure and function from the circadian clock protein KaiA of *Synechococcus elongatus*: A potential clock input mechanism. Proc Natl Acad Sci USA 99:15357–15362

Xu Y, Piston DW, Johnson CH (1999) A bioluminescence resonance energy transfer (BRET) system: application to interacting circadian clock proteins. Proc Natl Acad Sci USA 96:151–156

Ye S, Vakonakis I, Ioerger TR, LiWang AC, Sacchettini JC (2004) Crystal structure of circadian clock protein KaiA from *Synechococcus elongatus*. J Biol Chem 279:20511–20518

Zapf J, Sen U, Madhusudan, Hoch JA, Varughese KI (2000) A transient interaction between two phosphorelay proteins trapped in a crystal lattice reveals the mechanism of molecular recognition and phosphotransfer in signal transduction. Structure 8:851–862

Zhang X, Dong G, Golden SS (2006) The *pseudo*-receiver domain of CikA regulates the cyanobacterial circadian input pathway. Mol Microbiol 60:658–668

Chapter 7
Structural Aspects of the Cyanobacterial KaiABC Circadian Clock

Martin Egli and Phoebe L. Stewart

Abstract The KaiABC circadian clock in the cyanobacterium *Synechococcus elongatus* can be reconstituted in vitro from three proteins in the presence of ATP. This oscillator displays the pertinent features of circadian rhythms including a self-sustained 24-h period and temperature compensation. At every phase of the cycle there is a mixture of types of Kai complexes and the proportions of the various types are oscillating. The KaiC protein is an auto-kinase and auto-phosphatase whose phosphorylation levels oscillate over the daily period whereby KaiA and KaiB interact with KaiC to increase and decrease its phosphorylation, respectively. This chapter provides an overview of the three-dimensional (3D) structural characterization of Kai proteins and the understanding of the KaiA–KaiC interaction gained by NMR and 3D electron microscopy (EM). Despite impressive advances in the structural realm, many open questions remain regarding the control of KaiC phosphorylation by KaiA and KaiB and conformational changes accompanying the transition between the hypo and hyper phosphorylated states of KaiC.

7.1 Introduction

Recent research has shown that the cyanobacterial circadian clock is able to function without de novo synthesis of clock gene mRNAs and the proteins encoded by them, and accurate determination of the period is achieved without transcriptional/translational feedback (Tomita et al. 2005). In the model organism *Synechococcus elongatus* PCC 7942 there exists a minimal timing loop in vivo that functions without transcription and translation and exhibits temperature compensation. Even more remarkably, it was found that the circadian clock in *S. elongatus* can be fully

M. Egli(✉)
Department of Biochemistry, School of Medicine, Vanderbilt University, Nashville, TN 37232, USA, e-mail: martin.egli@vanderbilt.edu

P.L. Stewart
Department of Molecular Physiology and Biophysics, School of Medicine, Vanderbilt University, Nashville, TN 37232, USA

J.L. Ditty et al. (eds.), *Bacterial Circadian Programs.*

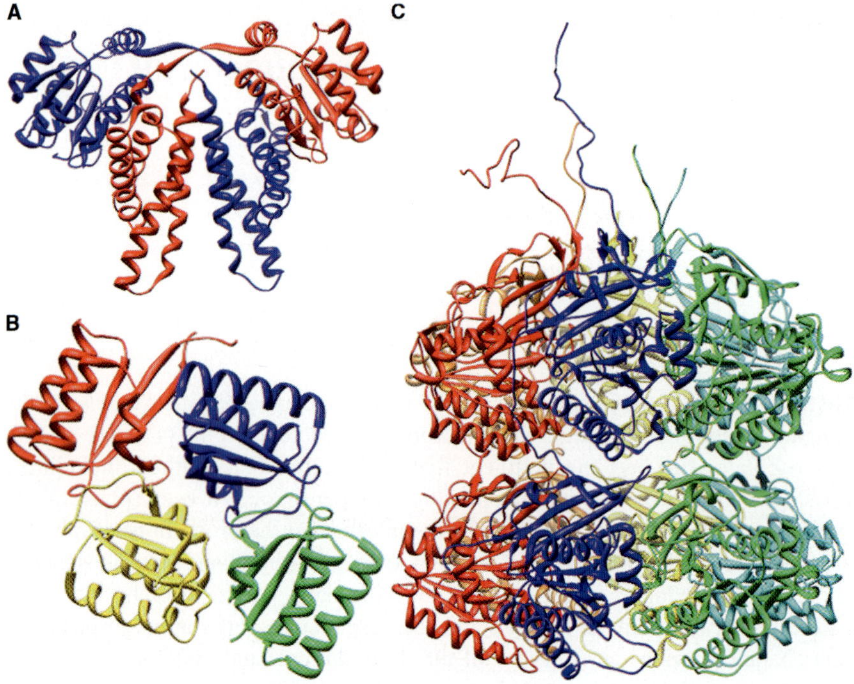

Fig. 7.1 Structures of the cyanobacterial clock proteins KaiA, KaiB, and KaiC. **A** The crystal structure of the *Synechococcus elongatus* KaiA dimer, molecular mass ~64 kDa (PDB-ID 1R8J; Ye et al. 2004). **B** The crystal structure of the *Synechocystis* KaiB tetramer, molecular mass ~48 kDa (PDB-ID 1WWJ; Hitomi et al. 2005). **C** The crystal structure of the *Synechococcus elongatus* KaiC hexamer with extended C-termini, molecular mass ~360 kDa (PDB-ID 2GBL; Pattanayek et al. 2006). Each subunit of the multimeric proteins (KaiA, KaiB, KaiC) is colored differently. Molecular graphics image produced with the UCSF Chimera package (Pettersen et al. 2004)

reconstituted in vitro by the three proteins KaiA, KaiB and KaiC in the presence of ATP (Nakajima et al. 2005). These discoveries render the KaiABC timekeeper a unique target for biochemical and biophysical analyses. Three-dimensional (3D) structures of full-length versions of the KaiA, KaiB and KaiC proteins from different cyanobacterial strains became available in 2004 (for reviews, see Golden 2004; Johnson and Egli 2004; Egli et al. 2007; Fig. 7.1).

The first structure to be determined was that of the N-terminal pseudo-receiver domain of KaiA from *S. elongatus* (Williams et al. 2002; Table 7.1). Afterward, the structure of the KaiC protein was initially characterized by negative-stain electron microscopy (EM) studies that revealed a homo-hexameric particle with a central opening (Hayashi et al. 2003; Mori et al. 2003). Subsequently, a crystal structure of full-length KaiA from *S. elongatus* exposed a domain-swapped dimer with three different dimer interfaces (Ye et al. 2004). One of these connects the N-terminal receiver domain with the C-terminal KaiC-interacting domain (Fig. 7.1A). Further KaiA structures include those of the C-terminal dimerization and KaiC-interacting

Table 7.1 Three-dimensional structures of protein KaiA

Protein	Construct	Organism	Technique	Reference	PDB identity code[a]
KaiA	N-terminal domain	PCC 7942 *Synechococcus elongatus*	NMR	Williams et al. (2002)	1M2E
KaiA	Full length	*S. elongatus*	X-ray	Ye et al. (2004)	1R8J
KaiA	Full length	PCC 7120 *Anabaena*	X-ray	Garces et al. (2004)	1R5Q
KaiA	C-terminal domain	*Thermosynechococcus elongatus* BP-1	X-ray	Uzumaki et al. (2004)	1V2Z
KaiA	C-terminal domain	*T. elongatus*	NMR	Vakonakis et al. (2004a)	1Q6A

[a]http://www.rcsb.org (Berman et al. 2000)

Table 7.2 Three-dimensional structures of protein KaiB

Protein	Construct	Organism	Technique	Reference	PDB identity code
KaiB	Full length	*Anabaena*	X-ray	Garces et al. (2004)	1R5P
KaiB	Full length	PCC 6803 *Synechocystis*	X-ray	Hitomi et al. (2005)	1WWJ
KaiB	Full length (T64C mutant)	*T. elongatus*	X-ray	Iwase et al. (2005)	1VL
KaiB	Full length (wild type)	*T. elongatus*	X-ray	Pattanayek et al. (2008)	2QKE

domain of KaiA from *Thermosynechococcus elongatus* BP-1, determined separately by X-ray crystallography (Uzumaki et al. 2004) and NMR (Vakonakis et al. 2004a), and the crystal structure of the C terminal domain of KaiA from the cyanobacterium *Anabaena* PCC 7120 (Garces et al. 2004; Table 7.1).

The crystal structure of full-length *Anabaena* KaiB revealed a thioredoxin-like fold (Garces et al. 2004; Fig. 7.1B; Table 7.2). In this structure KaiB was found to be a dimer with a loop region in the monomer contributing the majority of the interactions between the two subunits. In the crystal structures of KaiB from *Synechocystis* PCC 6803 (Hitomi et al. 2005) and *T. elongatus* (Iwase et al. 2005), the protein forms a tetramer with a positively charged perimeter, a negatively charged center and a zipper of aromatic rings important for oligomerization. There is evidence based on mutational data that supports the importance of the tetrameric state of KaiB (Hitomi et al. 2005) and the C-terminal acidic region (Iwase et al. 2005) for proper clock function.

The crystal structure of the full-length KaiC protein from *S. elongatus* was determined in one of our laboratories (Pattanayek et al. 2004). As expected, based on earlier EM results, the central and largest protein from the cyanobacterial clock exists in the form of a homo-hexamer with a central pore. Its shape resembles a

Table 7.3 Three-dimensional structures of protein KaiC

Protein	Construct	Organism	Technique	Reference	PDB identity code
KaiC	Full length	*S. elongatus*	X-ray	Pattanayek et al. (2004)	1TF7
KaiC	Full length; the structure is based on the same crystallographic data as 1TF7, but phosphate groups were added to S431 (in four subunits) and T432 (in six subunits)	*S. elongatus*	X-ray	Xu et al. (2004)	1U9I
KaiC	Full length; the structure is based on the same crystallographic data as 1TF7, but the C-terminal tails of subunits were extended to varying degrees, and for two of them all residues up to S519 were added	*S. elongatus*	X-ray	Pattanayek et al. (2006)	2GBL

double-doughnut, in which the N-terminal CI and the C-terminal CII halves constitute the lower and the upper rings, respectively (Fig. 7.1C; Table 7.3). Twelve ATP molecules are bound between the interfaces of CI and CII domains of monomers. The key phosphorylation sites S431 and T432 were independently identified through their crystal structure (Xu et al. 2004) and from a mass spectrometric approach (Nishiwaki et al. 2004); and they map exclusively to the CII half of the protein. In the crystal structure, all six T432 sites were phosphorylated, whereas four of the S431 sites were phosphorylated. The phosphorylation occurs across subunits and, when S431 is phosphorylated, the hydroxyl group of T426 forms a hydrogen bond to that phosphate group (Xu et al. 2004). We took this as evidence that T426 represents a third possible phosphorylation site. These three residues (T426, S431, T432), when mutated to alanine individually, abolish rhythmicity and the triple mutant (T426/S431/T432→A) is no longer phosphorylatable.

An NMR structure of the complex of a KaiC peptide with the C-terminal domain of KaiA showed that the C-terminal peptide of KaiC interacts with KaiA (Vakonakis and LiWang 2004; Table 7.4). However there were suggestions from yeast two-hybrid studies that KaiA might also bind to the linker region of KaiC between the CI and CII domains (Taniguchi et al. 2001). A 3D EM study of the KaiA–KaiC complex with full-length proteins combined with an analysis of KaiC truncation mutants showed that KaiA binds exclusively to the CII half of KaiC (Pattanayek et al. 2006), suggesting that the previous observation of KaiC binding to the linker region might have been an artifact of yeast two-hybrid interaction methodology.

The past several years have also witnessed a flurry of functional advances in the cyanobacterial circadian clock field (Kageyama et al. 2006; Ito et al. 2007; Mori

Table 7.4 Three-dimensional structures of KaiA–KaiC complexes

Protein	Construct	Organism	Technique	Reference	PDB identity code
KaiA/KaiC	C-terminal KaiC peptide (amino acids 488–518), in complex with the C-terminal domain of KaiA	*T. elongatus*	NMR	Vakonakis and LiWang (2004)	1SUY
KaiA/KaiC	Full length KaiA and KaiC; the complex was modeled using coordinates of the crystal structures of KaiC (2GBL) and KaiA (1R8J) and the KaiC peptide–KaiA complex (1SUY) from NMR	*T. elongatus*	EM	Pattanayek et al. (2006)	–

et al. 2007; Rust et al. 2007; Nishiwaki et al. 2007; Terauchi et al. 2007). Despite the wealth of recent structural and functional data on the KaiABC clock, mysteries regarding its inner workings still remain (Golden et al. 2007). This chapter provides a detailed account of the anatomy of the *S. elongatus* KaiC hexamer based on the crystal structure obtained at 2.8 Å resolution (Pattanayek et al. 2004). We also summarize insights obtained regarding the KaiA–KaiC interaction based on our own studies, combining X-ray, crystallography, EM and modeling (Pattanayek et al. 2006; Mori et al. 2007).

7.2 Overall Structure of KaiC

The crystal structure of *S. elongatus* KaiC determined at 2.8 Å resolution revealed a hexamer in the form of a double doughnut with a constricted waist region and overall dimensions of ca. 100 × 100 Å (Fig. 7.1C; Pattanayek et al. 2004). The central channel of the hexamer is ca. 20 Å wide on average, but the channel is constricted at the CII side by six arginine residues (Fig. 7.2). The two hexameric rings, CI and CII, have similar overall shapes, but the CII side of the hexamer differs in that it has protruding C-terminal peptides (Fig. 7.2C). In the initial 3D structural model, the last 20 residues of individual KaiC subunits were missing because the C-terminal regions of individual KaiC molecules exhibit considerable conformational flexibility, resulting in poorly defined electron density. The discovery that the C-terminal residues of KaiC are crucial for KaiA binding (Vakonakis and LiWang 2004) prompted us to a carefully inspect the density above the C-terminal dome and incorporate into the model individual C-terminal tails of various lengths (Fig. 7.1C). For two of the subunits all C-terminal residues, including S519, were built into the electron density (Pattanayek et al. 2006). Twelve ATP molecules are bound between individual subunits, six each in the N- and C-terminal

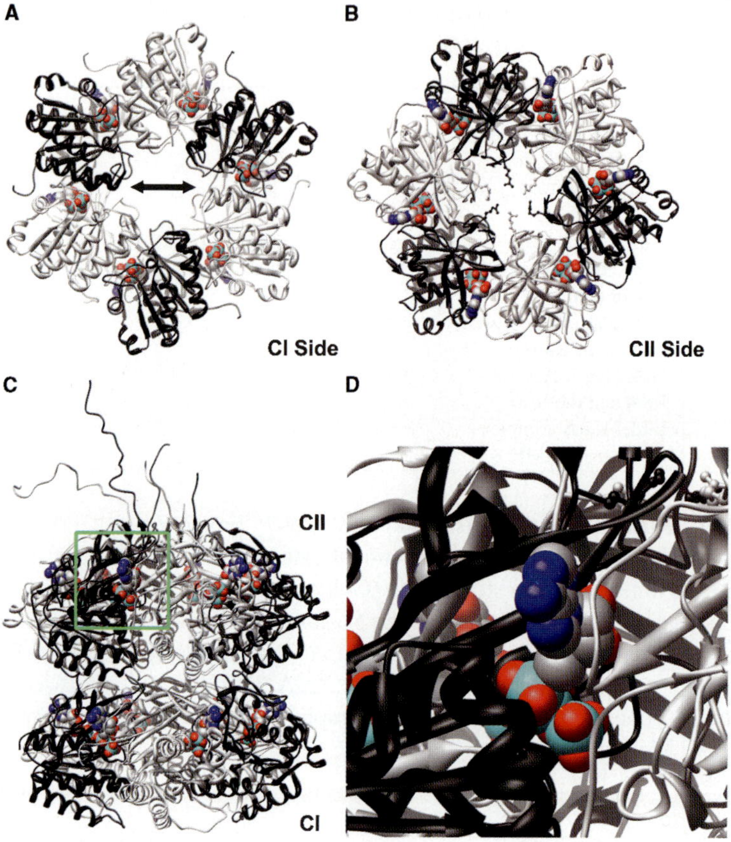

Fig. 7.2 Twelve ATP binding sites in KaiC. **A** There are six ATP binding sites between subunits in the ring of KaiC CI domains. The average diameter of the central channel is 20 Å, as indicated by the *double arrow*. View is from the CI surface of the hexamer. **B** There are an additional six ATP binding sites between subunits in the ring of KaiC CII domains. The side-chains of Arg488, which constrict the opening of the central channel on the CII side, are shown. View is from the CII surface of the hexamer. **C** Side view of the KaiC hexamer. The CI and CII surfaces are indicated. **D** Enlarged view of boxed region in part C showing one ATP molecule between two CII domains with the tip of the adenine base (*blue* and *white*) most accessible and the three phosphate groups (*cyan* and *red*) buried within the protein. ATP molecules are shown colored by element type and alternately KaiC subunits are shown in *gray* and *black* (PDB-ID 2GBL)

rings. ATP molecules are almost completely buried in the space between neighboring subunits; only an edge of the adenine base is exposed on the surface of the hexamer (Fig. 7.2D).

The KaiC protein is the product of a gene duplication (Ishiura et al. 1998) and the 3D-structure mirrors characteristics at the gene and primary sequence levels in that the KaiC monomer exhibits a two-domain fold (Fig. 7.3B). The N-terminal CI and C-terminal CII domains are arranged in a serial fashion and are related by a translation of 42 Å and a rotation of 15°. They adopt fairly similar core structures with a root mean square (r.m.s.) deviation of 2.45 Å, based on 208 matching

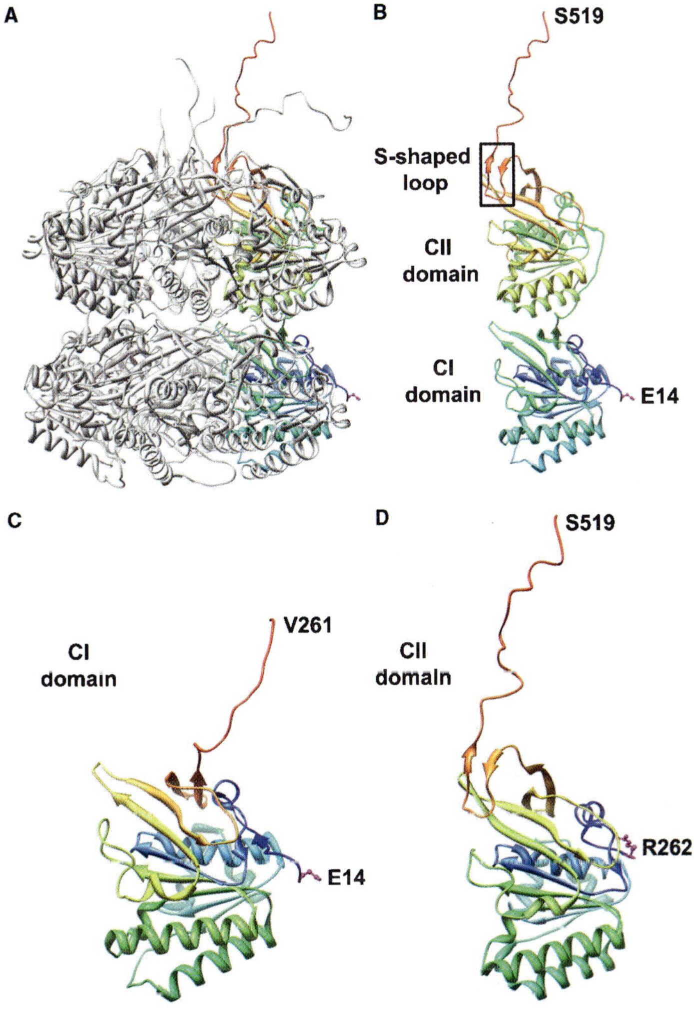

Fig. 7.3 KaiC is the product of gene duplication. **A** The KaiC hexamer with one of the six subunits (chain A) shown *rainbow colored*, N-terminus in *magenta*, and C-terminus in *red*. **B** KaiC chain A with the CI and CII domains indicated and the S-shaped loop (aa485–497) boxed. The N-terminal 13 residues, which extend from the side of the CI domain, are missing due to disorder. Residue E14 (*magenta*) is shown with its side-chain in a ball-and-stick representation. Residue S519 is the C-terminal residue of *S. elongatus* KaiC and is modeled in two of the six chains (A, F; PDB-ID 2GBL). **C** CI domain (aa14–261) including the extended linker between domains (*red*). The N-terminal residue, E14, is shown with its side-chain (*magenta*). **D** CII domain (aa262–519) including the flexible C-terminal tail (*red*). The N-terminal residue, R262, is shown with its side-chain (*magenta*)

Cα pairs. Notable deviations in the structures of CI and CII are found in their N-terminal and C-terminal regions. In the case of CI, the N-terminal region protrudes from the outer side of the domain and the positions of the first 14 residues were not resolved in electron density maps. Whereas the C-terminal tails of CII jut out from the dome region, following an S-shaped loop that borders on the channel exit, the C-terminal portion of CI links the two domains (Fig. 7.3D). Following a short β-strand that is part of the waist, the linker winds up on the outside of the CII domain and enters it near the border of the dome (Fig. 7.3C). The existence of this lever-like arrangement suggests an inherent flexibility in the relative orientation of the CI and CII domains. The covalent linkage between CI and CII is required for proper function of the clock. Using the full-length KaiC protein from *T. elongatus* and separately expressed CI and CII domains, it was shown that the combination of the two domains in the absence of the linker led to a drastic reduction in the thermodynamic stability relative to that of the wild-type protein (Hayashi et al. 2006).

As expected from comparisons of the primary sequences that implicated KaiCI and KaiCII as members of the DnaB/RecA superfamily of proteins (Leipe et al. 2000), the structures of the CI and CII cores display similarity to the folds adopted by DNA helicases (Pattanayek et al. 2004). However, comparison of the CI and CII hexameric rings with the hexamers of helicases demonstrates that the similarities are much closer at the monomer level, i.e., ring diameter and locations of the ATP binding cleft differ considerably in some cases. Moreover, none of the helicases features covalently linked rings as seen with KaiC. Unexpectedly, based on sequence alignments, F1 ATPase, a single ring comprised of a trimer of αβ-heterodimers (Abrahams et al. 1994), was found to exhibit the closest structural similarity to the CI and CII hexameric rings (Pattanayek et al. 2004).

7.3 ATP Binding and Relative Stability of the CI and CII Hexamers

ATP molecules are bound between individual subunits in both the CI and CII halves of the KaiC hexamer (Fig. 7.2). Note that the crystal structure features the more slowly hydrolyzing ATPγS analog. In both the CI and CII domains of KaiC, ATP forms specific interactions with the two neighboring subunits. In the CI half contacts to ATP phosphate groups from one monomer include the conserved P-loop amino acids T50, K52 and T53. Residues S89, K232 and D241 form hydrogen bonds to the nucleobase. Residues from the second monomer stabilizing ATP comprise K224 and R226 that contact the γ-phosphate group and H230 that is engaged in a hydrogen bond to the 2′-hydroxyl group of the ribose moiety. The observed binding mode involving P-loop residue K52 is in line with the earlier observation that this lysine is indispensable for ATP binding (Nishiwaki et al. 2000). In addition, the specific interactions made to the nucleobase portion of ATP help rationalize the preference for ATP over GTP by KaiCI. Interestingly, these interactions between CI residues and adenine atoms are missing in the CII half: no direct contacts between amino acid side-chains and the nucleobase exist there. This provides a rationalization why KaiCII is unable to discriminate to a significant degree between ATP and GTP (Nishiwaki et al. 2000; Mori and Johnson 2001). Another difference between the CI and CII halves concerns the effect of the mutation of the P-loop lysine: the KaiCI K52H mutant triggered a complete disruption of the rhythm whereas mutation of the corresponding CII K294 to histidine resulted in a long-period phenotype (Nishiwaki et al. 2000). The structure reveals that, unlike K52, K294 does not engage in a direct contact to the ATP γ-phosphate. Instead of K294 another lysine, K457, forms a salt bridge to that phosphate group. In contrast, threonines from the Walker A motif (T53 in CI and T295 in CII) interact with the γ-phosphate of ATP directly or via Mg^{2+} (T295) in both halves, thus explaining the arrhythmic phenotype of T53A and T295A mutants (Mori and Johnson 2001).

The crystal structure revealed different binding modes for ATP in the CI and CII halves. The binding interface in CI is specific for ATP (due to direct interactions with the nucleobase), but a conformational disorder of the γ-phosphate group in CI evident in electron density maps supports the notion that this portion of ATP is rather loosely bound. Conversely, the CII-binding interface manifests only weak restraints of the nucleobase portion, whereas the γ-phosphate is tightly gripped by surrounding residues as well as by Mg^{2+}. That these observations regarding differences between ATP binding and subunit interfaces in the CI and CII halves are not simply artifacts of the crystal structure is corroborated by thermodynamic data that indicate that the CI ring is more stable than the CII ring. This finding is essential as it implies different functions of the CI and CII rings.

7.4 Subunit Interface and Phosphorylation

KaiC is an auto-kinase that phosphorylates serines and threonines as well as an auto-phosphatase, and it exhibits both of these functions in vitro and in vivo (Nishiwaki et al. 2000; Iwasaki et al. 2002; Xu et al. 2003) and clock speed is correlated with the level of phosphorylation (Xu et al. 2003). Expression in *Escherichia coli* and purification of the KaiC protein from *S. elongatus* are always carried out in the presence of ATP. Therefore, the resultant protein is a mixture of the phospho- and dephospho-forms as judged by SDS-PAGE analysis. Although ATP was replaced by ATPγS prior to crystallization, the crystal structure could thus be expected to disclose phosphorylation sites. Following refinement of the crystal structure, inspection of difference Fourier electron density maps revealed peaks in the vicinity of two residues, T432 and S431, that are consistent with the presence of phosphate groups (Pattanayek et al. 2004; Xu et al. 2004; Fig. 7.4). In the crystal structure, the threonine is phosphorylated in all six subunits and serine is phosphorylated in four subunits. A second threonine, T426, is located in the immediate vicinity of S431 and its hydroxyl group forms a hydrogen bond to the phosphate of the latter. However, T426 does not seem to become phosphorylated.

Individual T432A and S431A mutations and also the T426A mutation alter KaiC phosphorylation in vivo (Xu et al. 2004). Both the T432A and T426A mutations lead to a significant reduction in the amount of phosphorylated KaiC. In contrast, the S431A mutation increases the ratio of phospho-KaiC to dephospho-KaiC. The S431A/T426A double mutant displays phosphorylation that is similar to that of wild-type KaiC and the triple mutant T432A/S431A/T426A shows no sign of phosphorylation. This latter observation and the lack of phosphorylation with the T432A/S431A double

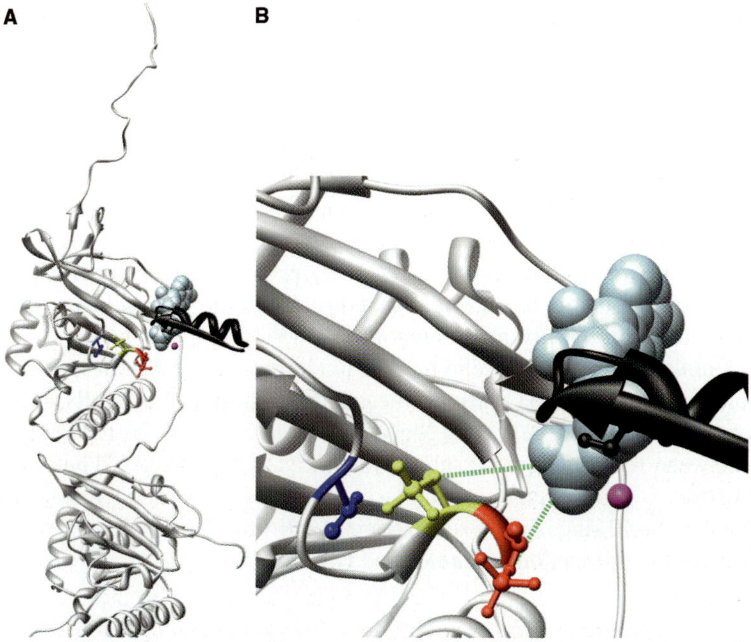

Fig. 7.4 Phosphorylation sites in KaiC. **A** KaiC chain A (*gray*) with a short portion of the neighboring chain F (*black*), including the P-loop (aa288–295; PDB-ID 2GBL). Also shown are the ATPγS bound between the CII domains (*sky blue*) and the nearby Mg^{2+} ion (*magenta*). Chain A T426 (*blue*), phosphorylated S431 (*yellow*), phosphorylated T432 (*red*), and chain F K294 in the P-loop (*black*) are shown with their side-chains. **B** Enlarged view of the kinase active site with the distances between the S431 and T432 hydroxyl groups with the ATPγS–phosphate indicated by dashed lines (*green*). These distances are 8.2 Å and 7.1 Å, respectively

mutant (Nishiwaki et al. 2004) provide evidence that there are two main phosphorylation sites per KaiC subunit. The KaiCI domain seems to be devoid of phosphorylatable Ser and The residues. The individual T432A, S431A and T426A mutants abolish circadian rhythmicity and the double and triple mutants are also arrhythmic. The effect does not seem related to the inability of phosphorylation mutants to hexamerize as the mutations do not disrupt hexamer formation (Xu et al. 2004).

Residues T432, S431 and T426 are located in a loop region that connects two β-strands and the two latter residues face each other across the loop (Fig. 7.4). A Mg^{2+} ion coordinates to the β- and γ-phosphate groups of ATP and engages in additional inner-sphere contacts to the side-chains of residues T295, E318, E319 and D378 from a subunit adjacent to that carrying phosphorylated T432 and S431 residues. Hence, KaiC phosphorylation proceeds across the subunit interface and the presence of phosphate groups at T432 and S431 results in additional interactions between amino acids from neighboring subunits. Phosphorylation of T432 leads to new contacts to R385 and E318 from the adjacent subunit. In the case of S431 addition of a phosphate results in a hydrogen bond to H429 from the same subunit. This histidine in turn interacts with D427 from the adjacent subunit. An additional interaction of the S431 phosphate concerns the previously described hydrogen bond to the γ-hydroxyl group of T426, a contact that is absent in the dephospho-KaiC structure. Based on the analysis of the interactions at the subunit interfaces in the phospho-KaiC hexamer, it would appear that phosphorylation of T432 and S431 leads to tighter binding between adjacent CII domains.

Recently it was shown that the phosphorylation cycle of the KaiC protein entails four steps – T432 phosphorylation, S431 phosphorylation, T432 dephosphorylation and S431 dephosphorylation – and that the product of each step regulates the reaction in the next step (Nishiwaki et al. 2007; Rust et al. 2007). Complete phosphorylation of both T432 and S431 converts KaiC from an auto-kinase to an auto-phosphatase. The finding that T432 is the first site to be phosphorylated is consistent with the crystallographic data that revealed phosphorylated T432 residues in all six subunits (only four of the six S431 residues were phosphorylated). Although all T432 and S431 residues are relatively far removed from the γ-phosphate of ATP, the distances between the T432 hydroxyl groups and the γ-phosphates are shorter on average (7.1 Å for chain A) than those between the S431 hydroxyl groups and the γ-phosphates (8.2 Å for chain A; Fig. 7.4). Phosphorylation of T432 generates new contacts across subunit interfaces (i.e., the interaction to E318) that could in turn lead to increased phosphorylation of S431. The hexamer trapped in the crystal structure is likely representative of the hyperphosphorylated form of KaiC. The distances between the hydroxyl groups of both T432 and S431 and the γ-phosphate of ATP exceed by far the spacing consistent with an active form of the kinase. This points to a considerable plasticity of the interface between CII domains, a notion that is in line with the different functions of the KaiCI and KaiCII halves inferred above from divergent ATP-binding and thermodynamic stabilities as well as the fact that the CI half lacks phosphorylation sites. It is reasonable to view the CI ring as a structural platform with a relatively rigid interface between subunits. Conversely, the CII ring is composed of subunits with variable relative orientations – most likely the result of small conformational adjustments in the central linker region between CI and CII – that form the basis for the controlled step-by-step phosphorylation and dephosphorylation process with a concomitant transition of KaiC from an auto-kinase to an auto-phosphatase.

7.5 The KaiA–KaiC Interaction

7.5.1 Binding of the KaiA Dimer to the C-Terminal Tail of KaiC

Overexpression of KaiA results in enhancement of *kaiBC* promoter activity, while continuous high levels of KaiC result in repression of the *kaiBC* promoter (Ishiura et al. 1998). Both in vitro and in vivo, KaiA is an enhancer of KaiC phosphorylation and KaiB antagonizes the action of KaiA (Iwasaki et al. 2002; Williams et al. 2002; Kitayama et al. 2003; Xu et al. 2003). In the case of *T. elongatus* it was found that a single KaiA dimer is sufficient to upregulate the phosphorylation of a KaiC hexamer to saturated levels (Hayashi et al. 2004b), consistent with the higher abundance of KaiC in vivo relative to KaiA (Kitayama et al. 2003). It is noteworthy that the question whether KaiA actually increases phosphorylation or decreases dephosphorylation is still open at the moment. Early models had placed the KaiA–KaiC binding interface in the waist region of the KaiC double hexamer (Taniguchi et al. 2001; Vakonakis et al. 2004). The models by Taniguchi and coworkers relied on yeast two-hybrid screens of KaiA with fragments of KaiC. KaiA exhibited affinities to C-terminal fragments from CI and CII that were not drastically reduced compared to that for the full-length KaiC protein. Prior to the crystal structure of KaiC, it was believed that the CII domain might fold back onto the CI domain, resulting in a tail-to-tail orientation of the two halves. The C-terminal regions implicated in KaiA binding would then map to the central waist region. However, the crystal structure demonstrated that CI and CII are arranged head-to-tail (Fig. 7.1). Thus, the KaiA-binding regions based on the two-hybrid analysis would be located at the waist and the CII-terminal dome and therefore be far removed from one another. An alternative theoretical model had the refolded KaiA dimer inserted into the central channel of KaiC (Wang 2005). Neither model is consistent with more recent biochemical and structural data.

Vakonakis and LiWang (2004) determined the NMR solution structure of the complex between the C-terminal domain of *T. elongatus* KaiA (residues 180–283) and a 30-mer peptide (residues 488–518) derived from the C-terminus of *T. elongatus* KaiC (Table 7.4; see Chap. 6). Unlike the C-terminal region of KaiCII for which a specific interaction with the dimer of the C-terminal domain of KaiA was found, a KaiCI C-terminal peptide corresponding to residues 241–260 of *T. elongatus* KaiC when mixed with KaiA did not trigger any changes in the NMR spectra of the latter. The conformation of the dimerized C-terminal domain of KaiA does not fundamentally alter upon binding to two KaiC peptides. Thus, the overall fold of KaiA with the four α-helices organized into two antiparallel helix–loop–helix pairs is maintained in the complex (Fig. 7.5). KaiA molecules dimerize along the C-terminal half of the longest α-helix, primarily via coiled-coil hydrophobic interactions. The dimer interface is stabilized by additional hydrophobic interactions and an intersubunit salt bridge as well as two putative hydrogen bonds. The relatively wide angle of around 50° between pairs of antiparallel α-helices at the dimerization interface of KaiA opens up somewhat as a consequence of binding of KaiC peptides. Whereas the KaiA monomeric subunit is more or less unchanged by KaiC-peptide binding, the angle of dimerization changes between free and bound KaiA through a relative rotation around the dimerization interface. This rotation is due to the KaiC peptides inserting non-polar side-chains of residues L505 and A506 into the KaiA dimerization groove, thereby forming a hydrophobic cluster with side-chains of KaiA residues L233, H236, L264 and I265. It has been suggested that the KaiA–KaiC affinity can be modulated by changes in the dimerization geometry of the KaiA C-terminal domain (Vakonakis and LiWang 2004). The backbone r.m.s. deviation between the structures of free and bound KaiA dimers amounts to about 1.3 Å including all ordered residues. C-terminal KaiCII peptides bound to the KaiA dimer adopt an extended L-shaped conformation (Fig. 7.5).

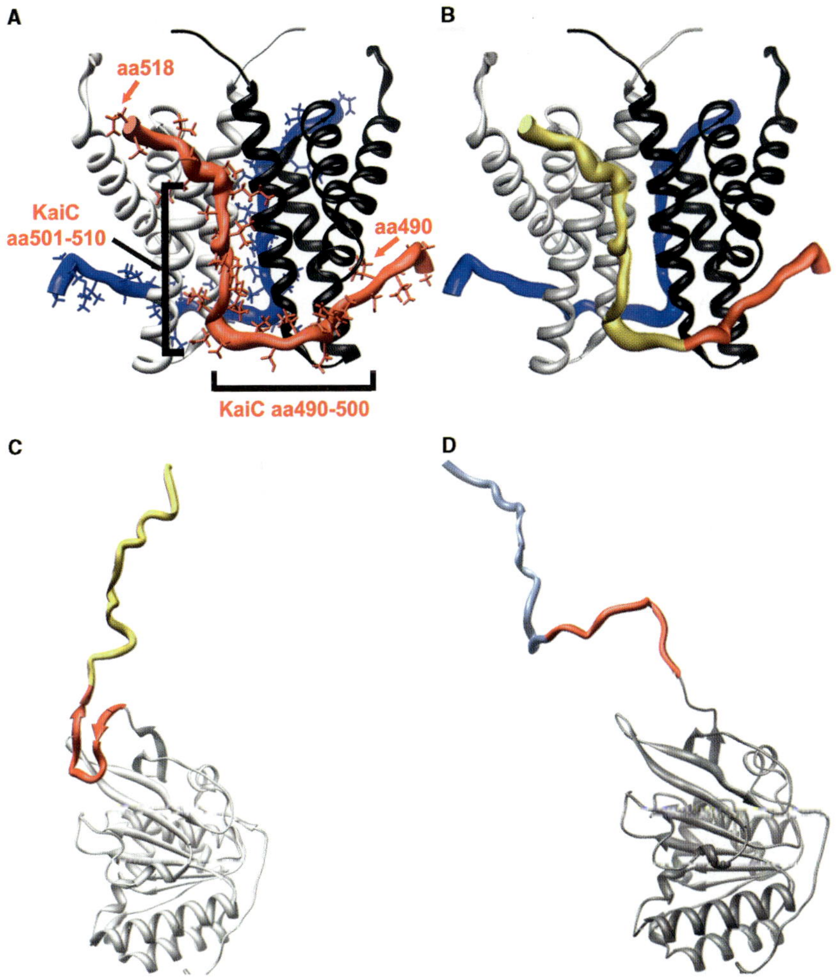

Fig. 7.5 Interaction of the KaiC C-terminal peptide with KaiA. **A** NMR structure of a complex with a C-terminal KaiC peptide (aa488–518) and the C-terminal domain of KaiA (PDB-ID 1SUY; Vakonakis and LiWang 2004). Two KaiC peptides are shown in *red* and *blue* and the KaiA dimer is shown in *gray* and *black*. There are two roughly perpendicular interaction regions in the KaiC peptide, which correspond to aa490–500 and aa501–510. The first interaction region crosses the apical helix-loop-helix of a KaiA subunit (*black*), while the second interaction region follows the groove between KaiA subunits. Note the KaiC residue numbers have been adjusted to correspond to the *S. elongatus* sequence. **B** NMR structure with the KaiC S-shaped loop residues (aa485–497) of one chain in *red*. **C** One KaiC subunit CII domain from the crystal structure (PDB-ID 2GBL) with the S-shaped loop in *red* and the remainder of the C-terminal tail in *yellow*. **D** One KaiC subunit CII domain modified to show the S-shaped loop in a modified pulled-out conformation with the remainder of the C-terminal tail (aa498–519) in *light blue* (Pattanayek et al. 2006)

Beginning at the N-terminus, peptides cross the helix–loop–helix pair of one KaiA C-terminal subunit and then, after a turn, follow the groove between KaiA subunits. Because individual KaiC peptides engage in extensive contacts to both KaiA subunits, KaiA dimerization is a prerequisite for KaiC binding. Binding of KaiC peptides involves a combination of hydrophobic, electrostatic and hydrogen-bonding interactions with KaiA (Vakonakis and LiWang 2004).

7.5.2 *Electron Microscopy Studies*

We carried out a negative-stain electron microscopic analysis of the *T. elongatus* KaiA–KaiC complex using full-length proteins (Pattanayek et al. 2006). Under the conditions used, only 1:1 KaiA dimer:KaiC hexamer complexes were observed. Although it is possible that two KaiA dimers are bound to KaiC (Hayashi et al. 2004b), this is unlikely to occur in vivo because the concentration of KaiC molecules in the cell far exceeds that of KaiA (Kitayama et al. 2003; Johnson and Egli 2004). And, as stated above, one KaiA dimer was sufficient to saturate KaiC phosphorylation. A 1:1 stoichiometry was subsequently also found for the KaiA–KaiC complex from *S. elongatus* in a time-dependent analysis of the interactions between Kai proteins employing negative-stain EM, native gels and fluorescence (Mori et al. 2007). Electron micrographs show the KaiA dimer protruding from the dome surface at one end of the KaiC hexamer particle (Fig. 7.6). Native PAGE of mixtures of KaiA either with wild-type KaiC or a C-terminal deletion mutant lacking the last 25 residues demonstrated that the lack of the C-terminal tail prevents binding by KaiA (Pattanayek et al. 2006). A C-terminal deletion in KaiC also abolishes rhythmicity in vivo, but not hexamerization in vitro. An EM study of KaiA mixed with a truncated form of KaiC also showed no evidence of complex formation without the C-terminal residues. These observations are consistent with the results obtained by Vakonakis and LiWang that the KaiA dimer specifically recognizes the C-terminal KaiCII peptide and indicates that KaiA probably contacts only the CII half.

Electron microscopy revealed that the KaiA dimer assumes various orientations vis-à-vis the CII dome surface (Fig. 7.6). In the crystal structure of full-length *S. elongatus* KaiC, C-terminal peptides from the six subunits exhibited various conformations, one of which resembles that of the model peptide in the NMR structure of the KaiA dimer–KaiC peptide complex. Given the flexibility of the C-terminal region the range of orientations observed for the KaiA dimer bound to KaiC is not surprising. However, EM images of the KaiA–KaiC complex also reveal that the KaiA dimer is at some distance (~35 Å) from the hexameric barrel of KaiC. This suggests that the KaiC S-shaped loop bordering on the channel at the surface of the CII dome may become unraveled and pulled out upon binding of KaiA (Fig. 7.5C, D). An EM reconstruction of the KaiA–KaiC complex reveals two plumes of weak density extending from one end of the KaiC hexameric barrel (Fig. 7.6B). These plumes extend in two directions and suggest that KaiA is not bound to KaiC in a single defined orientation, but rather KaiA occupies a variety of positions relative to the main barrel of KaiC. Using the EM reconstruction of the KaiA–KaiC complex as a guide, models of "tethered" and "engaged" KaiA–KaiC complexes were built (Fig. 7.7A, B; Pattanayek al. 2006). Given the limited resolution of

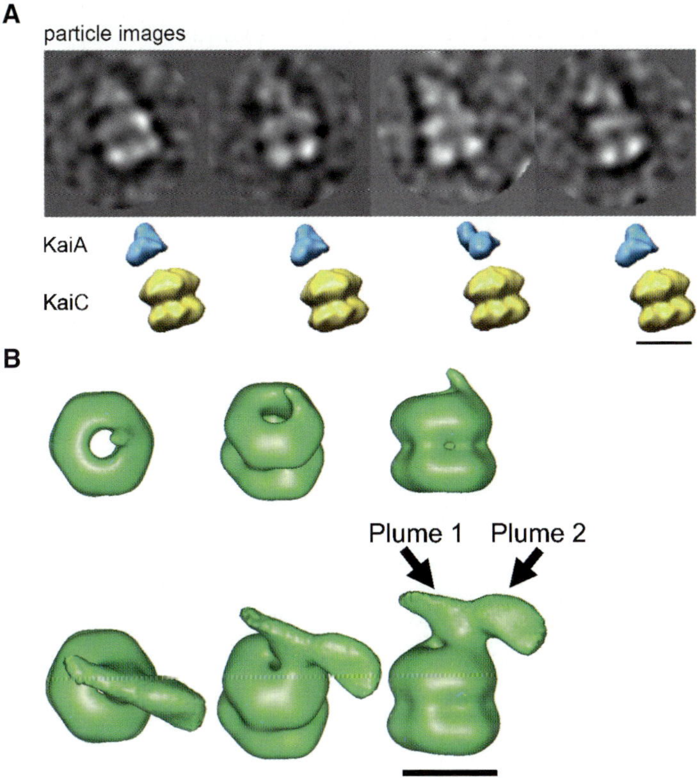

Fig. 7.6 Electron microscopy of KaiA–KaiC complex. **A** Individual particle images of negatively stained *T. elongatus* BP-1 KaiA–KaiC (*top row*) compared to 25-Å filtered representations of the KaiA (*blue*) and KaiC (*yellow*) crystal structures. The particle images are shown filtered to 20 Å resolution. **B** EM reconstruction of KaiA–KaiC based on ~4,000 negatively stained particle images. The resolution of the reconstruction is 24 Å. The reconstruction is shown in three views (0°, 45°, 90°) and with two isosurface values. At the lower isosurface two plumes of weak, diffuse KaiA density (indicated by *arrows*) connect to the KaiC hexameric barrel near the central channel. *Bars* 100 Å. Modified from Pattanayek et al. (2006)

~24 Å of the EM structure, a more detailed model of the KaiA–KaiC complex cannot be built. However it is clear that some region near the C-terminus of KaiC must serve as a flexible linker between the KaiA dimer and the KaiC hexameric barrel. On the basis of these models, we postulated that the engaged mode might enable a secondary contact between the apical loop of one KaiA monomer and the region between two KaiC subunits harboring ATP. A contact involving this region of KaiA is consistent with the effects of mutations of individual loop residues on the period of the clock (Ye et al. 2004). For example, KaiA residues could interact with KaiCII residues surrounding the ATP cleft and thus affect the intersubunit phosphorylation activity (Pattanayek et al 2004; Xu et al. 2004). Alternatively, covering the ATP cleft could simply enhance the residence time of ATP thus resulting in an enhancement

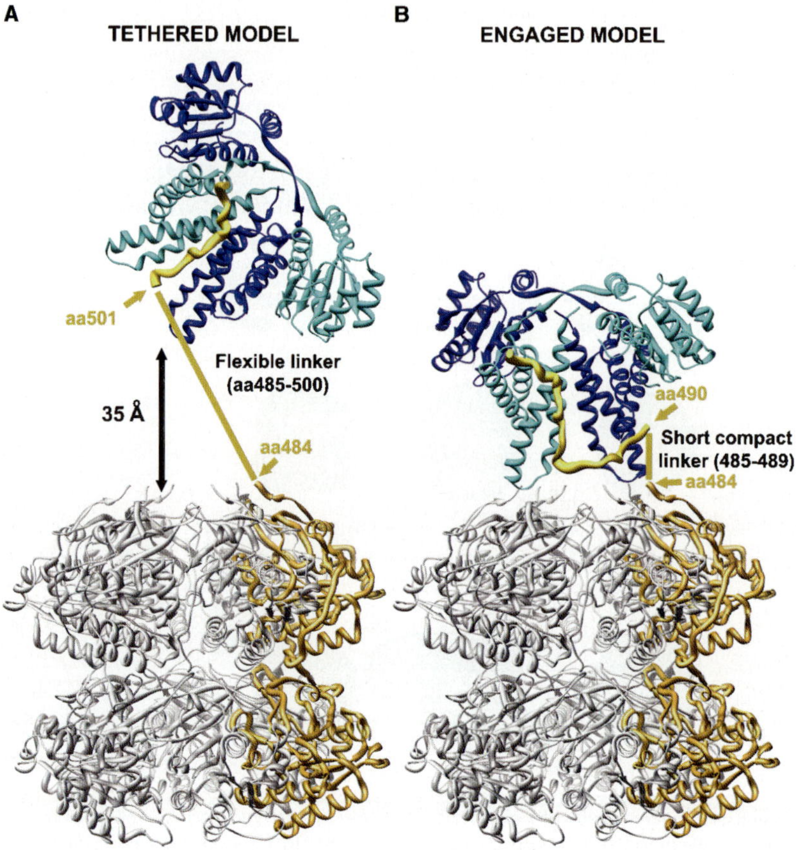

Fig. 7.7 Tethered and engaged models of the KaiA–KaiC complex. **A** The "tethered" model of the KaiA–KaiC complex with KaiC aa485–500 forming an extended flexible linker between the KaiA dimer and the hexameric barrel of KaiC. In this model KaiA is ~35 Å above the KaiC hexameric barrel, in agreement with EM images of the KaiA–KaiC complex. Also note that, in this model, both the KaiC S-shaped loop (aa485–497) and one of the two KaiC interaction regions (aa490–500) have been extended to form the linker. **B** The "engaged" model of the KaiA–KaiC complex with KaiC aa 485–489 forming a short compact linker. For clarity the S-shaped loops have been removed from all six chains of KaiC. Modified from Pattanayek et al. (2006)

of phosphorylation. This would be consistent with direct measurements of the dephosphorylation rate of turnover (Xu et al. 2003).

7.6 Summary and Outlook

A decade after the *kaiA*, *kaiB* and *kaiC* genes were shown to be essential for proper circadian function in the model organism *S. elongatus* (Ishiura et al. 1998), the KaiABC clock has now become the best characterized clock system at the molecular level. This is to some extent due to the remarkable finding that a circadian oscillator can be reconstituted in vitro from the KaiA, KaiB and KaiC proteins in the presence of ATP (Nakajima et al. 2005). Major advances have been made in terms of the functional and structural characterization of the clock in recent years. A possible view of the cyanobacterial circadian clock as consisting of single KaiC particles associating with KaiA and KaiB to different extents over the daily cycle has given way to a model based on a dynamic equilibrium entailing different types of complexes whose concentrations oscillate with a period of ca. 24 h (Kageyama et al. 2006; Mori et al. 2007). Thus, free KaiC hexamers coexist with KaiCs bound to KaiA or KaiB or both. KaiC in these complexes exhibits alternative phosphorylation states and during a single 24 h cycle, KaiC progresses from the hypo- to the hyper-phosphorylated and back to the hypo-phosphorylated state.

Three-dimensional structures for all three proteins have been available for some time but high-resolution structures of complexes are still elusive. The crystal structure of the KaiC hexamer provided a wealth of information on the architecture and conformational underpinnings of the central cog (Pattanayek et al. 2004). Phosphorylation sites were readily discernible in electron density maps (Xu et al. 2004) and the T432 and S431 sites (Nishiwaki et al. 2007; Rust et al. 2007) were phosphorylated in the crystal structure of KaiC. The structure and subsequent analyses of KaiA–KaiC interactions using solution NMR (Vakonakis and LiWang 2004), biochemical (Hayashi et al. 2004a, b) and hybrid structural approaches including EM (Pattanayek et al. 2006) also demonstrated different functions of the CI and CII domains of KaiC. Thus, the CI hexamer serves as a structural platform and the CII hexamer is conformationally more flexible and harbors all phosphorylation sites as well as the kinase and phosphatase activities.

Although the combined structural and functional data have provided important insights on the possible mechanisms of the control of KaiC phosphorylation by KaiA and the kinase activity of KaiC, there is an urgent need to gain a better understanding of the conformational changes in KaiC that underlie the switch from the kinase (endpoint hyper-phosphorylated state) to the phosphatase activity (endpoint hypo-phosphorylated state; Nishiwaki et al. 2007). Although it is now reasonably clear that both KaiA and KaiB interact with the C-terminal CII domains (see also our recent paper on the model of the binary KaiB–KaiC complex; Pattanayek et al. 2008), the lack of high-resolution structures of binary and ternary complexes of KaiC and structures of KaiC hexamers or mutants trapped in different phosphorylation states constitutes a bottleneck on the road to a complete mechanistic dissection of the KaiABC clock. Crystal structure determinations for KaiCs of various phosphorylation states and a range of complexes are important near-term goals of research concerning the *S. elongatus* KaiABC clock. A further problem of central importance concerns the molecular basis of temperature compensation. Because the in vitro KaiABC timer is temperature compensated, it stands to reason that combined biochemical and biophysical analyses will eventually uncover the molecular origins of this salient property of all circadian clocks. Long-term goals also concern a structural characterization of the interactions between the minimal components of the circadian oscillator and mediators of input (i.e., the histidine kinase CikA;

Schmitz et al. 2000) and output signals (i.e., the histidine kinase SasA; Iwasaki et al. 2000), the latter pathway including those that regulate the general transcription mechanism (Tomita et al. 2005).

References

Abrahams JP, Leslie AGW, Lutter R, Walker JE (1994) Structure at 2.8 Å resolution of F1 ATPase from bovine heart mitochondria. Nature 370:621–628

Berman HM, Westbrook J, Feng Z, Gilliland G, Bhat TN, Weissig H, Shindyalov IN, Bourne PE (2000) The protein data bank. Nucleic Acids Res 28:235–242

Egli M, Pattanayek R, Pattanayek S (2007) Protein–protein interactions in the cyanobacterial KaiABC circadian clock. In: Boeyens JCA, Ogilvie JF (eds) Models, mysteries, and magic of molecules; proceedings of the INDABA-5 conference, Kruger National Park, South Africa, August 20–25, 2006. Springer, Dordrecht, pp 287–303

Garces RG, Wu N, Gillon W, Pai EF (2004) *Anabaena* circadian clock proteins KaiA and KaiB reveal potential common binding site to their partner KaiC. EMBO J 23:1688–1698

Golden SS (2004) Meshing the gears of the cyanobacterial circadian clock. Proc Natl Acad Sci USA 101:13697–13698

Golden SS, Cassone VM, LiWang A (2007) Shifting nanoscopic clock gears. Nat Struct Mol Biol 14:362–363

Hayashi F, Suzuki H, Iwase R, Uzumaki T, Miyake A, Shen J-R, Imada K, Furukawa Y, Yonekura K, Namba K, Ishiura M (2003) ATP-induced hexameric ring structure of the cyanobacterial circadian clock protein KaiC. Genes Cells 8:287–296

Hayashi F, Itoh N, Uzumaki T, Iwase R, Tsuchiya Y, Yamakawa H, Morishita M, Onai K, Itoh S, Ishiura M (2004a) Roles of two ATPase-motif-containing domains in cyanobacterial circadian clock protein KaiC. J Biol Chem 50:52331–52337

Hayashi F, Ito H, Fujita M, Iwase R, Uzumaki T, Ishiura M (2004b) Stoichiometric interactions between cyanobacterial clock proteins KaiA and KaiC. Biochem Biophys Res Comm 316:195–202

Hayashi F, Iwase R, Uzumaki T, Ishiura M (2006) Hexamerization by the N-terminal domain and intersubunit phosphorylation by the C-terminal domain of cyanobacterial circadian clock protein KaiC. Biochem Biophys Res Comm 318:864–872

Hitomi K, Oyama T, Han S, Arvai AS, Getzoff ED (2005) Tetrameric architecture of the circadian clock protein KaiB: a novel interface for intermolecular interactions and its impact on the circadian rhythm. J Biol Chem 280:18643–18650

Ishiura M, Kutsuna S, Aoki S, Iwasaki H, Andersson CR, Tanabe A, Golden SS, Johnson CH, Kondo T (1998) Expression of a gene cluster *kaiABC* as a circadian feedback process in cyanobacteria. Science 281:1519–1523

Ito H, Kageyama H, Mutsuda M, Nakajima M, Oyama T, Kondo T (2007) Autonomous synchronization of the circadian KaiC phosphorylation rhythm. Nat Struct Mol Biol 14:1084–1088

Iwasaki H, Williams SB, Kitayama, Y, Ishiura M, Golden SS, Kondo T (2000) A KaiC-interacting sensory histidine kinase, SasA, necessary to sustain robust circadian oscillation in cyanobacteria. Cell 101:223–233

Iwasaki H, Nishiwaki T, Kitayama Y, Nakajima M, Kondo T (2002) KaiA-stimulated KaiC phosphorylation in circadian timing loops in cyanobacteria. Proc Natl Acad Sci USA 99:15788–15793

Iwase R, Imada K, Hayashi F, Uzumaki T, Morishita M, Onai K, Furukawa Y, Namba K, Ishiura M (2005) Functionally important substructures of circadian clock protein KaiB in a unique tetramer complex. J Biol Chem 280:43141–43149

Johnson CH, Egli M (2004) Visualizing a biological clockwork's cogs. Nat Struct Mol Biol 11:584–585

Kageyama H, Nishiwaki T, Nakajima M, Iwasaki H, Oyama T, Kondo T (2006) Cyanobacterial circadian pacemaker: Kai protein complex dynamics in the KaiC phosphorylation cycle in vitro. Mol Cell 23:161–171

Kitayama Y, Iwasaki H, Nishiwaki T, Kondo T (2003) KaiB functions as an attenuator of KaiC phosphorylation in the cyanobacterial circadian clock system. EMBO J 22:1–8

Leipe DD, Aravind L, Grishin NV, Koonin EV (2000) The bacterial replicative helicase DnaB evolved from a RecA duplication. Genome Res 10:5–16

Mori T, Johnson CH (2001) Circadian programming in cyanobacteria. Semin Cell Dev Biol 12:271–278

Mori T, Saveliev SV, Xu Y, Stafford WF, Cox MM, Inman RB, Johnson CH (2002) Circadian clock protein KaiC forms ATP-dependent hexameric rings and binds DNA. Proc Natl Acad Sci USA 99:17203–17208

Mori T, Williams DR, Byrne M, Qin X, Egli M, Mchaourab H, Stewart PL, Johnson CH (2007) Elucidating the ticking of an in vitro circadian clockwork. PLoS Biol 5:841–853

Nakajima M, Imai K, Ito H, Nishiwaki T, Murayama Y, Iwasaki H, Oyama T, Kondo T (2005) Reconstitution of circadian oscillation of cyanobacterial KaiC phosphorylation in vitro. Science 308:414–415

Nishiwaki T, Iwasaki H, Ishiura M, Kondo T (2000) Nucleotide binding and autophosphorylation of the clock protein KaiC as a circadian timing process of cyanobacteria. Proc Natl Acad Sci USA 97:495–499

Nishiwaki T, Satomi Y, Nakajima M, Lee C, Kiyohara R, Kageyama H, Kitayama Y, Temamoto M, Yamaguchi A, Hijikata A, Go M, Iwasaki H, Takao T, Kondo T (2004) Role of KaiC phosphorylation in the circadian clock system of *Synechococcus elongatus* PCC 7942. Proc Natl Acad Sci USA 101:13927–13932

Nishiwaki T, Satomi Y, Kitayama Y, Terauchi K, Kiyohara R, Takao T, Kondo T (2007) A sequential program of dual phosphorylation of KaiC as a basis for circadian rhythm in cyanobacteria. EMBO J 26:4029–4037

Pattanayek R, Wang J, Mori T, Xu Y, Johnson CH, Egli M (2004) Visualizing a circadian clock protein: crystal structure of KaiC and functional insights. Mol Cell 15:375–388

Pattanayek R, Williams DR, Pattanayek S, Xu Y, Mori T, Johnson CH, Stewart PL, Egli M (2006) Analysis of KaiA–KaiC protein interactions in the cyanobacterial circadian clock using hybrid structural methods. EMBO J 25:2017–2038

Pattanayek R, Williams DR, Pattanayek S, Mori T, Johnson CH, Stewart PL, Egli M (2008) Structural model of the circadian clock KaiB-KaiC complex and mechanism for modulation of KaiC phosphorylation. EMBO J 27:1767–1778

Pettersen EF (2004) UCSF chimera – a visualization system for exploratory research and analysis. J Comput Chem 25:1605–1612

Rust MJ, Markson JS, Lane WS, Fisher DS, O'Shea EK (2007) Ordered phosphorylation governs oscillation of a three-protein circadian clock. Science 318:809–812

Schmitz O, Katayama M, Williams SB, Kondo T, and Golden SS (2000) CikA, a bacteriophytochrome that resets the cyanobacterial circadian clock. Science 289:765–768

Taniguchi Y, Yamaguchi A, Hijikata A, Iwasaki H, Kamagata K, Ishiura M, Go M, Kondo T (2001) Two KaiA-binding domains of cyanobacterial circadian clock protein KaiC. FEBS Lett 496:86–90

Terauchi K, Kitayama Y, Nishiwaki T, Miwa K, Murayama Y, Oyama T, Kondo T (2007) The ATPase activity of KaiC determines the basic timing for circadian clock in cyanobacteria. Proc Natl Acad Sci USA 104:16377–16381

Tomita J, Nakajima M, Kondo T, Iwasaki H (2005) Circadian rhythm of KaiC phosphorylation without transcription-translation feedback. Science 307:251–254

Uzumaki T, Fujita M, Nakatsu T, Hayashi F, Shibata H, Itoh N, Kato H, Ishiura M (2004) Crystal structure of the C-terminal clock-oscillator domain of the cyanobacterial KaiA protein. Nat Struct Mol Biol 11:623–631

Vakonakis I, LiWang AC (2004) Structure of the C-terminal domain of the clock protein KaiA in complex with a KaiC-derived peptide: implications for KaiC regulation. Proc Natl Acad Sci USA 101:10925–10930

Vakonakis I, Sun J, Wu T, Holzenburg A, Golden SS, LiWang AC (2004a) NMR structure of the KaiC-interacting C-terminal domain of KaiA, a circadian clock protein: Implications for the KaiA–KaiC Interaction. Proc Natl Acad Sci USA 101:1479–1484

Vakonakis I, Klewer DA, Williams SB, Golden SS, LiWang AC (2004b) Structure of the N-terminal domain of the circadian clock-associated histidine kinase SasA. J Mol Biol 342:9–17

Wang J (2005) Recent cyanobacterial Kai protein structures suggest a rotary clock. Structure 13:735–741

Williams SB, Vakonakis I, Golden SS, LiWang AC (2002) Structure and function from the circadian clock protein KaiA of *Synechococcus elongatus*: a potential clock input mechanism. Proc Natl Acad Sci USA 99:15357–15362

Xu Y, Mori T, Johnson CH (2003) Cyanobacterial circadian clockwork: roles of KaiA, KaiB, and the *kaiBC* promoter in regulating KaiC. EMBO J 22:2117–2126

Xu Y, Mori T, Pattanayek R, Pattanayek S, Egli M, Johnson CH (2004) Identification of key phosphorylation sites in the circadian clock protein KaiC by crystallographic and mutagenetic analyses. Proc Natl Acad Sci USA 101:13933–13938

Ye S, Vakonakis I, Ioerger TR, LiWang AC, Sacchettini JC (2004) Crystal structure of circadian clock protein KaiA from *Synechococcus elongatus*. J Biol Chem 279:20511–20518

Chapter 8
Mechanisms for Entraining the Cyanobacterial Circadian Clock System with the Environment

Shannon R. Mackey, Jayna L. Ditty, Gil Zeidner, You Chen, and Susan S. Golden

Abstract The importance of resetting one's internal biological clock is most obvious when traveling across multiple time zones. Just as humans have mechanisms that allow recovery from jet lag, the cyanobacterial circadian system also possesses pathways that transduce external stimuli to the central pacemaker. The daily synchronization of the endogenous circadian rhythm with the environment allows the clock to control downstream processes at the time of day most beneficial to that organism. Here, the strategies used by *Synechococcus elongatus* to coordinate its internal cycle with daily cues are described.

8.1 Introduction

The coordination of metabolic processes – notably photosynthesis – with the presence of sunlight is essential for the obligate photoautotrophic cyanobacterium *Synechococcus elongatus* PCC 7942, because light is its only energy source. The anticipation of, and preparation for, the predictable daily cycles of sunrise and sunset is controlled by an internal biological clock (Pittendrigh 1981). The cyanobacteria are among a growing number of organisms that have been shown to utilize intrinsic biological clock systems to regulate rhythmic gene expression, as well as metabolic and/or behavioral processes. These rhythmic processes/behaviors are

S.R. Mackey(✉)
Department of Biology, St. Ambrose University, Davenport, IA 52803, USA,
e-mail: MackeyShannonR@sau.edu

J.L. Ditty
Department of Biology, The University of St. Thomas, St. Paul, MN 55105, USA,
e-mail: jlditty@stthomas.edu

G. Zeidner, Y. Chen, S.S. Golden
Department of Biology and Center for Research on Biological Clocks, Texas A&M University, College Station, TX 77843, USA, e-mails: gzeidner@mail.bio.tamu.edu, ychen@syntheticgenomics.com, sgolden@tamu.edu

J.L. Ditty et al. (eds.), *Bacterial Circadian Programs.*

not simply driven by the light-to-dark transition, but are instead regulated by environmental cues indirectly through the resetting of an endogenous circadian rhythm. Although the internal biological rhythm maintains a near 24-h periodicity, it remains in synchrony with the world in which it exists through interpretation of external factors.

8.1.1 Entraining Agents

The molecular and physiological processes exhibited by organisms that possess biological clocks are directly controlled by the timing (period and phasing) of their internal rhythms, which indirectly reflects that of environmental rhythmic conditions (Dunlap et al. 2004; Koukkari and Sothern 2006). The central oscillator (or pacemaker) of a circadian system generates and maintains near 24-h time. Because the periodicity of the oscillator is rarely an exact 24-h measurement, the biological rhythm must be reset daily in order to avoid discrepancy between the internal rhythm and that of the environment. This entrainment of the internal biological clock results in a period of the organism's biological rhythm whose average value is equal to that of the environmental entraining stimuli. As a result of entrainment, a stable phase relationship exists between the entrained (internal) oscillation and that of the entraining factor, such that clock-controlled processes occur at appropriate times each day (Johnson et al. 2004). Any external stimulus that can entrain the internal oscillation is termed a zeitgeber, or "time giver." In nature, there are numerous entraining agents that fluctuate over the daily cycle. The strongest and most obvious zeitgeber is the daily light/dark (LD) cycle, although temperature (Liu et al. 1998) and food availability (Damiola et al. 2000) can also serve to entrain biological rhythms.

8.1.2 Effects of Light Input on the Clock System

The effect of light on a circadian rhythm can result in entrainment by either continuous or discrete entrainment (Johnson et al. 2004). As the Earth rotates on its axis, the intensity of the sun changes during the course of the day, with lower intensities occurring at dawn and dusk as opposed to the high-intensity light exhibited during the afternoon hours. During continuous entrainment, gradual changes in the environment (including the changing intensity of light) modulate the period of the internal rhythm. In diurnal organisms such as *S. elongatus*, the circadian period decreases with increasing light intensity, reflective of a faster clock (Fig. 8.1A); the response is opposite in nocturnal organisms where the clock slows in response to higher light. This phenomenon is also known as Aschoff's Rule in recognition of Jürgen Aschoff, who originally described this property of circadian rhythms (Aschoff 1981).

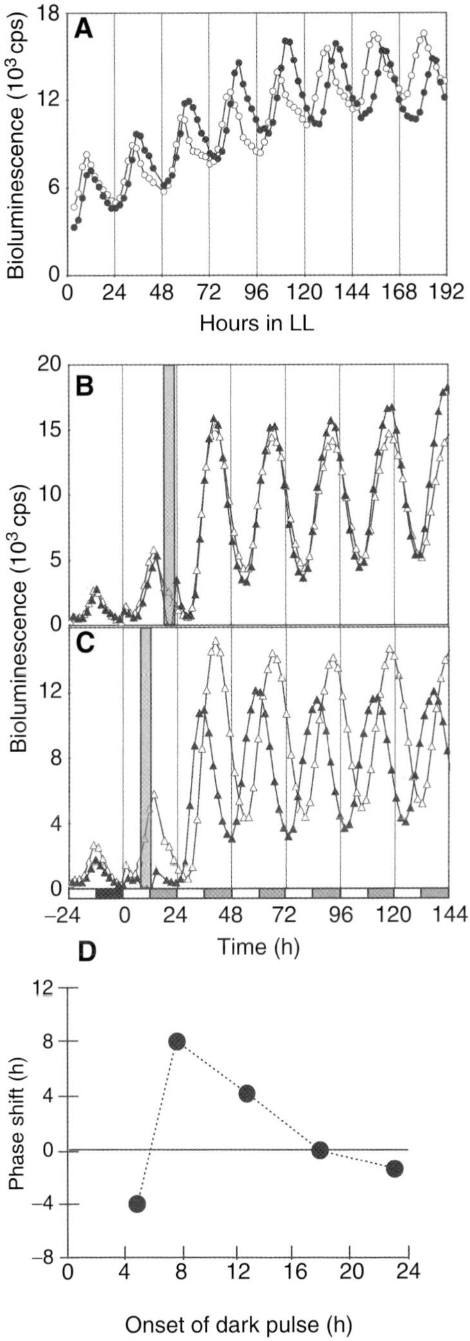

Fig. 8.1 Entrainment of the *Synechococcus elongatus* circadian system. **A** Bioluminescence in counts per second (*cps*) recorded from a P*psbAI*::*luxAB* reporter strain in LL conditions. Obeying Aschoff's Rule, the endogenous biological clock of this diurnal organism runs faster, i.e., shorter

Discrete entrainment describes the effect of the phase angle of the internal rhythm in response to abrupt environmental transitions. Each day, external signals adjust the phase of the circadian rhythm by advancing or delaying the rhythm so that the internal rhythm entrains to that of its environment. The adjustment that occurs is equal to the time difference between the free-running period (FRP) and that of the entraining cycle. The shift in phase of the internal rhythm in response to a particular zeitgeber can be measured over circadian time to produce a phase response curve.

In many eukaryotic model systems, a phase response curve is generated by maintaining an organism in constant darkness (DD) and subjecting it to brief light pulses; the corresponding phase response is measured after the return to DD (Johnson et al. 2004). *S. elongatus* relies on light as its source of energy to drive photosynthesis; therefore, maintaining these cultures in DD results in their metabolic quiescence and eventual demise. To measure the response of *S. elongatus* to external stimuli, cells are maintained in constant light (LL) and are subjected to 5-h pulses of darkness (Schmitz et al. 2000). This amount of time in the dark invokes a stable phase shift in this species without resetting the cells to dawn. A 5-h dark pulse given during the time in LL that corresponds to nighttime in an LD cycle (i.e., subjective night) does not cause a noticeable change in the phase of the rhythm (Fig. 8.1B) as compared to the phase shift exhibited (up to 8 h) when the dark pulse is given during the subjective day (Fig. 8.1C; Schmitz et al. 2000; Kiyohara et al. 2005).

In another cyanobacterial model system in which a circadian monitoring system has been developed, *Synechocystis* sp. strain PCC 6803, it may be more straightforward to study the effect of light pulses on the cyanobacterial clock (see Chap. 15). This strain has demonstrated a circadian rhythm in gene expression using a luciferase reporter (Kucho et al. 2005). In contrast to *Synechococcus elongatus*, *Synechocystis* sp. is capable of growing in the dark in the presence of glucose if provided with a brief 5-min light pulse each day, termed light-activated heterotrophic growth (LAHG; Anderson and McIntosh 1991). Use of *Synechocystis* sp. and the LAHG growth regimen may lead to the identification of proteins necessary for interpreting acute light cues as entraining agents for the circadian oscillator.

Fig. 8.1 (Continued) period, at higher light intensity [*open circles*; free-running period (FRP) = 24.1 h] than lower light intensity (*closed circles*; FRP = 24.9 h). **B**, **C** Bioluminescence from a P*kaiBC::luxAB* reporter strain in LL (*open triangles*) or in response to a 5-h dark pulse (*closed triangles*) at LL18 (**B**) or LL8 (**C**). All cells were synchronized by a LD12:12 cycle, as depicted on the *x*-axis as alternating *white* and *black* bars and negative time. During the first cycle of LL, samples were subjected to 5 h of darkness at the indicated time (depicted by the vertical *gray bar*) and then returned to LL for the duration of the bioluminescence measurements. Control samples, maintained in LL, were not subjected to a dark pulse. **D** A phase response curve can be generated by plotting the time during LL at which a 5-h dark pulse is begun (*x*-axis) and the value, in hours (*h*), of the phase shift that results from that dark pulse (*y*-axis). Note that, during the subjective night (LL12–24) when the cells would normally be in the dark of an LD cycle, there is little response to the dark pulse. In contrast, a 5-h dark pulse given during the subjective day (LL0–12) produces phase shifts of up to 8 h

8.2 The Kai Oscillator

To fully appreciate the manner by which the input pathways affect the internal timing mechanism, a brief review of the central components that make up the *Synechococcus elongatus* oscillator is necessary (for a complete review, see Williams 2007). The KaiA, KaiB, and KaiC proteins comprise the known circadian oscillator in *S. elongatus*. Deletion of any of the genes that encode these core oscillator components renders the clock arrhythmic; missense mutations result in altered period and/or phasing, or arrhythmia (Ishiura et al. 1998). The Kai proteins appear to be essential only for the circadian rhythm, because *kai* mutants grow as pure cultures at a rate similar to that of wild-type cells.

The *kaiA* gene is expressed from its own promoter, while *kaiB* and *kaiC* are arranged in an operon such that they are transcribed as a *kaiBC* dicistron (Ishiura et al. 1998). Both transcripts accumulate with a near 24-h rhythm in LD cycles or LL conditions. KaiB and KaiC protein levels fluctuate over the daily cycle, with peak levels occurring 4–6 h after the peak in mRNA, while KaiA accumulation is relatively constant, displaying only a low amplitude rhythm in overall levels (Xu et al. 2000). The *kai* genes and their protein products display regulatory feedback (Ishiura et al. 1998), like that of the oscillator components of the eukaryotic model systems (Bell-Pedersen et al. 2005). Overexpression of KaiA protein increases expression from the *kaiBC* promoter, while excess KaiC protein down-regulates its own promoter activity (Ishiura et al. 1998). The transcription/translation regulation in *S. elongatus* appears to reinforce the robustness of the rhythm rather than maintain it (Ditty et al. 2005). Instead, the generation of rhythms in *S. elongatus* stems from a post-translational oscillator (Nakajima et al. 2005; Tomita et al. 2005).

The Kai proteins undergo both homo- and heterotypic interactions throughout the circadian cycle (Kitayama et al. 2003; see Chaps. 6, 7). The net result of these interactions is the alteration of the phosphorylation state of the KaiC protein. KaiC forms an ATP-dependent homohexamer (Mori et al. 2002) that autophosphorylates on at least two adjacent residues (serine 431, threonine 432; Pattanayek et al. 2004; Xu et al. 2004); the series of phosphorylation events at these two amino acids occurs in a progressive manner like that of a well choreographed routine (Nishiwaki et al. 2007). The autokinase activity is enhanced through interactions with KaiA, whereas KaiB alleviates the positive effect of KaiA on KaiC autophosphorylation (Williams et al. 2002). These contradicting efforts result in a fluctuation of the phosphorylation state of KaiC in a circadian manner in vivo (Kitayama et al. 2003). In contrast to eukaryotic systems where progressive phosphorylation of core oscillator components leads to their rapid degradation (Mackey 2007), phosphorylated KaiC does not appear to be targeted for degradation (Iwasaki et al. 2002). Rather, the oscillation in KaiC phosphorylation can be maintained in vitro in the presence of only KaiA, KaiB, KaiC, and ATP with a temperature-compensated circadian rhythm (Nakajima et al. 2005; see Chap. 5). This in vitro "test tube oscillator" demonstrates the importance of the Kai-based post-translational timing circuit in

S. elongatus; however, an oscillator alone is not a functional clock. The action of this oscillator coupled with both input and output pathways comprise the entire cyanobacterial biological clock system that coordinates the independent Kai oscillation with the environmental day/night cycle.

8.3 Input Pathways to the Kai-Based Oscillator

Three proteins, CikA (circadian input kinase), LdpA (light-dependent period), and Pex (period extender), have been identified in *S. elongatus* as components of the input pathway of the circadian system (Kutsuna et al. 1998; Schmitz et al. 2000; Katayama et al. 2003). These input pathways to the Kai-based oscillator interpret environmental cues that alter the rhythm through slight modifications in period and abrupt changes in phase in order to keep the internal oscillations in synchrony with the external world. Many eukaryotic circadian systems have identified at least one photoreceptor that receives light information to entrain their clocks (Mackey 2007). In response to light, these photoreceptors directly influence the steady-state level of core oscillator components, which oftentimes results in rapid degradation of oscillator proteins. This abrupt change in the abundance of the core clock protein(s) alters the phase angle of the resulting rhythm.

In contrast, no true photoreceptor in *S. elongatus* has yet been associated with light signal transduction in the circadian system. Extensive forward genetic screens to look for phase-resetting mutants, in both the S. Golden (Texas A&M University) and T. Kondo laboratories (Nagoya University), have failed to identify photoreceptors that affect the clock. The availability of a genome sequence for *S. elongatus* has made it possible to identify potential photoreceptor genes by the predicted domains they encode. We have tested inactivation mutants for each of the genes shown in Fig. 8.2, all of which showed wild-type phase-resetting phenotypes like that shown in Fig. 8.3. The genes being examined are predicted to encode potential blue-light photoreceptors, including two LOV-domain proteins (*Synpcc7942_0188*, *Synpcc7942_1355*) and a predicted FAD-binding protein (*Synpcc7942_1867*), as well as three GAF-domain proteins (potential red-light receptors, *Synpcc7942_0858*, *1357*, *2534*). Figure 8.3 shows normal resetting of the wild-type strain and a mutant defective for both ORFs *0858* and *2534*, genes that encode the only two predicted GAF-containing proteins whose GAF domains clearly belong to the bilin-lyase subfamily of GAFs that are expected to bind a bilin chromophore (Wu and Lagarias 2000). *Synpcc7942_1357* is the closest homolog in *S. elongatus* PCC 7942 of Cph1, a bilin-containing protein from *Synechocystis* with known photoreceptor activity (Yeh et al. 1997). However, neither of the GAF domains in the ORF 1357 protein is a clear match to bilin lyase domains: one, marked with an asterisk in Fig. 8.2, lacks a residue for covalent linkage of a bilin; the other contains a cysteine that does not align well with bilin- binding orthologs (Wu and Lagarias 2000).

Although PCR assays of each of the individual mutants confirmed complete segregation of the mutant alleles, the recovery of some merodiploids from the

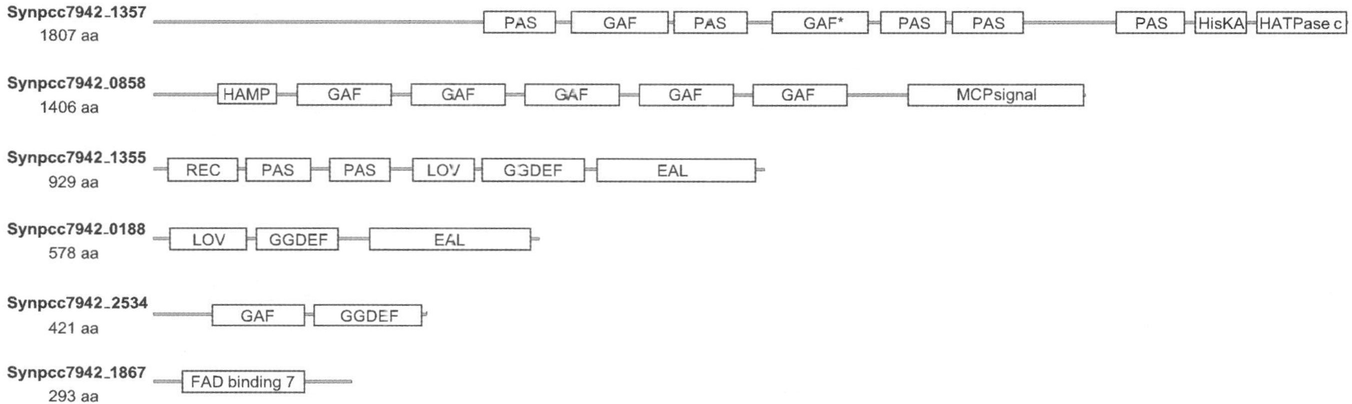

Fig. 8.2 Domain organization of putative photoreceptors of *S. elongatus*. Domains known as GAF are potential bilin-binding red photoreceptors; those designated as LOV and FAD binding 7 are potential blue photoreceptors. Other domains predicted for these proteins are not discussed in this chapter

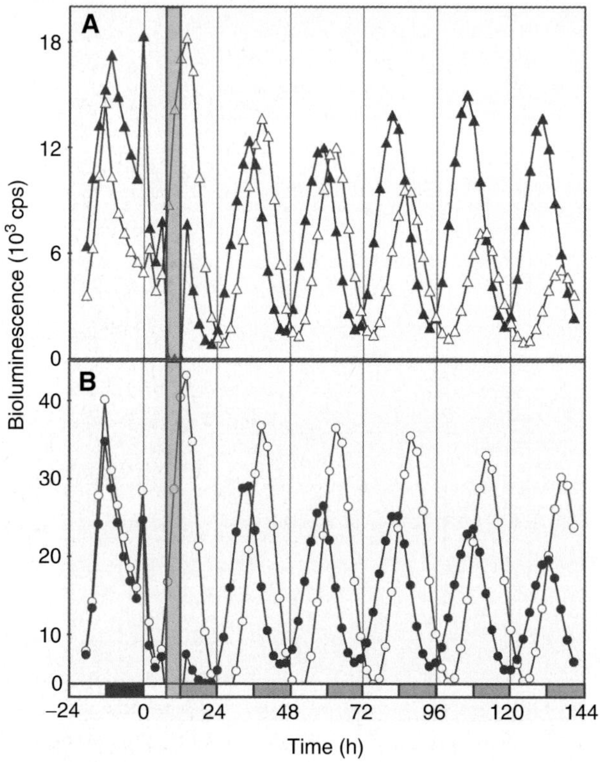

Fig. 8.3 Putative photoreceptor mutations do not affect phase response to dark pulses. Bioluminescence from a P*kaiBC::luxAB* reporter strain in a wild-type background (**A**) or after insertional inactivation of both *orf0858* and *orf2534* (**B**). All cells were synchronized by a LD12:12 cycle, as depicted on the *x*-axis as alternating *white* and *black* bars and negative time. During the first cycle of LL, samples (*closed symbols*) were subjected to 5 h of darkness at LL8 (depicted by the vertical *gray bar*) and then returned to LL for the duration of the bioluminescence measurements. Control samples (*open symbols*) were not subjected to a dark pulse

transformation to produce null mutants of *Synpcc7942_0188* suggests that loss of this locus is deleterious (Holtman et al. 2005; Clerico et al. 2007). Indeed, we were unable to obtain double mutants defective for both *Synpcc7942_0188* and *1355*. Thus, redundant functions of the LOV domain proteins in clock function cannot be ruled out. Overall the data suggest that, if any of these proteins is a blue- or red-light photoreceptor, it is involved in a process unrelated to the clock. For example, regulation of the family of *psbA* genes is known to be controlled specifically by blue and red light (Tsinoremas et al. 1994).

As demonstrated by the collective data described in the following sections, the cyanobacterial clock system may use the photosynthetic antenna as a megaphotoreceptor to detect environmental information in the form of fluctuations in cellular redox state (Ivleva et al. 2006).

8.3.1 CikA

The CikA protein is a major component of the input pathways that are involved in resetting the phase of the internal oscillation in response to the timing of acute external stimuli (i.e., discrete entrainment), such as a pulse of darkness. The *cikA* gene was originally identified in a luciferase reporter strain that displays slightly altered expression of the *psbAII* gene, which encodes a variant D1 protein of photosystem II (Schmitz et al. 2000). Additional analyses revealed that a *cikA* null mutant resets the phase of its rhythm very little (<2 h) in response to pulses of darkness throughout the circadian cycle as compared to wild-type cells that can reset their phase by up to 8 h in response to the same stimulus. In addition to the input defect, a *cikA* mutant strain exhibits a shortened circadian period (~22 h); this period alteration has an additive effect in *kai* period mutant backgrounds, which suggests that CikA and the Kai proteins are involved in non-overlapping functions of the clock (Schmitz et al. 2000). Additionally, cells that lack *cikA* display a defect in cell division that produces cells twice as long as their wild-type counterparts (Miyagishima et al. 2005). This combination of seemingly unconnected phenotypes suggests that CikA serves to link the circadian rhythm to fundamental cellular processes.

CikA contains a central histidine protein kinase (HPK) domain like that of sensor kinase proteins belonging to bacterial two-component signal transduction systems (Stock et al. 2000). The conserved histidine residue (H393) undergoes autophosphorylation both in vitro and in vivo; and this phosphorylation is essential for normal CikA function (Mutsuda et al. 2003). Precedence predicts that CikA, as a functional HPK, would transfer its phosphate group to a partner response regulator (RR) protein; however, no cognate RR has yet been identified through the use of numerous saturation mutagenesis and biochemical strategies. The modulation of autokinase activity, and therefore functionality, occurs through interactions with the GAF and pseudo-receiver (PsR) domains that flank the CikA HPK domain.

The presence of an N-terminal GAF domain initially suggested that CikA would act as a red/far-red photoreceptor by binding a bilin chromophore at that domain as occurs in other bacterial photoreceptors. This idea was appealing as it would have pinpointed CikA as a clock-related photoreceptor, but the CikA GAF does not contain the conserved cysteine amino acid residue necessary for bilin adduct formation; experiments designed to detect a covalently-bound bilin were negative (Mutsuda et al. 2003). One function attributed to the GAF domain is the activation of the autokinase activity of CikA in vitro; removal of the GAF results in a substantial decrease in CikA phosphorylation levels (Mutsuda et al. 2003). In vivo, a CikA variant that lacks the GAF domain cannot complement a *cikA* null strain (Zhang et al. 2006), which may be due to the lessened phosphorylation and hence lessened activity of the CikA protein.

The CikA PsR domain that lies downstream of the HPK region resembles that of a true RR in both structure and sequence, except that it lacks the conserved aspartic acid residue that would normally receive a phosphate group from its partner HPK. Experiments designed to visualize the transfer of a phosphate group from the HPK of CikA to its PsR in vitro did not show any phosphotransfer event (Mutsuda et al. 2003). The PsR domain has been implicated in repressing CikA autophosphorylation in vitro by physically blocking the histidine residue that would normally undergo autophosphorylation (Gao et al. 2007). CikA localizes to the pole of the cyanobacterial cell and the PsR domain is necessary for this localization to occur (Zhang et al. 2006). At the pole, it is predicted that the PsR and GAF domains interact with yet-to-be-identified protein partners that result in a conformational change in CikA to alter the phosphorylation and thus functionality of the protein. The identification of proteins that interact with CikA in a yeast two-hybrid system may help to pinpoint which protein partners play crucial roles in the different processes in which CikA is involved (Mackey et al. 2008).

CikA protein accumulation fluctuates during the circadian cycle with peak levels occurring during the subjective night (or during the dark of an LD cycle; Ivleva et al. 2006). The regulation of CikA accumulation, degradation, and function was further revealed through work on another input protein, LdpA (see Sect. 8.3.2). In a *cikA* null mutant the rhythmic output of bioluminescence continues to occur (albeit with a shortened period), yet the rhythm in KaiC phosphorylation is absent or occurs at a much lower amplitude than that of wild-type cells, such that there is a nearly equal abundance of phosphorylated and unphosphorylated KaiC protein throughout the circadian cycle (Ivleva et al. 2006). A similar phenotype in KaiC phosphorylation is exhibited by a mutant form of *kaiC* (*pr1*) (Kiyohara et al. 2005). This *kai* mutant strain displays a wild-type period of bioluminescence from luciferase reporters in LL conditions. Thus, post-translational modifications to the core clock components may play a more crucial role in the resetting of the oscillator rather than in the maintenance of circadian period in constant conditions.

8.3.2 LdpA

Unlike the phase-resetting phenotype exhibited by a *cikA* mutant, a strain that lacks *ldpA* can no longer modulate the period length of the *S. elongatus* rhythm in response to changing light intensities. In the absence of *ldpA*, cyanobacterial cells no longer obey Aschoff's Rule. In an *ldpA* mutant, the rhythm of bioluminescence produced by luciferase reporter fusions maintains its short period (22–23 h) – indicative of high light intensity – regardless of the actual light quantity (Katayama et al. 2003). In an *ldpA* null, the levels of CikA are at their trough level throughout the cycle regardless of light intensity, which corresponds to the high-light phenotype of a wild-type strain (Ivleva et al. 2005). Conversely, levels of KaiA protein are higher in an *ldpA* null strain than in wild-type cells (Ivleva et al. 2005), which may contribute to the short period of the mutant strain (see Sect. 8.3.3). Interestingly, although an *ldpA* null strain shows an increase in KaiA protein and a short period, there is no noticeable change in the level of P*kaiBC* expression, as would be predicted due to the fact that an increased level of KaiA has been shown to activate expression from the *kaiBC* promoter.

LdpA contains two iron–sulfur clusters that allow the protein to detect changes in the redox state of the cell as changes in light quantity (i.e., intensity). Producing an overall reduced state of the plastoquinone pool in the photosynthesis apparatus by the addition of the quinone analog DBMIB (2,5-dibromo-3-methyl-6-isopropyl-*p*-benzoquinone) results in rapid degradation of both the LdpA and CikA proteins (Ivleva et al. 2005). Degradation of CikA in the presence of DBMIB occurs more quickly when LdpA is present in the cell, and variants of CikA that lack the PsR domain are resistant to DBMIB-induced degradation. It is not yet known whether degradation of CikA is the normal response to changes in the plastoquinone pool in vivo. The PsR of CikA binds directly DBMIB (and other quinone analogs), suggesting that the CikA PsR binds plastoquinone in vivo; ligand binding is an unprecedented role for this type of protein domain (Ivleva et al. 2006). Although a distinction between sensing of the redox state and the level of quinone present cannot be made, these data support the role of CikA as a bridge between the circadian rhythm and essential cellular functions, such as metabolism.

LdpA co-purifies from *S. elongatus* cells with components of the oscillator (KaiA), the input pathway (CikA), and the output pathway (*Synechococcus* adaptive sensor, SasA; see Chap. 9; Ivleva et al. 2005). CikA co-purifies with both KaiA and KaiC in a large multimeric complex in vivo (Ivleva et al. 2006). The interactions among these proteins likely serve to transduce environmental information to the central oscillator. One potential mechanism of signal transduction to the Kai-based oscillator is through the N-terminal PsR domain of KaiA. Like the PsR of CikA, this domain of KaiA lacks the conserved aspartic acid necessary for phosphotransfer (Williams et al. 2002). Protein–protein interactions that occur between KaiA and members of the input pathways may cause conformational changes within KaiA to allow the C-terminal domain of KaiA access to KaiC. Interaction between the C-terminal domain of KaiA and KaiC activates the autokinase activity of KaiC (Williams et al. 2002; Kim et al. 2008) and may speed up or slow down the progression of the KaiC phosphorylation cycle depending upon the current phosphorylation state of KaiC when the stimulus is perceived.

8.3.3 *Pex*

A more direct route for external stimuli to alter the Kai-based oscillator may exist by way of the Pex protein. The *pex* gene was originally identified through its ability to "complement" a short-period (22-h) mutant (Kutsuna et al. 1998), which was later shown to be the result of a mutation in the *kaiC* gene. This apparent complementation resulted not because Pex is a core component of the clock's oscillator, but rather because circadian period lengthening occurs when an extra copy of the *pex* gene is present. Cells that lack a functional *pex* gene display a 1-h period shortening of their circadian rhythm, while overexpression of *pex* causes a dose-dependent period lengthening up to 3 h in LL conditions. Pex is placed in the input pathway to the oscillator because it is necessary for delaying the internal phase of the oscillation during multiple days of exposure to 12 h light/12 h dark (LD12:12; Takai et al. 2006). After being subjected to this LD cycle for 8 days, cells that lack *pex* show an advanced phase as compared to wild-type cells, while cells that overexpress Pex protein during those LD cycles display a more delayed phase than that of the wild type upon entering LL conditions.

The mechanism by which Pex functions to effect the period and phase of the rhythm is likely through its repression of *kaiA* expression. Structural analyses of the Pex protein depict a winged-helix motif that forms a homodimer (Arita et al. 2007). Pex accumulates in a circadian pattern with peak levels during the subjective night (or dark phase of an LD cycle; Takai et al. 2006). Pex binds to a negative regulator sequence in the promoter of the *kaiA* gene (Arita et al. 2007), which makes Pex the first protein to be identified as directly regulating expression of one of the *S. elongatus* core oscillator components. The period effects that result from Pex overexpression are dependent upon its ability to bind to the promoter because overexpression of a Pex variant that cannot bind upstream of *kaiA* does not alter circadian period (Arita et al. 2007). The circadian period phenotypes of *pex* inactivation or overexpression can be mimicked by altering the expression levels of *kaiA* mRNA (Kutsuna et al. 2007). Increasing expression of *kaiA* by an inducible promoter causes a dose-dependent period shortening, which is similar to the *pex* null in which repression of *kaiA* expression is lifted. Knocking down the expression of *kaiA* by expressing an anti-sense construct increases the rhythm of bioluminescence to a period similar to that seen when Pex is overexpressed. It should be noted that specific *cis* elements in the *kai* promoters, including the Pex binding site upstream of *kaiA*, are not required for rhythmic transcription or the maintenance of rhythmic processes. Expression of either *kaiA* or the *kaiBC* operon from a heterologous *Escherichia coli* promoter (Xu et al. 2003) or an *S. elongatus* promoter that drives expression 12-h out of phase from that of the wild-type *kai* locus (Ditty et al. 2005) produces circadian output of luciferase like that of wild type. Therefore, the role of Pex in the transcriptional/translational feedback loop is probably an auxiliary one.

8.3.4 Additional Input

Short-term overexpression of clock-related genes and their corresponding protein products can alter the phase of peak luciferase expression in subsequent cycles like that of an abrupt environmental cue. When the levels of *kaiBC* mRNA are on the rise during the subjective day, a pulse of elevated *kaiC* expression causes a phase advance. A short pulse of *kaiC* overexpression during the subjective night, when *kaiBC* levels are declining, causes a phase delay. The magnitude of the phase shift is proportional to the phase difference between the time when the pulse was given and the time when *kaiBC* levels would normally be at their maximum (Ishiura et al. 1998; Xu et al. 2000).

The direction of the phase shifts that result from short pulses of SasA overexpression differs from those seen when KaiC is overproduced (Iwasaki et al. 2000).

Short pulses of SasA cause phase delays when *sasA* levels are on the rise (subjective day) and advances in the oscillation when *sasA* levels are decreasing (subjective night). This discrepancy between the effect of excess KaiC or SasA shows that the function of SasA, though clock-related, is distinct and separate from that of the Kai complex. These results also suggest that SasA function includes some degree of feedback on the input pathways or the oscillator to direct the phase of the oscillation.

Mutations in the *kaiC* gene not only affect the innate oscillatory behavior but some also alter the response of the resulting oscillator to input stimuli. As mentioned in Sect. 8.3.1, the *kaiC* (*pr1*) mutant displays wild-type rhythms of luciferase expression in LL, but is incapable of shifting the phase of its rhythm in response to 5-h dark pulses throughout the circadian cycle (Kiyohara et al. 2005). Another mutant, *kaiC22a*, is more sensitive to changes in light intensity than that of the wild type with respect to modulation of circadian period (Katayama et al. 2003). This *kaiC* allele results in a cyanobacterial strain whose periodicity of the circadian rhythm varies by 3 h over the same light gradient that causes only a 1-h period difference in wild-type cells. Taken together, these data suggest that KaiC is involved in both the discrete and continuous entrainment of the internal circadian rhythm that it, along with KaiA and KaiB, generates.

8.4 Concluding Remarks

Despite the great detail by which the Kai-based oscillator has been described, both genetically and biochemically (see Chap. 5), there is relatively little information with regard to the mechanisms that allow this internal oscillation to be entrained to daily cycles. There are likely multiple pathways that interpret the very different external cues of light, temperature, and other yet-to-be-identified zeitgebers for the cyanobacterial system.

The ability of an *S. elongatus* cell to respond to the 5-h pulses of darkness may be related to changes in chromosome topology in addition to the input proteins described. In the biological clock system of rodents, the core oscillator protein CLOCK functions as a histone acetyltransferase that is involved in the remodeling of chromatin as a means of regulating gene expression in a circadian manner (Doi et al. 2006). The cyanobacterial chromosome has also been shown to oscillate such that there are times during the circadian cycle when the chromosome is compacted and others when it is diffuse (Smith and Williams 2006). Strains that are incapable of resetting their gene expression rhythm in response to 5-h dark pulses are also unable to fully condense their nucleoid during that same time period; wild-type cells require only 5 h to display a dark-induced chromosome compaction (see Chap. 10).

Of the three core oscillator components, KaiA appears to be the portal through which environmental information reaches the inner workings of the clock. Expression of the *kaiA* gene is regulated directly by the input protein Pex (Arita et al. 2007); KaiA is associated in complexes with both the CikA (Ivleva et al. 2006) and LdpA (Ivleva et al. 2005) proteins. The protein–protein interactions that occur in these complexes may allow for the transduction of external signals to the PsR domain at the N-terminus of the KaiA protein, which is predicted to result in a conformational change in the C-terminal, KaiC-interacting region of KaiA (Williams et al. 2002) to alter the activation of KaiC autophosphorylation; this change in the post-translational modification to KaiC is likely to alter the phase of the resulting rhythm.

Together, components of the sensory input pathways and the circadian oscillator ensure that *S. elongatus* has a reliably ticking clock that is in synchrony with a reliably rhythmic world. With this robust and elegant timepiece, the organism tunes its physiology and metabolism to do everything "in due time."

References

Anderson SL, McIntosh L (1991) Light-activated heterotrophic growth of the cyanobacterium *Synechocystis* sp. strain PCC 6803: a blue-light-requiring process. J Bacteriol 173:2761–2767

Arita K, Hashimoto H, Igari K, Akaboshi M, Kutsuna S, Sato M, Shimizu T (2007) Structural and biochemical characterization of a cyanobacterium circadian clock-modifier protein. J Biol Chem 282:1128–1135

Aschoff J (1981) Freerunning and entrained circadian rhythms. In: Aschoff J (ed) Handbook of behavioral neurobiology: biological rhythms. Plenum, New York, pp 81–93

Bell-Pedersen D, Cassone VM, Earnest DJ, Golden SS, Hardin PE, Thomas TL, Zoran MJ (2005) Circadian rhythms from multiple oscillators: lessons from diverse organisms. Nat Rev Genet 6:544–556

Clerico EM, Ditty JL, Golden SS (2007) Specialized techniques for site-directed mutagenesis in cyanobacteria. In: Rosato E (ed) Methods in molecular biology, Humana, Totowa, pp 155–172

Damiola F, Le Minh N, Preitner N, Kornmann B, Fleury-Olela F, Schibler U (2000) Restricted feeding uncouples circadian oscillators in peripheral tissues from the central pacemaker in the suprachiasmatic nucleus. Genes Dev 14:2950–2961

Ditty JL, Canales SR, Anderson BE, Williams SB, Golden SS (2005) Stability of the *Synechococcus elongatus* PCC 7942 circadian clock under directed anti-phase expression of the *kai* genes. Microbiology 151:2605–2613

Doi M, Hirayama J, Sassone-Corsi P (2006) Circadian regulator CLOCK is a histone acetyltransferase. Cell 125:497–508

Dunlap JC, Loros JJ, DeCoursey PJ (eds) (2004) Chronobiology: biological timekeeping. Sinauer, Sunderland, Mass

Gao T, Zhang X, Ivleva NB, Golden SS, LiWang A (2007) NMR structure of the *pseudo*-receiver domain of CikA. Protein Sci 16:465–475

Holtman CK, Chen Y, Sandoval P, Gonzales A, Nalty MS, Thomas TL, Youderian P, Golden SS (2005) High-throughput functional analysis of the *Synechococcus elongatus* PCC 7942 genome. DNA Res 12:103–115

Ishiura M, Kutsuna S, Aoki S, Iwasaki H, Andersson CR, Tanabe A, Golden SS, Johnson CH, Kondo T (1998) Expression of a gene cluster *kaiABC* as a circadian feedback process in cyanobacteria. Science 281:1519–1523

Ivleva NB, Bramlett MR, Lindahl PA, Golden SS (2005) LdpA: a component of the circadian clock senses redox state of the cell. EMBO J 24:1202–1210

Ivleva NB, Gao T, LiWang AC, Golden SS (2006) Quinone sensing by the circadian input kinase of the cyanobacterial circadian clock. Proc Natl Acad Sci USA 103:17468–17473

Iwasaki H, Williams SB, Kitayama Y, Ishiura M, Golden SS, Kondo T (2000) A KaiC-interacting sensory histidine kinase, SasA, necessary to sustain robust circadian oscillation in cyanobacteria. Cell 101:223–233

Iwasaki H, Nishiwaki T, Kitayama Y, Nakajima M, Kondo T (2002) KaiA-stimulated KaiC phosphorylation in circadian timing loops in cyanobacteria. Proc Natl Acad Sci USA 99:15788–15793

Johnson CH, Elliott J, Foster R, Honma K-I, Kronauer R (2004) Fundamental properties of circadian rhythms. In: Dunlap JC, Loros JJ, DeCoursey PJ (eds) Chronobiology: biological timekeeping. Sinauer, Sunderland, Mass., pp 67–84

Katayama M, Kondo T, Xiong J, Golden SS (2003) *ldpA* encodes an iron-sulfur protein involved in light-dependent modulation of the circadian period in the cyanobacterium *Synechococcus elongatus* PCC 7942. J Bacteriol 185:1415–1422

Kim Y-I, Dong G, Carruthers C, Golden SS, LiWang A (2008) The day/night switch in KaiC, a central oscillator component of the circadian clock of cyanobacteria. Proc Natl Acad Sci USA 105:12825–12830

Kitayama Y, Iwasaki H, Nishiwaki T, Kondo T (2003) KaiB functions as an attenuator of KaiC phosphorylation in the cyanobacteria circadian clock system. EMBO J 22:1–8

Kiyohara YB, Katayama M, Kondo T (2005) A novel mutation in *kaiC* affects resetting of the cyanobacterial circadian clock. J Bacteriol 187:2559–2564

Koukkari WL, Sothern RB (eds) (2006) Introducing biological rhythms. Springer, Heidelberg

Kucho K, Aoki K, Itoh S, Ishiura M (2005) Improvement of the bioluminescence reporter system for real-time monitoring of circadian rhythms in the cyanobacterium *Synechocystis* sp. strain PCC 6803. Genes Genet Syst 80:19–23

Kutsuna S, Kondo T, Aoki S, Ishiura M (1998) A period-extender gene, *pex*, that extends the period of the circadian clock in the cyanobacterium *Synechococcus* sp. strain PCC 7942. J Bacteriol 180:2167–2174

Kutsuna S, Kondo T, Ikegami H, Uzumaki T, Katayama M, Ishiura M (2007) The circadian clock-related gene pex regulates a negative cis element in the *kaiA* promoter region. J Bacteriol 189:7690–7696

Liu Y, Merrow M, Loros JJ, Dunlap JC (1998) How temperature changes reset a circadian oscillator. Science 281:825–829

Mackey SR (2007) Biological rhythms workshop IA: molecular basis of rhythms generation. Cold Spring Harb Symp Quant Biol 72:7–19

Mackey SR, Choi JS, Kitayama Y, Iwasaki H, Dong G, Golden SS (2008) Proteins found in a CikA interaction assay link the circadian clock, metabolism, and cell division in *Synechococcus elongatus*. J Bacteriol 190:3738–3746

Miyagishima SY, Wolk CP, Osteryoung KW (2005) Identification of cyanobacterial cell division genes by comparative and mutational analyses. Mol Microbiol 56:126–143

Mori T, Saveliev SV, Xu Y, Stafford WF, Cox MM, Inman RB, Johnson CH (2002) Circadian clock protein KaiC forms ATP-dependent hexameric rings and binds DNA. Proc Natl Acad Sci USA 99:17203–17208

Mutsuda M, Michel KP, Zhang X, Montgomery BL, Golden SS (2003) Biochemical properties of CikA, an unusual phytochrome-like histidine protein kinase that resets the circadian clock in *Synechococcus elongatus* PCC 7942. J Biol Chem 278:19102–19110

Nakajima M, Imai K, Ito H, Nishiwaki T, Murayama Y, Iwasaki H, Oyama T, Kondo T (2005) Reconstitution of circadian oscillation of cyanobacterial KaiC phosphorylation in vitro. Science 308:414–415

Nishiwaki T, Satomi Y, Kitayama Y, Terauchi K, Kiyohara R, Takao T, Kondo T (2007) A sequential program of dual phosphorylation of KaiC as a basis for circadian rhythm in cyanobacteria. EMBO J 26:4029–4037

Pattanayek R, Wang J, Mori T, Xu Y, Johnson CH, Egli M (2004) Visualizing a circadian clock protein: crystal structure of KaiC and functional insights. Mol Cell 15:375–388

Pittendrigh CS (1981) Circadian systems: general perspective and entrainment. In: Aschoff J (ed) Handbook of behavioral neurobiology: biological rhythms. Plenum, New York, pp 57–80, 95–124

Schmitz O, Katayama M, Williams SB, Kondo T, Golden SS (2000) CikA, a bacteriophytochrome that resets the cyanobacterial circadian clock. Science 289:765–768

Smith RM, Williams SB (2006) Circadian rhythms in gene transcription imparted by chromosome compaction in the cyanobacterium *Synechococcus elongatus*. Proc Natl Acad Sci USA 103:8564–8569

Stock AM, Robinson VL, Goudreau PN (2000) Two-component signal transduction. Annu Rev Biochem 69:183–215

Takai N, Ikeuchi S, Manabe K, Kutsuna S (2006) Expression of the circadian clock-related gene *pex* in cyanobacteria increases in darkness and is required to delay the clock. J Biol Rhythms 21:235–244

Tomita J, Nakajima M, Kondo T, Iwasaki H (2005) No transcription–translation feedback in circadian rhythm of KaiC phosphorylation. Science 307:251–254

Tsinoremas NF, Schaefer MR, Golden SS (1994) Blue and red light reversibly control *psbA* expression in the cyanobacterium *Synechococcus* sp. strain PCC 7942. J Biol Chem 269:16143–16147

Williams SB (2007) A circadian timing mechanism in the cyanobacteria. Adv Microb Physiol 52:229–296

Williams SB, Vakonakis I, Golden SS, LiWang AC (2002) Structure and function from the circadian clock protein KaiA of *Synechococcus elongatus*: a potential clock input mechanism. Proc Natl Acad Sci USA 99:15357–15362

Wu SH, Lagarias JC (2000) Defining the bilin lyase domain: lessons from the extended phytochrome superfamily. Biochemistry 39:13487–13495

Xu Y, Mori T, Johnson CH (2000) Circadian clock-protein expression in cyanobacteria: rhythms and phase setting. EMBO J 19:3349–3357

Xu Y, Mori T, Johnson CH (2003) Cyanobacterial circadian clockwork: roles of KaiA, KaiB and the *kaiBC* promoter in regulating KaiC. EMBO J 22:2117–2126

Xu Y, Mori T, Pattanayek R, Pattanayek S, Egli M, Johnson CH (2004) Identification of key phosphorylation sites in the circadian clock protein KaiC by crystallographic and mutagenetic analyses. Proc Natl Acad Sci USA 101:13933–13938

Yeh KC, Wu SH, Murphy JT, Lagarias JC (1997) A cyanobacterial phytochrome two-component light sensory system. Science 277:1505–1508

Zhang X, Dong G, Golden SS (2006) The *pseudo*-receiver domain of CikA regulates cyanobacterial circadian input. Mol Microbiol 60:658–668

Chapter 9
Factors Involved in Transcriptional Output from the Kai-Protein-Based Circadian Oscillator

Hideo Iwasaki

Abstract In the cyanobacterium *Synechococcus elongatus* PCC 7942, the central oscillator of the circadian system consists of three genes (*kaiA, kaiB, kaiC*) and their protein products. In the presence of ATP, the interactions among these proteins drive a temperature-compensated circadian rhythm of KaiC phosphorylation in vitro. The temporal information from this protein-based chemical oscillator regulates expression from essentially all gene promoters in vivo. Some insights have been reported that describe key factors involved in the transduction of temporal information from the central oscillator via circadian output pathways. Moreover, recent studies have demonstrated that the rhythm in gene transcription is sustained, even in the absence of rhythmic KaiC phosphorylation, suggesting that several interconnected output pathways are integrated into the robust circadian oscillatory system in cyanobacteria.

9.1 Introduction

In the cyanobacterium *Synechococcus elongatus* PCC 7942, the central oscillator of the circadian system consists of three genes (*kaiA*, *kaiB*, *kaiC*) and their protein products (Ishiura et al. 1998). Although KaiA and KaiC display positive and negative regulatory feedback, respectively, upon the *kaiBC* promoter (Fig. 9.1), no transcription/translation-based feedback process is necessary to drive a temperature-compensated circadian rhythm of KaiC phosphorylation in vitro (Nakajima et al. 2005), which suggests that this feedback is not necessary for the maintenance of circadian rhythmicity. Nevertheless, the protein-based chemical oscillator regulates expression from essentially all gene promoters in vivo. Although the functional relevance of transcriptional/translational feedback to the rhythmic expression of clock genes in

H. Iwasaki
Department of Electrical Engineering & Bioscience, Waseda University; and PRESTO, Japan Science & Technology Agency, 2-2 Wakamatsu-cho, Shinjuku, Tokyo 162-8480, Japan,
e-mail: hideo-iwasaki@waseda.jp

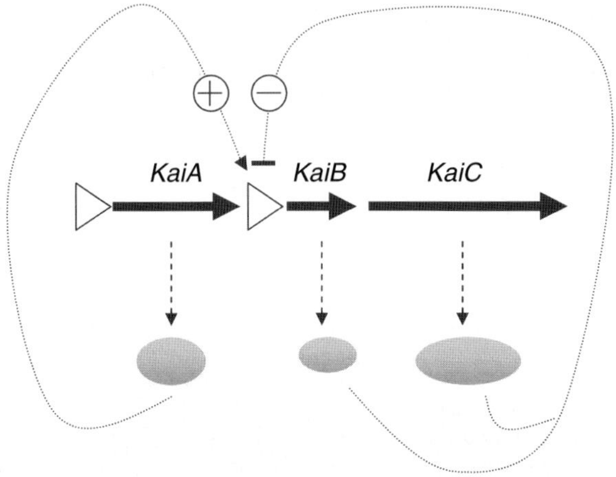

Fig. 9.1 The transcription/translation feedback model for the Kai-based clock proposed by Ishiura et al. (1998). KaiA and KaiC were interpreted to be an activator and a repressor for *kaiBC* transcription, respectively

the *S. elongatus* circadian system remains largely unknown, some insights have been reported that describe key factors involved in the transduction of temporal information from the central oscillator via circadian output pathways. Moreover, recent studies have demonstrated that the rhythm in gene transcription is sustained, even in the absence of rhythmic KaiC phosphorylation, which suggests that several interconnected output pathways are integrated into the robust circadian oscillatory system in cyanobacteria.

DNA microarray experiments have revealed the numbers of genes that are regulated by the circadian clock in many model organisms (for reviews, see Duffield et al. 2003; Hayes et al. 2005). Before these DNA microarray techniques became popular, a clock-controlled, genome-wide expression profile was examined in *S. elongatus* PCC 7942 with an elegant "promoter-trap" experiment (Fig. 9.2). Liu et al. (1995) generated a library of *S. elongatus* genomic DNA fragments fused to a promoterless luciferase gene set, *luxAB*, and introduced the reporter cassette into the *S. elongatus* chromosome by single homologous recombination. Strikingly, all (~800) clones that produced bioluminescence exhibited circadian rhythms in reporter (i.e., promoter) activity, although the phases of peak expression differed among them (Liu et al. 1995). This result suggested that the clock in *S. elongatus* regulates the expression of almost all genes. As expected, such rhythms were no longer observed in the *kaiABC*-null mutant strains for any tested promoters (Nakahira et al. 2004). Importantly, even a minimal promoter from *Escherichia coli* fused to *luxAB* was able to drive circadian bioluminescence rhythms in *S. elongatus* (Katayama et al. 1999). Based upon these data, it is plausible that the Kai-based oscillator regulates gene expression in a global, general mechanism of transcriptional regulation. Two possibilities are proposed to explain genome-wide circadian control (Fig. 9.3). One is based on a transcriptional network with cannonical transcriptional

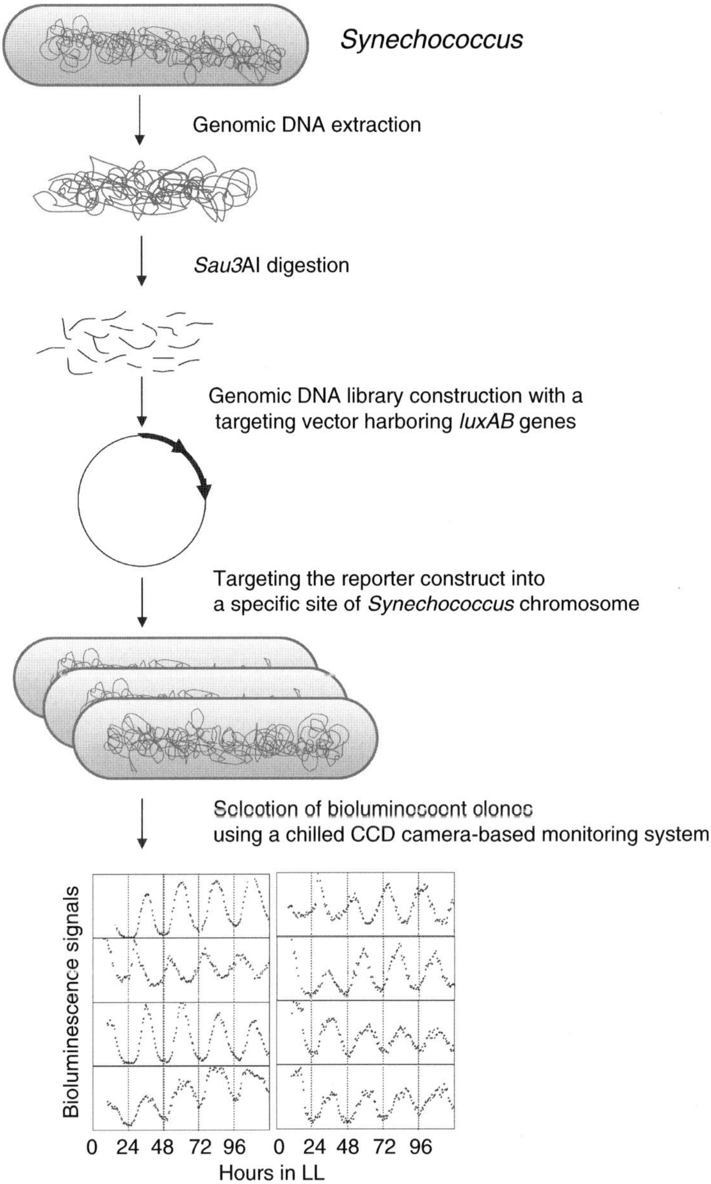

Fig. 9.2 Random promoter-trap experiments. Small fragments of *Sau*3AI-digested *Synechococcus elongatus* genomic DNA were fused to the promoterless luciferase (*luxAB*) gene set and introduced into wild-type *Synechococcus* strains. Brightly luminescent clones, which were expected to contain some promoter–reporter fusions, were selected with a cooled CCD camera. The time course of the bioluminescence of each clone was then monitored under continuous light (LL) conditions after 12 h darkness to reset the clock. Most of the bioluminescence peaked around the subjective dusk, whereas a subset of clones peaked around the subjective dawn

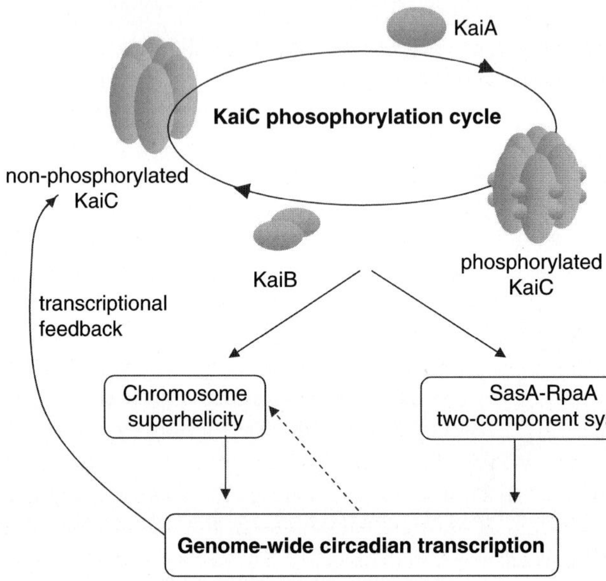

Fig. 9.3 Possible mechanisms for the genome-wide expression rhythms based on the post-translational oscillation model proposed by Tomita et al. (2005) and Nakajima et al. (2005). Alternating KaiC phosphorylation states due to enzymatic reaction networks among the Kai proteins are proposed to be the core process for driving genome-wide circadian transcription rhythms. Temporal information would be transmitted from some forms of KaiC protein in a circadian time-dependent manner to basic transcription machinery via the KaiC-associating His-to-Asp regulatory system composed of the SasA sensory histidine kinase and RpaA response regulator. In addition, temporal information may be transmitted to transcriptional processes via circadian modulation of chromosomal DNA superhelicity (Smith and Williams 2006). In either case, the post-translational chemical oscillator drives the circadian transcription cycle, including the clock genes themselves, to form the secondary transcriptional feedback loop under subjective light. Under subjective darkness, the enzymatic oscillator and the transcriptional circuit are disconnected, while the post-translational oscillator keeps time in the absence of transcriptional feedback

regulatory proteins, such as sigma factors and His-to-Asp two-component regulatory system proteins (see Sect. 9.2). An alternative possibility is the circadian control of DNA topology by the Kai-based clock (Mori and Johnson 2001; Smith and Williams 2006; Woelfle et al. 2007). Because the latter case is described in detail in Chap. 10, only the former molecular model is summarized here.

9.2 KaiC-Associating His-to-Asp Two-Component Regulatory Factors, SasA and RpaA

A yeast two-hybrid screen identified a KaiC-associating histidine protein kinase, SasA (*Synechococcus adaptive sensor A*; Iwasaki et al. 2000). Histidine kinases are factors in bacterial His-to-Asp two-component signal transduction system (for a review, see Stock et al. 2000). Generally, a histidine kinase receives a protein-specific ligand or an external signal, such as light, temperature, or osmolarity, at the amino-terminal input sensory domain, which activates or inactivates the autophosphorylation activity at a conserved histidine residue in the C-terminal domain. The phosphate group is transferred from the sensor kinase to a partner response regulator protein, which leads to a response to the original stimulus.

The presumed amino-terminal sensory domain of SasA shares sequence similarity to KaiB (Iwasaki et al. 2000); it is this amino-terminus of SasA that is sufficient for interacting with KaiC. The genetic inactivation of the *sasA* gene and a mutation in the autophosphorylating histidine residue of *sasA* each dramatically lowered the levels of *kaiBC* mRNA as well as KaiB and KaiC proteins (Iwasaki et al. 2000). The *sasA*-null strain also displays a shortened period length and an attenuated amplitude of the circadian transcriptional rhythms of all gene promoters tested, some of which became completely arrhythmic. The continuous overexpression of *sasA* eliminated such rhythms, whereas the temporal and transient overinduction of SasA induced a stable shift in the phase of gene expression rhythms (Iwasaki et al. 2000). In addition, KaiC has been shown in vitro to elevate the rate of the autophosphorylation of SasA (Takai et al. 2006). These observations suggest that as yet unknown, time-dependent form(s) of KaiC interact with SasA to regulate its enzymatic activity and thus elevate the expression of the *kaiBC* gene via the activity of a cognate response regulator of SasA.

To identify the partner response regulator of SasA, Takai et al. (2006) inactivated 24 genes, each of which was predicted to encode a response regulator in the *S. elongatus* genome. Of the resulting null strains, one mutant displayed a phenotype similar to that of the *sasA*-null mutant. The strain of interest harbored an inactivation in an OmpR-type DNA binding response regulator protein gene, named *rpaA* (*r*egulator of *p*hycobilisome-*a*ssociated *A*). Nullification of *rpaA* dramatically reduced *kaiBC* transcription and caused the arrhythmic expression of all the genes tested (Takai et al. 2006). In the presence of KaiC, a phosphotransfer reaction was demonstrated in vitro from SasA to RpaA. Moreover, the phosphotransfer activity from SasA to RpaA changed dramatically in vitro, depending on the circadian state of a coexisting Kai protein complex (Takai et al. 2006). Thus, the SasA-RpaA two-component system likely functions in the positive limb of clock gene expression and mediates temporal information from the Kai-based oscillator to key downstream regulators of genome-wide transcriptional control in cyanobacteria. An as yet unknown target gene(s), which is driven by RpaA, could encode such factors, such as RNA polymerase subunits (core and/or sigma factors; see Sect. 9.4), factors affecting DNA compaction (see Chap. 10), or transcription factors themselves.

SasA is not essential to drive basic oscillations because some residual unstable rhythms remain in a *sasA*-null background, specifically under conditions of constant dim light (Iwasaki et al. 2000); such residual rhythms are much less evident in the *rpaA* strain (Takai et al. 2006). The SasA-RpaA two-component system seems to be an activator of *kaiBC* gene expression by forming a secondary positive feedback loop to sustain the robust rhythmicity of the Kai-based posttranslational oscillator. Interestingly, the disruption of *sasA* or *rpaA*, but not that of *kaiABC*, impairs the growth of *Synechococcus* under light/dark (LD) cycles. Therefore, the SasA-RpaA system has two-fold importance for growth in a diurnal environment: it is necessary to sustain robust circadian rhythms and to adapt to LD alternations (Iwasaki et al. 2000; Takai et al. 2006).

9.3 LabA: a Possible Mediator of the Negative Limb of the Cyanobacterial Circadian System

The overexpression of *kaiC* dramatically suppresses its own (*kaiBC*) expression, which led Ishiura et al. (1998) to propose the initial transcription/translation-feedback model for *kaiBC* expression (Fig. 9.4). This negative effect is not *kaiBC*-specific because *kaiBC* overexpression suppressed the magnitude of the expression to the trough level for all the genes tested (Nakahira et al. 2004). To identify the factors involved in the negative limb of the circadian regulatory circuit, Taniguchi et al.

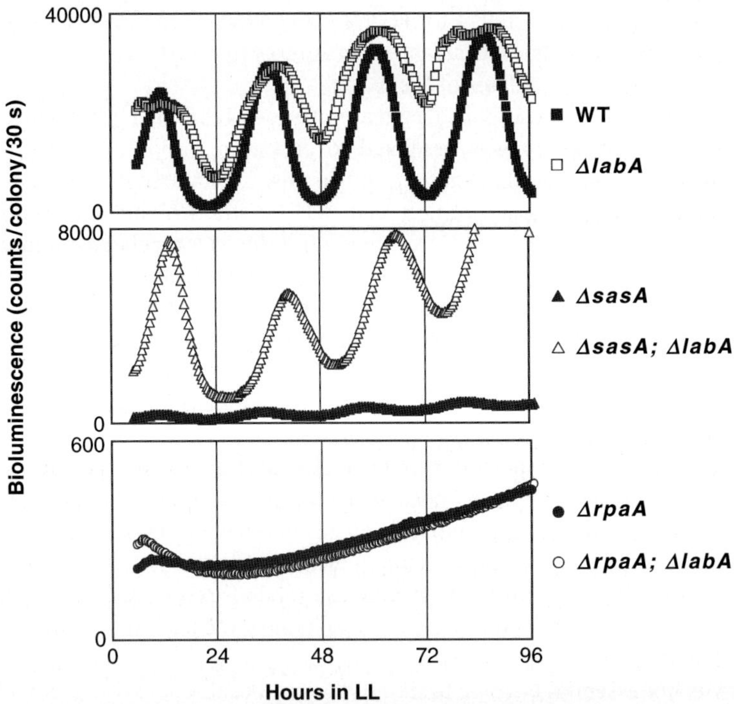

Fig. 9.4 Suppression of the *sasA* phenotype by an additional mutation in *labA*. Effects of *labA*-inactivation in wild-type (*WT*), *sasA*-null, or *rpaA*-null strains on *kaiBC* promoter activity monitored by a firefly luciferase reporter (P*kaiBC::luc*). Bioluminescence rhythms of the cells were measured under continuous light (*LL*) conditions after 12 h darkness in the presence of 0.5 mM luciferin. Modified from Taniguchi et al. (2007)

(2007) isolated a mutant impaired in *kaiC*-mediated transcriptional suppression, and identified the causal gene, *labA* (*l*ow-*a*mplitude, and *b*right *A*). The LabA protein is widely conserved, not only in cyanobacteria but also in various other bacteria, including some Archaea; however, no identifiable functional motif occurs in the amino acid sequence.

Genetic inactivation of *labA* attenuated the amplitude of *kaiBC* expression, with a highly elevated trough level; however, the KaiC protein continued to exhibit a circadian phosphorylation rhythm (Taniguchi et al. 2007). Additionally, in a strain that lacks *labA*, *kaiC* over-induction failed to completely suppress *kaiBC* expression as is seen in a wild-type background. It is known that the overexpression of either *sasA* or *rpaA* also suppresses the activity of the *kaiBC* promoter (Iwasaki et al. 2000; Takai et al. 2006). Taniguchi et al. (2007) tested whether the *sasA/rpaA* negative effect is modulated by LabA function. Interestingly, the repression of the *kaiBC* promoter by *sasA* overexpression was attenuated in *labA*-null strains, whereas the effect of *rpaA* overexpression was less attenuated. More importantly, a *labA/sasA* double mutant strains suppressed most of the characteristic *sasA*-null phenotypes, including short period, very low-amplitude rhythm under constant light (LL) conditions, and slow growth phenotype under LD conditions, whereas *labA* inactivation failed to suppress the *rpaA*-null phenotype (Fig. 9.4, Taniguchi et al. 2007). These results imply that temporal information from the Kai-based protein oscillator diverges into a SasA-dependent, subjective-day-specific positive pathway to regulate gene expression during the subjective day and a LabA-mediated, subjective-night-specific negative pathway to regulate gene expression during the subjective night, both of which converge to RpaA function to generate robust circadian gene expression rhythms.

9.4 Sigma Factors Are Involved in Circadian Output Control

Because RpaA contains a putative DNA-binding motif, its role is probably to regulate the transcriptional activity of as yet unknown target genes by associating with their promoter regions. Preliminary analysis suggested that RpaA does not bind to the *kaiBC* promoter. This is not surprising because, even when the *kaiBC* genes and *kaiC* gene were driven by an *E. coli*-derived inducible promoter, instead of by their own promoters, normal circadian rhythms were restored in *kaiBC*- or *kaiC*-null mutants, respectively (Xu et al. 2003; Nakahira et al. 2004), which indicates that the specific control of the *kaiBC* promoter itself is not likely to be essential in circadian control. Rather, RpaA might affect the expression of some regulatory genes that affect further downstream gene expression more extensively, such as RNA polymerase subunits and factors involved in DNA topology.

Cyanobacteria are unusual among bacteria because in addition to the essential, principal sigma factor, RpoD1, they also have multiple sigma-70-like group 2 sigma factors, which are not essential for growth. Golden and coworkers have extensively analyzed the functional relevance of group 2 sigma factors to circadian

Table 9.1 Phenotypes of clock-related mutant strains. *Notes*: *1* results from P*trc::kaiA* strains in the presence of 15 μM IPTG; arrhythmic in the presence of more than 30 μM IPTG; *2* rhythmic only under dim light conditions; arrhythmic under standard LL conditions; *3* arrhythmic under both standard and dim LL conditions

	kaiBC transcription	KaiC level	KaiC phosphorylation	*psbAI* transcription	Notes	References
Wild type (LL)	Rhythmic	Rhythmic	Rhythmic	Rhythmic		Ishiura et al. (1998)
Wild type (DD)	None	Constant	Rhythmic	None		Tomita et al. (2005)
In vitro	None	Constant	Rhythmic	None		Nakajima et al. (2005)
ΔkaiA (LL)	Downregulated, arrhythmic	Downregulated, arrhythmic	Arrhythmic	Arrhythmic		Ishiura et al. (1998)
ΔkaiB (LL)	Arrhythmic	Arrhythmic	Arrhythmic	Arrhythmic		Ishiura et al. (1998)
ΔkaiC (LL)	Arrhythmic	None	None	Arrhythmic		Ishiura et al. (1998)
kaiA overexpressor (LL)	Upregulated, low amplitude	Upregulated, low amplitude	Hyperphosphorylated, arrhythmic	Low amplitude	1	Kitayama et al. (2008)
kaiC overexpressor (LL)	Downregulated, no rhythm	Upregulated, arrhythmic	Arrhythmic	Arrhythmic		Ishiura et al. (1998)
ΔsasA (LL)	Downregulated, short period, low amplitude	Downregulated, arrhythmic	Hyperphosphorylated, arrhythmic	Almost arrhythmic	2	Iwasaki et al. (2000)
ΔrpaA (LL)	Arrhythmic	Arrhythmic	Arrhythmic		3	Takai et al. (2006)
ΔlabA (LL)	Upregulated, low amplitude	Upregulated, low amplitude	Rhythmic			Taniguchi et al. (2007)
ΔlabA; *ΔsasA* (LL)	Rhythmic	Rhythmic	Rhythmic			Taniguchi et al. (2007)
ΔlabA; *ΔrpaA* (LL)	Arrhythmic	Arrhythmic	Arrhythmic		3	Taniguchi et al. (2007)
ΔrpoD2 (LL)	Rhythmic			Low amplitude		Nair et al. (2002), Tsinoremas et al. (2005)
ΔrpoD3; *ΔrpoD4* (LL)	Short period					Nair et al. (2002)
ΔsigC (LL)	Rhythmic			Long period		Nair et al. (2002)

output control (Tsinoremas et al. 1995; Nair et al. 2002). Initially, they found that the inactivation of a sigma factor gene, *rpoD2*, lowered the amplitude of the circadian rhythm in the expression of a subset of genes, including that of *psbAI* (Tsinoremas et al. 1995), and slightly affected the phase of the transcriptional rhythm observed in various promoter activities (Nair et al. 2002). The inactivation of *rpoD3*, *rpoD4*, or *sigC* also differentially modulated the gene expression rhythms (Table 9.1), suggesting that these sigma factors are not entirely redundant in circadian output control. The most startling result reported was that the inactivation of *sigC* lengthened the period of the *psbAI::luxAB* bioluminescence rhythm, whereas it did not affect the rhythms of the *kaiBC* or *purF* promoter activities (Nair et al. 2002). This result suggests the coexistence of multiple timing circuits, which are somehow connected via a SigC protein function. Because the rhythmic transcription of *psbAI*, *kaiBC*, and *purF* requires the Kai protein functions, SigC may be a

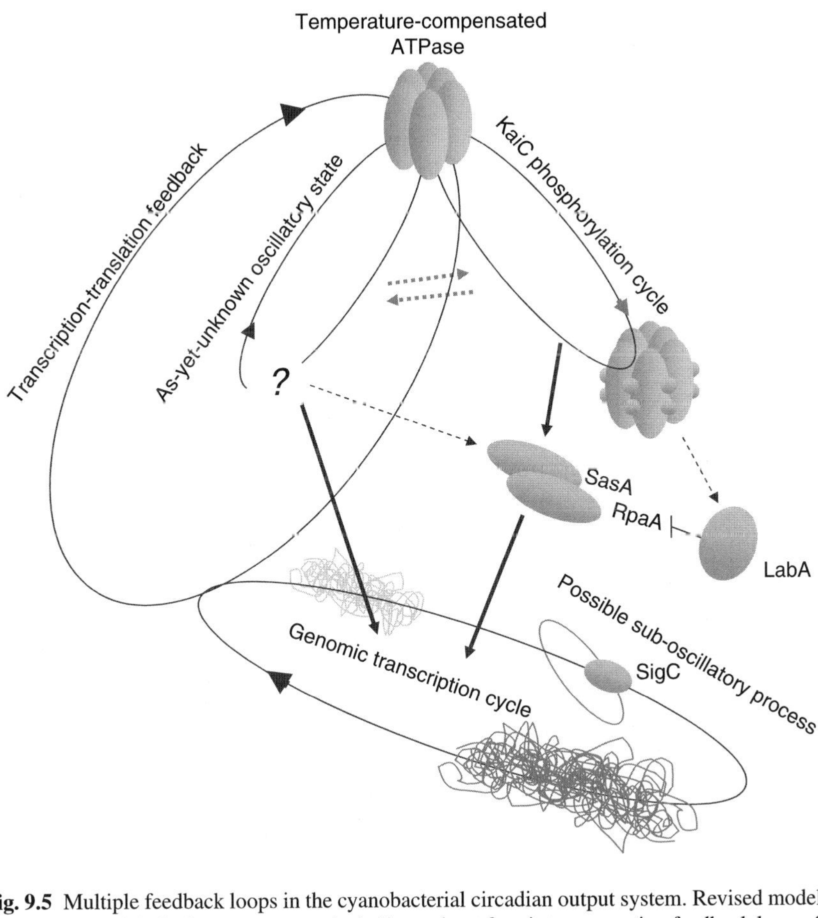

Fig. 9.5 Multiple feedback loops in the cyanobacterial circadian output system. Revised model of the cyanobacterial clock output system including at least four interconnecting feedback loops. See text for details

component regulating a suboscillatory process connected to the Kai-based central timing loop (Fig. 9.5). Experimental validation of this and alternative models to explain the period dissociation in the *sigC* mutant is yet to be addressed.

9.5 Reconsideration of Secondary Transcription/ Translation Feedback Control in the Cyanobacterial Circadian System

The period length of the in vitro reconstituted KaiC phosphorylation cycle is temperature compensated, and the period in bioluminescence in particular *kai* mutants is consistent with the period in KaiC phosphorylation cycle in vitro using the same mutant variants of KaiC (Nakajima et al. 2005). Therefore, it was concluded that this post-translational oscillation is the core of the temporal integration of the cyanobacterial circadian system. However, Kitayama et al. (2008) recently demonstrated that even in the absence of the rhythmic KaiC phosphorylation cycle some *kai*-mutant strains exhibited temperature-compensated transcription/translation rhythms. When *kaiA* was overexpressed under the control of an inducible promoter, the transcription of the *kaiBC* operon and the magnitude of KaiC phosphorylation were constitutively elevated, whereas *kaiBC* promoter activity, as monitored by a luciferase reporter, maintained its circadian oscillation. Moreover, the rhythms in *kaiBC* expression and KaiB and KaiC accumulation were observed even after two phosphorylation sites of KaiC (Ser 431, Thr 432) were replaced with Glu to mimic a constitutively phosphorylated form. Therefore, the KaiC phosphorylation cycle per se is not likely to be an essential process for Kai-dependent temperature-compensated transcriptional rhythms (Kitayama et al. 2008). Retrospectively, this situation is reminiscent of the residual low-amplitude, unstable transcription cycle in a *sasA*-null mutant strain, in which extremely down-regulated KaiC protein was constitutively phosphorylated (Iwasaki et al. 2000; Takai et al. 2006).

Interestingly, wild-type *S. elongatus* strains exhibit robust oscillations at 20°C, whereas the in vivo *kaiA*-overexpressor strain that lacks the KaiC phosphorylation cycle and the in vitro reconstituted KaiC phosphorylation cycle that lacks transcriptional feedback were not rhythmic at the same temperature condition (Kitayama et al. 2008). Thus, both the KaiC phosphorylation cycle and the KaiC-dependent transcriptional cycle are important in the integration of the robust oscillation system in cyanobacteria (Kitayama et al. 2008). How are these two processes related? Terauchi et al. (2007) demonstrated that the KaiC protein has an extremely weak but temperature-compensated ATPase activity, which correlates well with the period length of the circadian rhythm. Because this ATPase activity is the simplest reaction that is directly associated with basic circadian properties, it may independently regulate both the KaiC phosphorylation cycle and the transcriptional cycle.

In summary, there is much accumulated information on the coexistence of at least four types of feedback loops in the circadian system (Fig. 9.5): (i) the post- translational KaiC phosphorylation cycle, (ii) the Kai-based transcription/translation feedback loop, (iii) the SasA-RpaA-based feedback loop (circadian amplifier), and (iv) the possible SigC-based suboscillatory circuit. The detailed molecular mechanism of all these circuits and the relationships among them remain to be resolved.

Acknowledgements H.I. acknowledges his long-term collaboration with the Takao Kondo laboratory (Nagoya University) and the Susan Golden laboratory (Texas A&M University). This study was supported in part by Grants-in-Aid from the Ministry of Education, Culture, Sports, Science, and Technology of Japan (20370072, 19657019) and grants from the Uehara Memorial Foundation, the Mitsubishi Foundation, and the Asahi-Glass Foundation to H.I.

References

Duffield GE (2003) DNA microarray analyses of circadian timing: the genomic basis of biological time. J Neuroendocrinol 15:991–1002

Hayes KR, Baggs JE, Hogenesch JB (2005) Circadian clocks are seeing the systems biology light. Genome Biol 6:219

Ishiura M, Kutsuna S, Aoki S, Iwasaki H, Andersson CR, Tanabe A, Golden SS, Johnson CH, Kondo T (1998) Expression of a gene cluster *kaiABC* as a circadian feedback process in cyanobacteria. Science 281:1519–1523

Iwasaki H, Williams SB, Kitayama Y, Ishiura M, Golden SS, Kondo T (2000) A KaiC-interacting sensory histidine kinase, SasA, necessary to sustain robust circadian oscillation in cyanobacteria. Cell 101:223–233

Katayama M, Tsinoremas NF, Kondo T, Golden SS (1999) *cpmA*, a gene involved in an output pathway of the cyanobacterial circadian system. J Bacteriol 181:3516–3524

Kitayama Y, Nishiwaki T, Terauchi K, Kondo T (2008) Dual KaiC-based oscillations constitute the circadian system of cyanobacteria. Genes Dev 22:1513–1521

Liu Y, Tsinoremas NF, Johnson CH, Lebedeva NV, Golden SS, Ishiura M, Kondo T (1995) Circadian orchestration of gene expression in cyanobacteria. Genes Dev 9:1469–1478

Mori T, Johnson CH (2001) Circadian programming in cyanobacteria. Semin Cell Dev Biol 12:271–278

Nair U, Ditty JL, Min H, Golden SS (2002) Roles for sigma factors in global circadian regulation of the cyanobacterial genome. J Bacteriol 184:3530–3538

Nakahira Y, Katayama M, Miyashita H, Kutsuna S, Iwasaki H, Oyama T, Kondo T (2004) Global gene repression by KaiC as a master process of prokaryotic circadian system. Proc Natl Acad Sci USA 101:881–885

Nakajima M, Imai K, Ito H, Nishiwaki T, Murayama Y, Iwasaki H, Oyama T, Kondo T (2005) Reconstitution of circadian oscillation of cyanobacterial KaiC phosphorylation in vitro. Science 308:414–415

Smith RM,Williams SB (2006) Circadian rhythms in gene transcription imparted by chromosome compaction in the cyanobacterium *Synechococcus elongatus*. Proc Natl Acad Sci USA 103:8564–8569

Stock AM, Robinson VL, Goudreau PN (2000) Two-component signal transduction. Annu Rev Biochem 69:183–215

Takai N, Nakajima M, Oyama T, Kito R, Sugita C, Sugita M, Kondo T, Iwasaki H (2006) A KaiC-associating SasA–RpaA two-component regulatory system as a major circadian timing mediator in cyanobacteria. Proc Natl Acad Sci USA 103:12109–12114

Taniguchi Y, Katayama M, Ito R, Takai N, Kondo T, Oyama T (2007) *labA*: a novel gene required for negative feedback regulation of the cyanobacterial circadian clock protein KaiC. Genes Dev 21:60–70

Terauchi K, Kitayama Y, Nishiwaki T, Miwa K, Murayama Y, Oyama T, Kondo T (2007) ATPase activity of KaiC determines the basic timing for circadian clock of cyanobacteria. Proc Natl Acad Sci USA 104:16377–16381

Tomita J, Nakajima M, Kondo T, Iwasaki H (2005) No transcription–translation feedback in circadian rhythm of KaiC phosphorylation. Science 307:251–254

Tsinoremas NF, Ishiura M, Kondo T, Andersson CR, Tanaka K, Takahashi H, Johnson CH, Golden SS (1995) A sigma factor that modifies the circadian expression of a subset of genes in cyanobacteria. EMBO J 15:2488–2495

Woelfle MA, Xu Y, Qin X, Johnson CH (2007) Circadian rhythms of superhelical status of DNA in cyanobacteria. Proc Natl Acad Sci USA 104:18819–18824

Xu Y, Mori T, Johnson CH (2003) Cyanobacterial circadian clockwork: roles of KaiA, KaiB and the *kaiBC* promoter in regulating KaiC. EMBO J 22:2117–2126

Chapter 10
Chromosome Compaction: Output and Phase

Rachelle M. Smith and Stanly B. Williams

Abstract The precise mechanism by which the *Synechococcus elongatus* circadian clock conveys information that temporally regulates global gene expression is unknown. A recent study demonstrated a circadian clock-regulated chromosome compaction rhythm with a period that matches that of a concurrent gene expression rhythm. Further chromosome compaction study suggests a relationship between dark-induced chromosome compaction and phase shifts in gene expression rhythms. Gene expression rhythms are known to be phase-shifted in a circadian time dependent manner by a 5-h dark treatment. These dark treatments have also been shown to induce complete chromosome compaction. Consistent with this correlation, two mutant alleles, *cikA* and *kaiC20*, confer altered phase-response phenotypes and lack dark-induced compaction. This chapter serves as an argument that chromosome compaction has a role in circadian clock-regulated gene expression patterns and, in particular, that the chromosome compaction state likely determines phase angle.

10.1 Output from the Circadian Clock

The *Synechococcus elongatus* circadian clock regulates global patterns of gene expression (Liu et al. 1995). In recent years, data on the biochemical processes required for circadian rhythm generation and maintenance has accumulated; however, the mechanism by which the *S. elongatus* circadian clock regulates rhythmic patterns of gene expression still needs to be addressed. As discussed in Chap. 9, there are data showing interactions between the Kai clock and the SasA–RpaA two-component regulatory system. Once phosphorylated, the RpaA response regulator may bind DNA and regulate the transcription of target genes. These target genes

R.M. Smith
Divison of Biological Sciences, University of California San Diego, La Jolla, CA 92093, USA,
e-mail: rms005@ucsd.edu

S.B. Williams(✉)
Department of Biology, University of Utah, Salt Lake City, UT 84112, USA,
e-mail: williams@biology.utah.edu

J.L. Ditty et al. (eds.), *Bacterial Circadian Programs*.

likely include sigma factors, which would then drive transcription cascades by activating sets of genes that include additional sigma factors (Takai et al. 2006); however, DNA-binding activity by RpaA has not been reported.

As previously mentioned in Chap. 9, expression rhythms from all tested promoters are abolished in the absence of any one of the *kai* genes; however, in a *sasA* strain, there are still rhythmic gene expression patterns from the *kaiBC* promoter (Iwasaki et al. 2000). These data show that, even in the absence of the major output pathway component (SasA), circadian clock-regulated gene expression still occurs; therefore, the Kai clock/SasA/RpaA pathway is not solely responsible for regulating rhythmic patterns of gene expression. In fact, there are at least two major classes of clock-regulated gene-expression phase patterns (Liu et al. 1995). And interestingly, no functional cis- or trans-acting elements that control whether a gene is expressed in the predominant Class 1 (Fig. 10.1a) or the minor Class 2 phase have been identified (Min and Golden 2000; Min et al. 2004). Additionally, no circadian phenotype was seen in a Class 2, P*purF::luxAB* reporter background when several known genes encoding nonessential sigma factors were inactivated. These data suggest that these two gene expression classes are not simply governed by the availability of particular sigma factors (Nair et al. 2002). The following sections explore a role for chromosome compaction in circadian clock-regulated gene expression patterns.

Heterologous promoters from bacteria without circadian clock-regulated gene expression are, nonetheless, expressed rhythmically in *S. elongatus* (Ditty et al. 2003). An engineered, consensus-type *Escherichia coli* σ^{70} regulated promoter (P*conII*) drives rhythmic *luxAB* expression in Class 1 phase (Tsinoremas et al. 1996), whereas other *E. coli* promoters (such as those for the *fis* and *tyrT* genes) drive *luxAB* expression in the Class 2 phase (Min et al. 2004). Expression of both the *fis* and *tyrT* genes are known to be affected by DNA topology in *E. coli* (Steck et al. 1993) and depend strongly on DNA superhelical density (Schneider et al. 2000). When a DNA topology-dependent *E. coli fis* promoter *luxAB* fusion is expressed in *S. elongatus*, it displays Class 2 phasing; therefore, DNA topology may be involved in Class 2 gene expression phasing in *S. elongatus* (Min et al. 2004). Also, sequence analysis reveals TGGC repeats in the upstream regions of all characterized Class 2 genes in *S. elongatus*, (Min and Golden 2000; Min et al. 2004). These repeats may act in the same way as the GC-rich discriminators in *E. coli* (Min et al. 2004); within the *tyrT* and *fis* promoters from *E. coli*, GC-rich segments surround the core promoters and are important for regulation by DNA topology (Pemberton et al. 2000; Schneider et al. 2000). It may be that access for binding by RNA polymerase holoenzyme is determined by DNA topology at these promoters.

KaiC protein is known to interact with DNA and when KaiC is overproduced, gene expression levels are repressed on a global scale. Taken together, these data suggest that DNA structure is involved in gene expression regulation (Mori et al. 2002; Nakahira et al. 2004). Additionally, the *Chlamydomonas reinhardtii* chloroplast displays rhythmic fluctuations in DNA supercoiling during light/dark cycles (Salvador et al. 1998). The clear homology between chloroplasts and cyano-

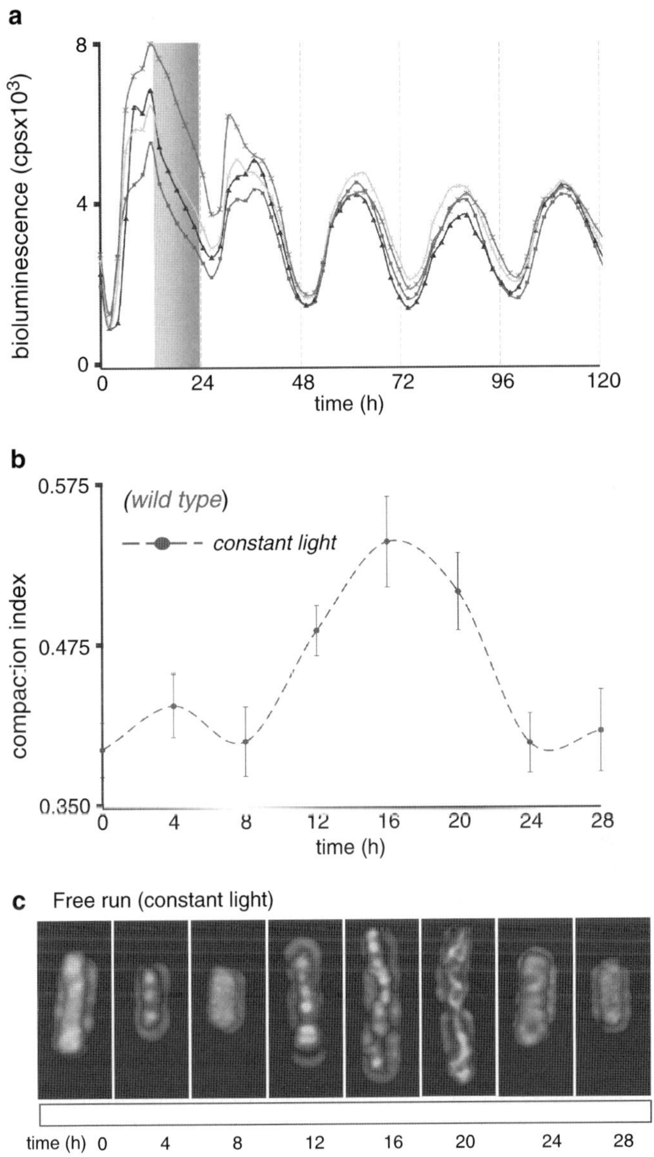

Fig. 10.1 Gene expression rhythms and chromosome compaction rhythms in *Synechococcus elongatus*. (**a**) Bioluminescence (counts per second, *cps*) from a P*kaiB::luc*⁺ reporter recorded over time (hours, *h*) in an otherwise wild-type *S. elongatus* strain. Four independent data sets are graphed. The *gray* box designates time without illumination during an entraining light/dark (LD) cycle. (**b**) Chromosome compaction under constant illumination in a wild-type *S. elongatus* strain. Mean compaction indices (CI values) from the images shown in (**c**) plotted as a function of time. Error bars indicate standard deviations. Samples for microscopy and subsequent quantification were taken after the cultures received two entraining LD12:12 cycles and one constant illumination cycle. Sampling began 24 h after constant illumination exposure and continued every fourth hour

bacteria has caused speculation that rhythmic fluctuations in DNA structure may also occur in cyanobacteria (Min et al. 2004). Together, these data have provoked thought that the circadian clock of *S. elongatus* may regulate DNA topology or chromosome arrangement dynamics as a way of regulating global gene expression patterns (Min et al. 2004; Smith and Williams 2006).

10.2 An *S. elongatus* Chromosome Compaction Rhythm

Recently an assay was developed to look at chromosome compaction in *S. elongatus* where live cells are treated with a fluorescent DNA-intercalating dye (DAPI) and visualized using deconvolution fluorescent microscopy. In the deconvolved images, obvious and striking changes in chromosome compaction are observed as a function of circadian time. The chromosome appears relatively diffuse during the day and compacted during the night. Quantification of the extent of chromosome compaction from these microscopy images yields a number called the compaction index (CI). Compacted states impart higher CI values than diffuse states; measured CI values are low during the day and high at night (Smith and Williams 2006). The chromosome compaction rhythm persists under constant illumination, which suggests that the process is under circadian clock-control (Fig. 10.1b, c). The compaction rhythm is also dependent on the *kai* genes but **is** independent of *sasA*. Additionally, in a mutant strain that harbors an allele of *kaiC* conferring a 14-h gene expression period, a chromosome compaction rhythm with a 14-h period is observed; the compaction rhythm matches concomitant gene expression rhythm periods. These data suggest that the *S. elongatus* circadian clock may regulate global gene expression, in part, by generating and maintaining a chromosome compaction rhythm (Smith and Williams 2006).

10.3 The Basics of Entrainment and Phase Resetting

A presumed role for the circadian clock is to assure that an organism's metabolism and, consequently, its behavior are both aptly timed with respect to environmental events, such as sunrise and sunset (Moore-Ede 1982). Circadian clocks entrain to environmental cycles such that even as day length varies (because of seasonal or migratory changes) the phase of the clock remains synchronized with the phase of any new-found environmental cycle. Light plays a pivotal role in circadian clock entrainment. Aschoff's Rule states that in diurnal organisms more intense light

Fig. 10.1 (Continued) for 28h. Notice that CI values are low during the subjective day, indicating a diffuse chromosomal state, and high during the subjective night, indicating a compacted chromosomal state. (**c**) Deconvolved fluorescence microscopy images (*gray*, cell autofluorescence; *white*, DAPI-stained DNA) of wild-type cells sampled at the indicated times under constant illumination. Images from the wild-type strain are representative of >99% of cells; cells are approximately 5 μm long

shortens the free-running period, whereas in nocturnal organisms more intense light lengthens the free-running period (Aschoff 1960). Entrainment compensates for the fact that free-running periods of circadian oscillators do not exactly match the period of the Earth's light/dark cycles (Moore-Ede 1982). The mechanism by which this entrainment occurs is a central question in contemporary circadian biology.

Two distinct models have been proposed to explain the method by which circadian clocks entrain: the continuous (phasic or nonparametric) model and the discrete (tonic or parametric) model (Daan 1977; Johnson 1999). The continuous model describes a circadian clock that continuously adjusts its cycle length to match that of the environmental cycle. This is accomplished by decreasing or increasing the free-running period in response to changes in light intensity (Aschoff 1960). The discrete model describes a circadian clock that is entrained by distinct environmental time cues, such as those produced by sunrise and sunset. Discrete light pulses or dark treatments in the laboratory are used to mimic these conditions (Pittendrigh 1964; Pittendrigh 1976; Johnson 1999). In fact, these discrete light pulses or dark treatments cause circadian rhythm phase shifts in organisms that are free-running in either constant darkness or constant light, respectively (Moore-Ede 1982).

Light is arguably the principal zeitgeber for circadian clock entrainment (Johnson 1999). Until recently, it was thought that the sensitivity of a circadian clock to light, or the absence of light, changed over the course of a circadian cycle (Smith 2008). This idea stems from the fact that a light pulse can induce a phase advance, no phase shift, or a phase delay in a circadian rhythm, depending upon when during the circadian cycle it is administered. Phase shifts are defined as the difference between the phase of a rhythm after exposure to a time cue and the phase of the free-running reference rhythm. The shifts can be expressed in terms of the difference in timing (hours) or the difference in phase angle (degrees). Phase-shifting data can be plotted by way of a phase-response curve, which is a method of graphing phase-shift data where the phase shift of a circadian rhythm is plotted as a function of the circadian time that the stimulus is given; examples of time-cue stimuli are light pulses, dark treatments and temperature pulses (DeCoursey 1959, 1960a, b; Pittendrigh 1960). Phase advances are designated with a positive sign, while phase delays are designated with a negative sign (Moore-Ede 1982).

10.4 Phase Shifting in Cyanobacterial Circadian Rhythms

The cyanobacterium *Synechocystis* sp. strain PCC 6803 exhibits circadian gene expression rhythms in constant darkness (Aoki et al. 1997). This strain grows heterotrophically on glucose in the dark for 6–8 days after exposure to a 15-min light pulse (see Chap. 15). Phase-response curves can be obtained by using cultures that are free-running in constant darkness, exposing them to a 3-h light treatment and then returning them to darkness. The resultant gene expression rhythm phase angle in the experimental culture is then compared to a control culture kept in constant darkness. The magnitude and direction of any phase shift is dependent upon when during the circadian cycle the light treatment was administered. Light treatments administered 8 h or 12 h after the onset of constant darkness induce phase delays. Phase advances are induced when light treatments are given 0 h or 20 h after being placed into constant darkness, whereas no shift is observed at 4 h or 16 h (Aoki et al. 1997).

Because *S. elongatus* is an obligate phototroph, phase-response curves are obtained with 5 h dark treatments administered while the organism is free-running under constant illumination (Kondo et al. 1993). *S. elongatus* gene expression rhythms are also phase shifted in a circadian time-dependent manner. When cultures free-running under constant illumination are subjected to 5 h dark treatments in the early to mid subjective day, considerable phase advances are achieved (roughly 6–8 h). Dark treatments given late in the subjective day or early in the subjective night show little or no phase response, whereas dark treatments administered in the late subjective night yield phase delays (approximately 3–5 h; Fig. 10.2a; Kondo et al. 1993). The resultant phase response curve is essentially the mirror image of that which is obtained using light treatments with *Synechocystis* sp. strain PCC 6803 free-running in constant darkness. Mutations in *S. elongatus* that cause altered phase-response phenotypes have been found in *kaiC* (Kiyohara et al. 2005), *cikA* (Schmitz et al. 2000; Kiyohara et al. 2005) and *smcA* genes (Smith 2008).

CikA was identified as a key component of the circadian clock input pathway (Schmitz et al. 2000). *S. elongatus cikA* null strains exhibit circadian rhythms in gene expression; however, the amplitudes are reduced and the circadian periods are shortened by approximately 2 h (Williams 2007). Arguably the most prominent circadian phenotype of a *cikA* strain is the lack of phase resetting (Fig. 10.2a; Schmitz et al. 2000). Subjecting a *cikA* strain to 5 h dark treatments yields little or no phase shift in gene expression rhythms, no matter when during the circadian cycle the dark treatment is administered. These data suggest that CikA is part of an input pathway that conveys environmental information to the circadian clock (Schmitz et al. 2000).

The *kaiC20* allele was identified in a genetic screen designed to isolate mutants that could not undergo circadian phase shifts in response to 5 h dark treatments (Kiyohara et al. 2005). The *kaiC20* mutant strain, also known as *pr1* (*p*hase-*r*esponse *1*), displays wild-type circadian gene expression rhythms; however, similar to a *cikA* strain, a *kaiC20* strain also has an altered response to 5 h dark treatments (Kiyohara et al. 2005). The *kaiC20* allele encodes the KaiC V422A protein (Kiyohara et al. 2005). The position of this mutation is near the two phosphorylation sites in KaiC (S431, T432; Nishiwaki et al. 2004); this mutation is also in a possible KaiA binding site (Taniguchi et al. 2001).

10.5 KaiC and Phase Determination

In wild-type cells, KaiC protein levels and phosphorylation state oscillate with circadian rhythmicity (Nishiwaki et al. 2000). It has been suggested that the KaiC phosphorylation state rhythm may determine the phase of the circadian oscillator

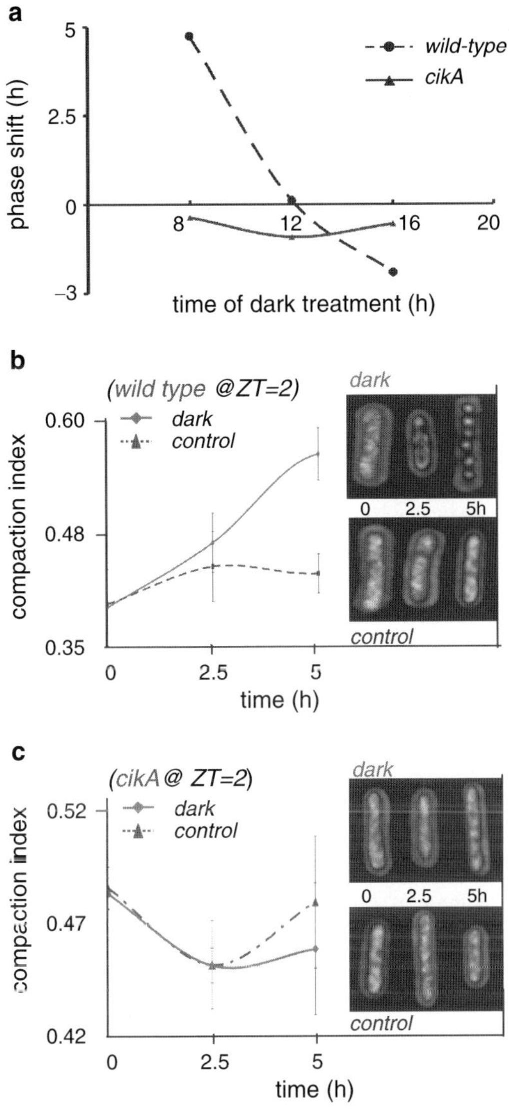

Fig. 10.2 Phase response and dark-induced compaction in wild-type and *cikA S. elongatus* strains. **(a)** Phase-response curve produced with 5 h dark treatments administered during constant illumination. Phase shifts in gene expression rhythms are plotted so that the amount of phase shift is shown as a function of the timing of the 5 h dark treatment. The *dashed line* represents the wild-type strain (*circles*). Note the large phase advance in the gene expression rhythms with 5 h dark treatments at ZT = 8, the absence of phase shifts at ZT = 12, and the phase delay at ZT = 16 in the wild-type strain. The *solid line* represents the *cikA* strain (*triangles*). Note that the *cikA* strain has an altered phase-response phenotype compared to the wild-type strain; 5 h dark treatments have little effect on phase no matter when during the circadian cycle they are given. **(b)** Dark-induced chromosome compaction in a wild-type *S. elongatus* strain. Mean compaction indices are plotted as a function of time. Error bars indicate standard deviation. Cultures were entrained with one

and thereby establish the phase of gene expression rhythms (Xu et al. 2000; Rust et al. 2007). However, neither KaiC levels nor KaiC phosphorylation state ratios change substantially during a 5 h dark treatment at ZT = 8 (Ivleva et al. 2006); recall that dark treatments given at this time lead to large phase advances in gene expression rhythms. Furthermore, the KaiC phosphorylation state rhythm is altered in at least two mutants (*cikA*, *kaiC20*), both of which show near wild-type gene expression patterns (Kiyohara et al. 2005; Ivleva et al. 2006). For example, a *cikA* strain exhibits relatively wild-type circadian rhythms in gene expression but the KaiC phosphorylation state rhythm is not wild type (Ivleva et al. 2006). Furthermore, in the *kaiC20* strain, the KaiC phosphorylation state rhythm is altered and the oscillation of KaiC protein level is abolished; however, this strain still has essentially wild-type circadian rhythms in gene expression (Kiyohara et al. 2005). These data make it difficult to model the KaiC phosphorylation state rhythm as a significant parameter in phase determination.

Several groups have independently determined the KaiC free-running phosphorylation state rhythm in vivo and it has been suggested that KaiC phosphorylation state determines the phase of the circadian oscillator; however, careful inspection of all these data reveals that the phosphorylation state of KaiC is not phase-defining (Iwasaki et al. 2002; Kageyama et al. 2003; Kitayama et al. 2003; Xu et al. 2003; Imai et al. 2004; Nakahira et al. 2004; Nishiwaki et al. 2004; Kiyohara et al. 2005; Nakajima et al. 2005; Tomita et al. 2005; Ivleva et al. 2006; Takai et al. 2006). Most reports in the literature show that KaiC is found in its unphosphorylated form during the early subjective day. In spite of this, close examination of immunoblotting data reveals that KaiC protein pools can be found completely unphosphorylated, completely phosphorylated, or at any stage in between these two states at any time in the circadian cycle (Iwasaki et al. 2002; Kageyama et al. 2003; Kitayama et al. 2003; Xu et al. 2003; Imai et al. 2004; Nakahira et al. 2004; Nishiwaki et al. 2004; Kiyohara et al. 2005; Nakajima et al. 2005; Tomita et al. 2005; Ivleva et al. 2006; Takai et al. 2006). In general, phosphorylated KaiC reaches peak levels somewhere between ZT = 12 and ZT = 20. Usually the ratio between phosphorylated and unphosphorylated KaiC drops by ZT = 24, but there are reports in the literature of KaiC still being found mostly in its phosphorylated form at ZT = 24 (Imai et al. 2004). Clearly, the phosphorylation state of KaiC in vivo cannot be considered the circadian phase predictor; therefore, there must be other factors involved in phase

---

Fig. 10.2 (Continued) LD12:12 cycle and then the experimental culture was placed in the dark at ZT = 2. The 5 h dark treatment induced full chromosome compaction. Compare the CI value at 5 h from the dark-placed experimental cultures (*solid line*) to that in the free-running, control cultures (*dashed line*). *Inset* Deconvolved fluorescence microscopy images (*gray*, cell autofluorescence; *white*, DAPI-stained DNA) of cells sampled at the indicated times. Images from the experimental culture are above those from the control culture. Cells are approximately 5 μm long. (**c**) Dark-induced chromosome compaction in a *cikA* strain; as in (**b**). Note that dark placement did not induce chromosome compaction; CI values stay similar between the free-running, control cultures (*dashed line*) and the experimental cultures (*solid line*). *Inset* Deconvolved fluorescence microscopy images are of the *cikA* strain sampled at the indicated times. Cells are around 5 μm long

determination of circadian gene expression rhythms in *S. elongatus*. The next section explores a role for chromosome compaction in the determination of phase.

10.6 The Role of Chromosome Compaction in Gene Expression Rhythm Phasing

Recently, a temporal correlation was observed between chromosome compaction kinetics and the stimulus required for phase response in *S. elongatus* (Smith 2008). The time in darkness required for stable, nontransient phase shifts in *S. elongatus* gene expression rhythms is 5 h. This 5 h dark treatment matches the time required to induce a diffuse chromosome into a fully compacted state (Fig. 10.2b). This compacted state is potentially restricting the access of RNA polymerase holoenzymes to particular gene promoters, which are sequestered within a compacted chromosome. This dark-induced chromosome compaction is independent of the circadian clock (Smith 2008). That is, the chromosome compacts in response to 5 h dark treatments at all times during the circadian cycle. Similar to gene expression rhythms, the chromosome compaction rhythm can be phase-shifted with 5 h dark treatments. The measured phase shifts are remarkably similar in direction and magnitude to those observed in the gene expression rhythms. As mentioned above, *cikA* and *kaiC20* strains, which retain rhythmic gene expression and chromosome compaction rhythms in free-running conditions, display altered phase response in gene expression. These nonphase responsive strains also do not undergo chromosome compaction in response to 5 h dark treatment (Fig. 10.2c), which demonstrates a relationship between dark-induced compaction and phase response.

The *smcA* gene encodes *S. elongatus* *s*tructural *m*aintenance of *c*hromosome protein *A*. These types of proteins are known to play significant roles in chromosome dynamics in both eukaryotes and prokaryotes (Hirano 1999). The *smcA3* allele of this gene was constructed by deleting most of the open reading frame and inserting an antibiotic resistance cassette. The *smcA3* strain lacks dark-induced chromosome compaction; similar to those of the *cikA* and the *kaiC20* strains, gene expression rhythm patterns in the *smcA3* strain also show a severely altered response to 5 h dark treatments (Smith 2008). In further support of a relationship between phase response and chromosome compaction, complete dark-induced chromosome compaction in the *kaiC14* strain (which displays a shortened gene expression rhythm period) only takes 2.5 h. Unlike wild-type strains, a phase response in gene expression can be elicited with a shortened 2.5 h dark treatment in a *kaiC14* strain. Additionally, overproduction of the KaiC protein has been shown to cause phase shifts in gene expression rhythms in a circadian time-dependent manner (Ishiura et al. 1998) and, interestingly, overproduction of KaiC also begets complete and total chromosome compaction.

A proposed model for phase response in *S. elongatus* is as follows: the Kai clock regulates a chromosome compaction rhythm, which may help drive circadian rhythms in global gene expression patterns. At any time in a circadian cycle when wild-type cultures are subjected to 5 h dark treatments, chromosomes fully compact and then slowly return to a diffuse state upon return to the light. Exposing free-running cultures to a dark treatment during the subjective day causes chromosome compaction at a time when they are normally diffuse. The chromosomes are induced into their potentially less transcriptionally accessible, compacted state, which presumably causes a decrease in gene expression. Exposing the cultures to light after dark treatment results in the chromosomes returning to a diffuse state earlier than if simply following a circadian clock directive; the chromosomes now return to a more transcriptionally accessible, diffuse state where presumably gene expression can commence again. The cells then continue with their newly phased chromosome compaction rhythm. The chromosome compaction rhythm is now phase-advanced compared to the compaction rhythm of cultures which did not undergo a dark treatment. The result is a phase advance in rhythmic gene expression patterns. Administration of 5 h dark treatments around subjective dusk is effectively equivalent to the typical, clock-driven chromosome compaction cycle since the chromosomes were already compacting when the dark treatment is given; few or no phase shift results are seen. If these dark treatments are given later in the subjective night, the compaction cycle is essentially prolonged because the chromosomes are kept in a compacted state longer than usual, which results in phase delays (Smith 2008).

The CikA protein plays some role in this dark-induced compaction process, perhaps by receiving environmental information that the cell recognizes as an absence of light and then relaying the information to both a chromosome compaction process and to the Kai-oscillator. This dark-induced chromosome compaction process is oscillator-independent and in a *ΔkaiC* strain compaction actually happens more quickly than in the wild-type strain; therefore, the clock may be gating the compaction process (Smith 2008). Seemingly, the KaiC20 protein is unable to receive information from CikA and therefore does not relay the lack-of-light information to cells that are in the dark. As a result, the Kai clock disallows the signal from CikA in the compaction process and does not initiate chromosome compaction until the normal circadian time.

10.7 Future Directions

Rhythmic chromosome compaction may not be necessary for gene expression rhythms; however, chromosome compaction seems to be important for the phasing of gene expression rhythms. It may be that large-scale compaction rhythms do not drive rhythmic gene expression, but instead, dark-induced compaction may be a mechanism for turning off gene transcription. The presence of a functional clock and the major output pathway components SasA and RpaA may be enough to drive rhythmic gene expression; and compaction might be part of another pathway that is used to shut down gene expression in response to environmental signals, such as darkness. Also, the chromosome compacts when KaiC is overproduced and this compaction corresponds with global repression in gene transcription rhythms (Ishiura et al. 1998). Futhermore, there are two other proteins, the *s*egregation and *c*ondensation *p*roteins ScpA and ScpB, which are known to interact with the Smc of *Bacillus subtilis*. The ScpA and ScpB proteins form a complex with Smc that is involved in chromosome structuring (Soppa et al. 2002). Potential *scpA* and *scpB* genes have been identified in *S. elongatus*. Strains harboring null alleles of these genes are currently being tested for gene expression rhythms and chromosome compaction rhythms. These alleles may need to be put in combination with one another to fully eliminate chromosome compaction.

The current chromosome compaction assay only detects large changes in chromosome structure. It may be that smaller changes are taking place which are not detectable by DAPI staining, such as changes in the level of supercoiling. Assays have been developed to determine the level of chromosomal DNA supercoiling in vivo in *E. coli* (Mojica and Higgins 1997). *S. elongatus* may display circadian rhythms in DNA supercoiling and gene expression rhythms may be dependent on supercoiling. A recent report demonstrated that the *S. elongatus* circadian clock regulates an in vivo rhythm of plasmid topological change. This plasmid DNA superhelical state rhythm persists under both light/dark and constant light conditions and is abolished in a *kaiC* background. It was also demonstrated that bioluminescent reporter fusions expressed on this plasmid drive rhythmic luciferase expression with the same period and phasing as when they are expressed in the chromosome (Woelfle et al. 2007). These data suggest that DNA supercoiling may play a role in rhythmic gene expression from plasmids in *S. elongatus*.

A way to directly test whether compaction represses transcription is to use a strain in which compaction can be induced. It has been suggested that Smc proteins affect chromosome compaction by affecting the level of DNA supercoiling (Lindow et al. 2002). In eukaryotes, an Smc-containing cohesion complex is able to constrain positive DNA supercoils in vitro (Kimura and Hirano 1997, 2000; Kimura et al. 1999; Hagstrom et al. 2002). In addition, the functional and structural analog of bacterial Smc proteins, the MukB protein of *E. coli*, affects the supercoiling of plasmid and chromosomal DNA in vivo (Hiraga et al. 1989, Niki et al. 1991; Weitao et al. 1999, 2000). The *B. subtilis* Smc also affects the supercoiling of plasmid DNA in vivo. It has been suggested that this protein affects compaction by stabilizing positive supercoils (Lindow et al. 2002). Other proteins involved in supercoiling include DNA gyrase (induces negative supercoils), topoisomerase I (relaxes negative supercoils) and topoisomerase IV (removes supercoils; Wang 1996; Levine et al. 1998; Lindow et al. 2002). The *smcA* gene could be overexpressed from an inducible plasmid and compaction levels could be assayed. Alternatively, cells may be depleted of topoisomerase I, which should increase the amount of negative supercoiling and thereby increase the level of chromosome compaction (DiNardo et al. 1982; Lindow et al. 2002). The *S. elongatus* genome contains potential topoisomerase-encoding genes.

The data discussed in this chapter suggests a role for chromosome structuring in the phase determination of circadian gene expression rhythms in *S. elongatus*.

References

Aoki S, Kondo T, Wada H, Ishiura M (1997) Circadian rhythm of the cyanobacterium *Synechocystis* sp. strain PCC 6803 in the dark. J Bacteriol 179:5751–5755

Aschoff J (1960) Exogenous and endogenous components in circadian rhythms. Cold Spring Harbor Symp Quant Biol 25:11–26

Daan S (1977) Tonic and phasic effects of light in the entrainment of circadian rhythms. Ann NY Acad Sci 290:51–59

DeCoursey PJ (1959) Daily activity rhythms in the flying squirrel, University of Wisconsin, Wisconsin

DeCoursey PJ (1960a) Daily light sensitivity rhythm in a rodent. Science 131:33–35

DeCoursey PJ (1960b) Phase control of activity in a rodent. Cold Spring Harbor Symp Quant Biol 25:49–55

DiNardo S, Voelkel KA, Sternglanz R, Reynolds AE, Wright A (1982) *Escherichia coli* DNA topoisomerase I mutants have compensatory mutations in DNA gyrase genes. Cell 31:43–51

Ditty JL, Williams SB, Golden SS (2003) A cyanobacterial circadian timing mechanism. Annu Rev Genet 37:513–543

Hagstrom KA, Holmes VF, Cozzarelli NR, Meyer BJ (2002) *C. elegans* condensin promotes mitotic chromosome architecture, centromere organization, and sister chromatid segregation during mitosis and meiosis. Genes Dev 16:729–742

Hiraga S, Niki H, Ogura T, Ichinose C, Mori H, Ezaki B, Jaffe A (1989) Chromosome partitioning in *Escherichia coli*: novel mutants producing anucleate cells. J Bacteriol 171:1496–1505

Hirano T (1999) SMC-mediated chromosome mechanics: a conserved scheme from bacteria to vertebrates? Genes Dev 13:11–19

Imai K, Nishiwaki T, Kondo T, Iwasaki H (2004) Circadian rhythms in the synthesis and degradation of a master clock protein KaiC in cyanobacteria. J Biol Chem 279:36534–36539

Ishiura M, Kutsuna S, Aoki S, Iwasaki H, Andersson CR, Tanabe A, Golden SS, Johnson CH, Kondo T (1998) Expression of a gene cluster *kaiABC* as a circadian feedback process in cyanobacteria. Science 281:1519–1523

Ivleva NB, Gao T, LiWang AC, Golden SS (2006) Quinone sensing by the circadian input kinase of the cyanobacterial circadian clock. Proc Natl Acad Sci USA 103:17468–17473

Iwasaki H, Williams SB, Kitayama Y, Ishiura M, Golden SS, Kondo T (2000) A KaiC-interacting sensory histidine kinase, SasA, necessary to sustain robust circadian oscillation in cyanobacteria. Cell 101:223–233

Iwasaki H, Nishiwaki T, Kitayama Y, Nakajima M, Kondo T (2002) KaiA-stimulated KaiC phosphorylation in circadian timing loops in cyanobacteria. Proc Natl Acad Sci USA 99:15788–15793

Johnson CH (1999) Forty years of PRCs – what have we learned? Chronobiol Int 16:711–743

Kageyama H, Kondo T, Iwasaki H (2003) Circadian formation of clock protein complexes by KaiA, KaiB, KaiC, and SasA in cyanobacteria. J Biol Chem 278:2388–2395

Kimura K, Hirano T (1997) ATP-dependent positive supercoiling of DNA by 13S condensin: a biochemical implication for chromosome condensation. Cell 90:625–634

Kimura K, Hirano T (2000) Dual roles of the 11S regulatory subcomplex in condensin functions. Proc Natl Acad Sci USA 97:11972–11977

Kimura K, Rybenkov VV, Crisona NJ, Hirano T, Cozzarelli NR (1999) 13S condensin actively reconfigures DNA by introducing global positive writhe: implications for chromosome condensation. Cell 98:239–248

Kitayama Y, Iwasaki H, Nishiwaki T, Kondo T (2003) KaiB functions as an attenuator of KaiC phosphorylation in the cyanobacterial circadian clock system. EMBO J 22:2127–2134

Kiyohara YB, Katayama M, Kondo T (2005) A novel mutation in *kaiC* affects resetting of the cyanobacterial circadian clock. J Bacteriol 187:2559–2564

Kondo T, Strayer CA, Kulkarni RD, Taylor W, Ishiura M, Golden SS, Johnson CH (1993) Circadian rhythms in prokaryotes: luciferase as a reporter of circadian gene expression in cyanobacteria. Proc Natl Acad Sci USA 90:5672–5676

Levine C, Hiasa H, Marians KJ (1998) DNA gyrase and topoisomerase IV: biochemical activities, physiological roles during chromosome replication, and drug sensitivities. Biochim Biophys Acta 1400:29–43

Lindow JC, Britton RA, Grossman AD (2002) Structural maintenance of chromosomes protein of *Bacillus subtilis* affects supercoiling in vivo. J Bacteriol 184:5317–5322

Liu Y, Tsinoremas NF, Johnson CH, Lebedeva NV, Golden SS, Ishiura M, Kondo T (1995) Circadian orchestration of gene expression in cyanobacteria. Genes Dev 9:1469–1478

Min H, Golden SS (2000) A new circadian class 2 gene, *opcA*, whose product is important for reductant production at night in *Synechococcus elongatus* PCC 7942. J Bacteriol 182:6214–6221

Min H, Liu Y, Johnson CH, Golden SS (2004) Phase determination of circadian gene expression in *Synechococcus elongatus* PCC 7942. J Biol Rhythms 19:103–112

Mojica FJ, Higgins CF (1997) In vivo supercoiling of plasmid and chromosomal DNA in an *Escherichia coli hns* mutant. J Bacteriol 179:3528–3533

Moore-Ede MC, Sulzman FM, Fuller CA (1982) The clocks that time us. Harvard University Press, Cambridge, Mass

Mori T, Saveliev SV, Xu Y, Stafford WF, Cox MM, Inman RB, Johnson CH (2002) Circadian clock protein KaiC forms ATP-dependent hexameric rings and binds DNA. Proc Natl Acad Sci USA 99:17203–17208

Nair U, Ditty JL, Min H, Golden SS (2002) Roles for sigma factors in global circadian regulation of the cyanobacterial genome. J Bacteriol 184:3530–3538

Nakahira Y, Katayama M, Miyashita H, Kutsuna S, Iwasaki H, Oyama T, Kondo T (2004) Global gene repression by KaiC as a master process of prokaryotic circadian system. Proc Natl Acad Sci USA 101:881–885

Nakajima M, Imai K, Ito H, Nishiwaki T, Murayama Y, Iwasaki H, Oyama T, Kondo T (2005) Reconstitution of circadian oscillation of cyanobacterial KaiC phosphorylation in vitro. Science 308:414–415

Niki H, Jaffe A, Imamura R, Ogura T, Hiraga S (1991) The new gene *mukB* codes for a 177 kd protein with coiled-coil domains involved in chromosome partitioning of *E. coli*. EMBO J 10:183–193

Nishiwaki T, Iwasaki H, Ishiura M, Kondo T (2000) Nucleotide binding and autophosphorylation of the clock protein KaiC as a circadian timing process of cyanobacteria. Proc Natl Acad Sci USA 97:495–499

Nishiwaki T, Satomi Y, Nakajima M, Lee C, Kiyohara R, Kageyama H, Kitayama Y, Temamoto M, Yamaguchi A, Hijikata A, Go M, Iwasaki H, Takao T, Kondo T (2004) Role of KaiC phosphorylation in the circadian clock system of *Synechococcus elongatus* PCC 7942. Proc Natl Acad Sci USA 101:13927–13932

Pemberton IK, Muskhelishvili G, Travers AA, Buckle M (2000) The G+C-rich discriminator region of the *tyrT* promoter antagonises the formation of stable preinitiation complexes. J Mol Biol 299:859–864

Pittendrigh CS (1960) Circadian rhythms and the circadian organization of living systems. Cold Spring Harbor Symp Quant Biol 25:159–182

Pittendrigh CS (1976) A functional anaysis of circadian pacemarkers in nocturnal rodents. IV. Entrainment: pacemaker as clock. J Comp Physiol 106:291–331

Pittendrigh CS, Minis DH (1964) The entrainment of circadian oscillations by light and their role as photoperiodic clocks. Am Nat 98:261–294

Rust MJ, Markson JS, Lane WS, Fisher DS, O'Shea EK (2007) Ordered phosphorylation governs oscillation of a three-protein circadian clock. Science 318:809–812

Salvador ML, Klein U, Bogorad L (1998) Endogenous fluctuations of DNA topology in the chloroplast of *Chlamydomonas reinhardtii*. Mol Cell Biol 18:7235–7242

Schmitz O, Katayama M, Williams SB, Kondo T, Golden SS (2000) CikA, a bacteriophytochrome that resets the cyanobacterial circadian clock. Science 289:765–768

Schneider R, Travers A, Muskhelishvili G (2000) The expression of the *Escherichia coli fis* gene is strongly dependent on the superhelical density of DNA. Mol Microbiol 38:167–175

Smith RM (2008) The role of chromosome compaction in phase determination of circadian gene expression rhythms in the cyanobacterium *Synechococcus elongatus*. Dissertation, University of Utah

Smith RM, Williams SB (2006) Circadian rhythms in gene transcription imparted by chromosome compaction in the cyanobacterium *Synechococcus elongatus*. Proc Natl Acad Sci USA 103:8564–8569

Soppa J, Kobayashi K, Noirot-Gros MF, Oesterhelt D, Ehrlich SD, Dervyn E, Ogasawara N, Moriya S (2002) Discovery of two novel families of proteins that are proposed to interact with prokaryotic SMC proteins, and characterization of the *Bacillus subtilis* family members ScpA and ScpB. Mol Microbiol 45:59–71

Steck TR, Franco RJ, Wang JY, Drlica K (1993) Topoisomerase mutations affect the relative abundance of many *Escherichia coli* proteins. Mol Microbiol 10:473–481

Takai N, Nakajima M, Oyama T, Kito R, Sugita C, Sugita M, Kondo T, Iwasaki H (2006) A KaiC-associating SasA-RpaA two-component regulatory system as a major circadian timing mediator in cyanobacteria. Proc Natl Acad Sci USA 103:12109–12114

Taniguchi Y, Yamaguchi A, Hijikata A, Iwasaki H, Kamagata K, Ishiura M, Go M, Kondo T (2001) Two KaiA-binding domains of cyanobacterial circadian clock protein KaiC. FEBS Lett 496:86–90

Tomita J, Nakajima M, Kondo T, Iwasaki H (2005) No transcription–translation feedback in circadian rhythm of KaiC phosphorylation. Science 307:251–254

Tsinoremas NF, Ishiura M, Kondo T, Andersson CR, Tanaka K, Takahashi H, Johnson CH, Golden SS (1996) A sigma factor that modifies the circadian expression of a subset of genes in cyanobacteria. EMBO J 15:2488–2495

Wang JC (1996) DNA topoisomerases. Annu Rev Biochem 65:635–692

Weitao T, Nordstrom K, Dasgupta S (1999) Mutual suppression of *mukB* and *seqA* phenotypes might arise from their opposing influences on the *Escherichia coli* nucleoid structure. Mol Microbiol 34:157–168

Weitao T, Nordstrom K, Dasgupta S (2000) *Escherichia coli* cell cycle control genes affect chromosome superhelicity. EMBO Rep 1:494–499

Williams SB (2007) A circadian timing mechanism in the cyanobacteria. Adv Microb Physiol 52:229–296

Woelfle MA, Xu Y, Qin X, Johnson CH (2007) Circadian rhythms of superhelical status of DNA in cyanobacteria. Proc Natl Acad Sci USA (in press)

Xu Y, Mori T, Johnson CH (2000) Circadian clock-protein expression in cyanobacteria: rhythms and phase setting. EMBO J 19:3349–3357

Xu Y, Mori T, Johnson CH (2003) Cyanobacterial circadian clockwork: roles of KaiA, KaiB and the *kaiBC* promoter in regulating KaiC. EMBO J 22:2117–2126

Chapter 11
Cell Division Cycles and Circadian Rhythms

Tetsuya Mori

Abstract Cell division cycles and circadian clocks are major periodic processes in living organisms. Circadian rhythms of cell division have been found in many eukaryotes and some prokaryotes. Circadian clocks gate cell division within discrete time windows. Among bacteria, circadian clocks have been found only in cyanobacteria. The freshwater unicellular cyanobacterium *Synechococcus elongatus* PCC 7942, which utilizes light as an energy source, grows and divides in the daytime and stops dividing during the night. When the cells are placed in continuous light conditions, the rhythmic occurrence of cell division continues with a period of ~24 h. Whether the cyanobacterial cells are rapidly growing or halted in their division cycle, the circadian clock appears to tick steadily and operates normally. This phenomenon implies an independence of circadian timekeeping from the cell division cycle. The mechanisms and significance of circadian control of the cell division cycle in cyanobacteria are discussed here.

11.1 Introduction

Self-reproduction is a fundamental feature of life. In cellular organisms, both unicellular and multicellular, cell division is the most basic process of reproduction. The period of the cell division cycle depends on many parameters such as nutrition, temperature, and developmental stage. Additionally, the generation of circadian rhythms is also an important process in living organisms. The planet Earth has continued to revolve on its axis since its formation about 4.5×10^9 years ago. The rotation of Earth results in daily environmental changes (e.g., light/dark, radiation, temperature, etc.) on its surface. Life on Earth has been exposed to more than 10^{12} cycles of day and night since its long history began. Such periodic environmental changes are predicted to have influenced many cellular and physiological processes,

T. Mori
Department of Biological Sciences, Vanderbilt University, VU Station B 35–1634, 2301 Vanderbilt Place, Nashville, TN 37235–1634, USA, e-mail: tetsuya.mori@vanderbilt.edu

J.L. Ditty et al. (eds.), *Bacterial Circadian Programs*.

including that of cell division. It is reasonable that the long course of evolution developed internal timekeeping systems that coordinate the biochemical and physiological processes with the daily changes of the environment. Daily rhythms (circadian rhythms) in the cell division cycle are found in both prokaryotes and eukaryotes. This chapter serves as an overview of the interactions between the cell division cycle and circadian clocks in unicellular eukaryotes and bacteria.

11.2 Circadian Control of Cell Division Cycles in Unicellular Eukaryotic Organisms

Circadian control of the timing of cell division has been reported in many organisms (Edmunds 1989). The first demonstration of a circadian cell division cycle rhythm was presented in the marine dinoflagellate *Gonyaulax polyedra* by Sweeney and Hastings 50 years ago (Sweeney and Hastings 1958). Cell division of this unicellular organism can be monitored either by counting recently divided daughter cells, which remain attached for about 30 min after cell division occurs and are easily distinguished, or simply by measuring the increase of cell numbers in cultures. When this marine dinoflagellate is placed in cycles of 12 h light and 12 h dark (LD12:12), the cells divide near the end of the 12-h dark period. When the cells are transferred to continuous light (LL) after being in LD12:12, the cells continued the near 24-h periodicity of cell division. The rhythmicity persists for at least 14 days under continuous dim light (2,152 lux; 1 lux = 1 cd m^2). The period of the cell division rhythm remained near 24 h in cultures that displayed different doubling times at different temperatures (Sweeney and Hastings 1958). These observations clearly indicate that the periodicity in the timing of cell division is under the control of an internal clock and is not a coincidence of simple cell cycle synchronization in a population of cells that each divide with a generation time of ~24 h. Later, the dynamic nature of circadian cell division in *Gonyaulax* was analyzed in detail by Honma and Hastings (1989). They hypothesized that the circadian clock gates cell division independently of the metabolism to proceed in a restricted circadian phase.

Since the initial finding of the cell division rhythm in *Gonyaulax*, circadian control of cell division has been reported and studied in many other eukaryotic unicellular organisms (Edmunds 1988). Edmunds and his colleagues extensively investigated the circadian regulation of cell division in *Euglena glacilis* (Edmunds 1966, 1988; Edmunds and Funch 1969; Jarrett and Edmunds 1970; Edmunds and Adams 1981). This plastid-containing, freshwater, unicellular flagellate can grow either photoautotrophically or chemoheterotrophically. The process of mitosis is restricted to a window of time between the end of the day and the end of the night (Fig. 11.1). The rhythm persists for more than one week in LL (photoautotroph), continuous darkness (DD; chemoheterotroph), or high frequency LD cycles such as LD1:1 or LD3:3.

Temperature compensation, entrainment by LD cycles, phase shifting by light pulses, and other fundamental properties of circadian rhythms have been demonstrated in *Euglena*, as well as the independence of the circadian oscillator from the

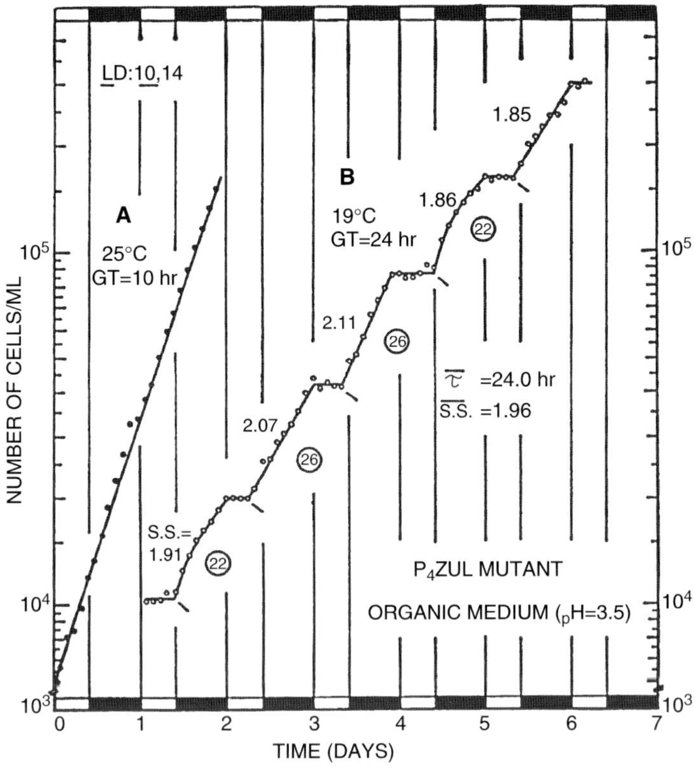

Fig. 11.1 Population growth of a photosynthetic mutant (P4ZUL) of *Euglena gracilis* var. *bacillaris* strain Z (Pringsheim) grown on defined heterotrophic medium in LD10:14. *Curve A* Exponential increase in cell number (generation time, GT, of 10h) at 25°C. *Curve B* Entrainment of the cell division rhythm at 19°C. Step size is indicated for successive division bursts; i.e., ratio of (cells ml^{-1} after a division burst):(cells ml^{-1} immediately preceding the onset of division in culture). The period of the rhythm is also given in hours (encircled to the right of each burst). The average period (τ) of the rhythm in population was 24.0h and the average step size (*ss*) was 1.96, yielding a GT of about 24h. From Jarrett and Edmunds (1970); reprinted with permission of the American Association for the Advancement of Science

cell division cycle (Edmunds 1988). *Euglena* is a cobalamin auxotroph, meaning that deprivation of vitamin B_{12} completely blocks cell division and suppresses the corresponding rhythm in *Euglena* (Bré et al. 1981). When vitamin B_{12} is added to the medium to release the cells from the temporary inhibition of cell division, the rhythm of cell division restarts in phase with a non-deprivation control. These results suggest that the circadian clock functions even when the cell division cycle is arrested in *Euglena* (Bré et al. 1981). Conversely, *Euglena* utilizes lactate as a carbon source. While an addition of lactate to the medium accelerates the cell division cycle to establish a generation time of 8–10h, a pulse of lactate to a free-running culture does not alter the phase of the cell division rhythm restored after the pulse (Jarrett and Edmunds 1970). In either case, these observations suggest independence of the circadian oscillator from the cell division cycle in *Euglena*.

Other eukaryotic unicellular organisms in which circadian control of cell division has been reported include *Tetrahymena* (Wille and Ehret 1968), *Chlamydomonas* (Bruce and Bruce 1981; Goto and Johnson 1995), and *Paramecium* (Barnett 1969). Recently, circadian control of cell division and expression of many cell division related genes, including cyclins and cyclin-dependent kinases, was demonstrated in the smallest photosynthetic eukaryote, *Ostreococcus tauri*, whose 12.5-Mb compact genome contains homologs of the plant clock genes *TOC1* and *CCA1* (Moulager et al. 2007; Bouget et al. 2008). Under constant dim light (15 μmol quanta cm^{-2} s^{-1}), *Ostreococcus* cells divide once per day at the beginning of the subjective night. The cells commit to G1 phase and are not arrested in S or G2/M phases, suggesting that the circadian clock controls the G1/S transitions.

Increasing evidence from studies of circadian rhythms in eukaryotic unicellular organisms (especially *Gonyaulax*, *Euglena*, *Tetrahymena*) in the 1950s to 1980s led to an empirical rule called the "circadian–infradian rule" or "GET effect" (GET stands for ***G**onyaulax*, ***E**uglena*, ***T**etrahymena*). In general, the rule stated that only cells dividing once per day (circadian growth mode) or more slowly (infradian growth modes) express circadian rhythms; any overt circadian rhythmicity could be suppressed or abolished in a cell population of eukaryotic microorganisms when the overall generation time was less than 24 h (ultradian growth mode; Ehret and Wille 1970; Ehret 1980). Figure 11.1 represents an example of the GET effect (Jarrett and Edmunds 1970). At a lower temperature (19°C), a population of a photosynthetic mutant of *Euglena gracillis* grows relatively slowly with an average generation time of 24 h, and the circadian rhythm in cell division can be entrained in LD10:14. However, when the culture is placed at a higher temperature (25°C), the population grows faster, with an average generation time of 10 h. In addition, no cell division rhythm is observed, even during entrainment conditions under LD cycles (Fig. 11.1). The circadian–infradian rule postulates the circadian oscillator to be either (1) cycling with the same period as the cell division cycle, (2) running with an approximately 24-h circadian period but not controlling the cell division cycle and other cellular processes, or (3) non-functional. In any case, the rule implied an interdependency between the cell division cycle and the circadian clock, such that when cells are growing rapidly and dividing more than once per day, they are unable to maintain a circadian oscillation or no longer employ the circadian control to achieve higher fitness. Although some exceptional observations had been reported (Chisholm et al. 1980; Edmunds 1989), the circadian–infradian rule was generally accepted until recently.

Molecular models of circadian clocks in eukaryotes have been proposed based on negative feedback regulation in expression of circadian clock genes (Hardin et al. 1990; Aronson et al. 1994). In those models, negative regulation of clock gene expression by its own gene product is thought to be essential. Many experimental data indicate that the time-delayed feedback could be achieved through protein–protein interactions, protein modifications and nuclear transport of the clock proteins (Harmer et al. 2001). It would therefore be expected that the processes of cell division would affect circadian time keeping. Therefore, the circadian–infradian rule had seemed to fit with molecular models of circadian clock mechanism in eukaryotic organisms. However, a recent report (Nagoshi et al. 2004) in mammalian tissue cultures (NIH 3T3 mouse fibroblasts) indicated that the cell division processes do not have much effect on the circadian timekeeping. In contrast to previous reports, this newer finding suggests that the effect of cell division on the circadian clockwork is negligible in the mammalian circadian clockwork.

11.3 Circadian Control of Cell Division in Bacteria

The existence of circadian rhythms in bacteria was first claimed by Halberg and Conner (1961), who analyzed data originally published by Rogers and Greenbank in 1930 on the growth of *Escherchia coli* in liquid culture. In the 1930 article, Rogers and Greenbank were interested in the growth of bacteria in "animal bodies" and attempted to simulate the conditions inside an animal body in the laboratory. They used a long glass tube (7 mm diameter, ca. 15 m long) filled with broth medium to serve as a model "intestine". *E. coli* cells were inoculated at one of the end of the long tube, and the extension of a liquid colony of the bacteria to the other end of the tube was monitored continuously at a constant temperature of 30°C for 175 h (lighting conditions were not described in detail). The original authors concluded that the liquid colonies of *E. coli* grew continuously in the tube but the growth rates varied (Rogers and Greenbank 1930). Thirty years latter, Halberg and Conner (1961) intensively analyzed Rogers and Greenbank's data, which were very noisy and did not exhibit obvious daily rhythms. Their periodogram and power spectra analyses of the data on growth (possibly cell division) and/or motility of *E. coli* estimated a period of approximately 20.5 h in the intermittent growth of the bacterial culture and suggested the existence of a circadian rhythm in cell growth (or cell division) and/or motility in bacteria.

Ten years later, an extensive study to search for biological rhythms in prokaryotic organisms was conducted by Sturtevant (1973a, b). She repeated Rogers and Greenbank's experiment in *E. coli* (Sturtevant 1973a) and extended her search for circadian rhythms of bacterial growth rates in *Klebsiella pneumoniae*, a gram-negative rod-shaped bacterium, using a continuous culture system under laboratory-controlled environmental conditions (Sturtevant 1973b). A continuous culture of *K. pneumoniae* was maintained for 310 h at a constant temperature of 37.1°C (±0.05°C) and optical density of the culture was measured at 510 nm every 15 or 30 min. She concluded that the growth of the bacteria showed a highly significant circadian fluctuation with a period of 24.1 h, as determined by a least-square cosine curve fitting method, even though the rhythm was not detected easily by visual (macroscopic) inspection of the data (Sturtevant 1973b).

Did those data and analyses indicate the existence of a circadian rhythm, possibly of circadian control of cell division, in bacteria? The hypothesis was controversial in the field of circadian biology (Halberg and Conner 1961; Halberg and Cornélissen 1991; Johnson et al. 1996; Halberg et al. 2003). A typical argument was that no positive experimental data had come out after their publications and supported their claim. Perhaps the bacterial strains, when used by others for repeating or following up the original experiments, had lost the circadian clock due to long maintenance of stock cultures in laboratory environments? Until the mid-1980s, it was believed by many that circadian rhythms, including the circadian control of cell division, were limited to eukaryotes (Johnson et al. 1996; Halberg et al. 2003).

As described in the Chap. 3, circadian rhythms in bacteria were first "persuasively" discovered in cyanobacteria (Grobbelaar et al. 1986; Mitsui et al. 1986; Huang et al. 1990; Chen et al. 1991). Shortly after the discovery of circadian rhythms in *Synechococcus* RF-1 (Grobbelaar et al. 1986), Sweeny and Borgese (1989) reported a circadian rhythm in cell division in the marine cyanobacterium *Synechococcus* WH7803. After being exposed to cycles of LD12:12, experimental cultures of the marine cyanobacterium were placed under continuous dim light (2 μE m^{-2} s^{-1}) and constant temperature. Cell division occurred mostly during subjective night for at least four cycles. The period of this division rhythm was approximately 24 h; the period remained nearly constant at different temperatures (16, 20, 22°C) and the calculated temperature coefficient (Q_{10}) was about 1.15, which demonstrated temperature compensation of the rhythm. The generation times at the differential temperatures were determined to be 200, 135, and 55 h, respectively, implying that cell growth rate does not affect the periodicity of cell division. Those experimental results strongly indicated that cell division is under the control of the circadian clock in *Synechococcus* WH7803 (Sweeny and Borgese 1989).

After the development of genetic tools to study circadian rhythms in *S. elongatus* PCC 7942 (Kondo et al. 1993) and the isolation of circadian clock mutants (Kondo et al. 1994), we decided to investigate the relationship or interconnection between the cell division cycle and the circadian clock in *S. elongatus*. Under LD12:12 cycles, this photosynthetic, single fission, unicellular cyanobacterium divides in the light phase and stops dividing in the dark phase (Mori et al. 1996). After imposing LD12:12 cycles, the cultures were placed in LL (45 μE m^{-2} s^{-1} white light). In LL, the AMC149 cells (wild-type strain carrying a bacterial luciferase reporter) continued to divide rhythmically. Most cells divide in the subjective day and stop dividing in the early subjective night (Fig. 11.2). The period of this cell division rhythm is approximately 24 h and is independent of the growth rate of the populations (Mori and Johnson 2001) or the temperature at which they are grown (unpublished data). In contrast, cell division of the clock-null strain (Δ*kaiC*), in which the essential circadian clock gene *kaiC* (Ishiura et al. 1998) is genetically deleted, does not continue to cycle in LL conditions although it exhibits almost identical growth patterns in LD12:12 as compared to those of the wild type (Fig. 11.2). These data indicate that the cell division cycle in *S. elongatus* is under control of the circadian oscillator, which is composed of *kai* gene products.

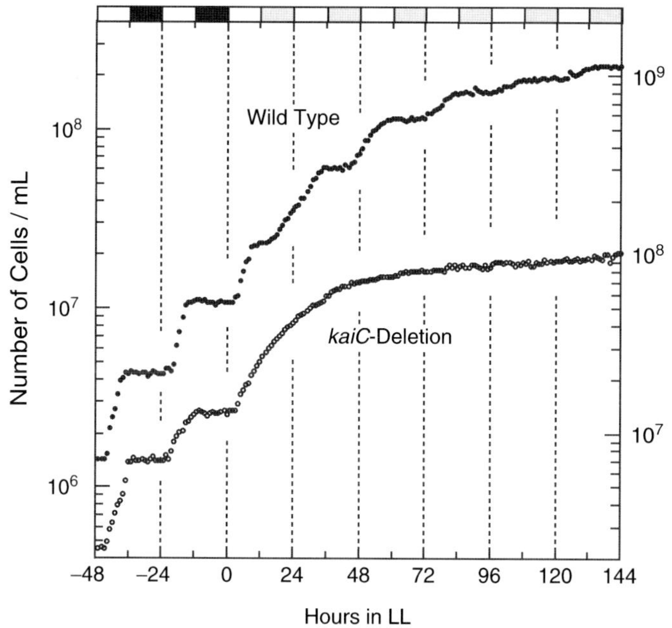

Fig. 11.2 Circadian rhythm of cell division in batch cultures of *S. elongatus* PCC 7942. Cell number data for the wild-type AMC149 strain (•) and a clock-null Δ*kaiC* strain (◦). The wild-type and Δ*kaiC* strains were grown in LD12:12 and transferred to LL (45 μE m^{-2} s^{-1}) at time zero. The last two LD cycles preceding LL are illustrated by the bars on the upper abscissa (*white* light, *black* darkness, *gray* subjective night phases of LL). The left ordinate is for the wild-type strain, and the right ordinate is for the Δ*kaiC* strain. From Mori and Johnson (2001); reprinted and adapted with permission of the American Society of Microbiology

11.4. Independence of the Circadian Clock from the Cell Division Cycle in Cyanobacteria

Despite the fact that the circadian–infradian rule was widely accepted in eukaryotic organisms about 10 years ago, it was unknown whether the rule could be generalized to the bacterial circadian system. To address this question, occurrence of cell division and expression from luciferase reporters were monitored in a culture of *S. elongatus* that was continuously diluted with fresh medium to maintain an approximately equal cell density with an average doubling time of less than 24 h. Even in rapidly growing culture, cyanobacteria exhibited circadian rhythms of cell division and gene expression (Mori et al. 1996; Figs. 11.3, 11.4). The cells slowed or stopped dividing in the early subjective night, although the growth rate (determined by the optical density of the culture) and DNA synthesis in the population were fairly constant (Fig. 11.3B). Cell division began again in the late subjective night and continued through the subjective day. This rhythm continued for the duration that the continuous culture was maintained – at least six cycles. The periods of the cell division rhythms were essentially the same in both batch cultures and

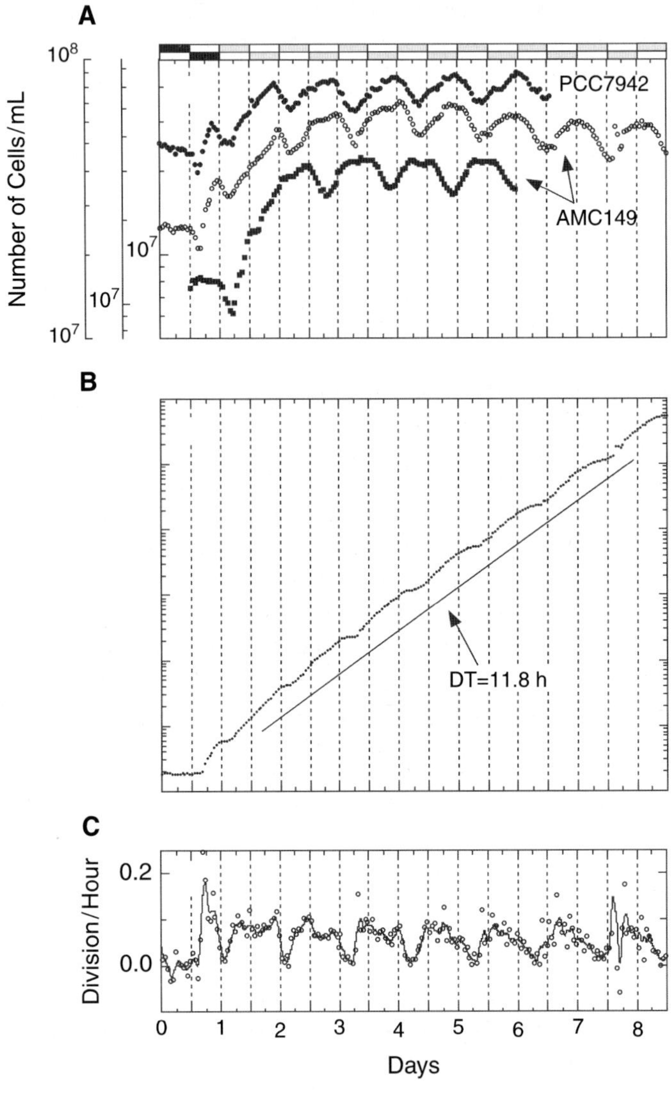

Fig. 11.3 Cell division rhythms in continuously diluted cultures of *S. elongatus* PCC 7942. **A** Cell number data for PCC 7942 and AMC149 cultures. The uppermost trace is for wild-type *S. elongatus* PCC 7942; the two AMC149 traces are from cultures that were previously entrained to LD cycles that were 12 h out of phase with each other compared with laboratory clock time (Central Standard Time). Abscissa: the last LD cycle preceding LL is illustrated by bars on the upper abscissa (*upper bar* for top and middle traces, *lower bar* for lowest trace; *white* light, *black* dark, *gray* subjective night phases of LL). Ordinates: the leftmost ordinate is for the bottom trace, the middle ordinate is for the middle trace, and the rightmost ordinate is for the top trace. After the entrainment, the cultures were released into LL and continuously diluted. The periods of the cultures estimated by the maximum entropy method were 24.0 h (*upper trace*), 25.2 h (*middle trace*), and 24.2 h (*lower trace*). **B** The data in (**A**) for the middle trace are replotted as a cumulative increase in cell number; i.e., a logistic growth curve, calculated from the rate of dilution and

continuous cultures, as well as in cells whose doubling times ranged between 9h and 90h. These observations suggest that the organism has circadian phases that "allow" and "forbid" cell division even in rapidly growing conditions (ultradian mode). In *S. elongatus,* we hypothesized that in LL the organism always proceeds with cell growth cycle (Asato 1984, 2003) depending on the availability of resources such as light. However, the circadian clock prohibits the cells from dividing at some specific phases of the circadian cycle (Mori et al. 1996; Mori and Johnson 2000).

Expression of a luciferase reporter under the control of the *psbAI* promoter (encoding form I of D1 protein of photosystem II) also exhibited a circadian rhythm in bioluminescence (strain AMC149) in rapidly growing cyanobacterial cells that were grown in continuous culture (Mori et al. 1996); peak expression occurred near the end of subjective day or early subjective night. Additionally, in the early stages of batch liquid cultures and solid cultures, bioluminescence patterns matched a model that could be made when assuming that cells proliferate exponentially with generation times of 7–10h and each single cell exhibits a circadian rhythm of *psbAI* promoter in a cosine fashion (Kondo et al. 1997). These observations indicate that the circadian timekeeping system is completely independent from the cell division cycle in *S. elongatus* and is not perturbed by cellular events associated with cell division, such as DNA replication, chromosome segregation, or cytokinesis. These data contradict the circadian–infradian rule that was developed to describe the interface between the cell division cycle and the circadian timekeeping system in eukaryotes and suggest that regulation of the cyanobacterial circadian system on cell division is fundamentally different from that in eukaryotes.

The circadian regulation of gene expression was also monitored in cells that did not undergo cytokinesis (Mori and Johnson 2001). Overexpression of the bacterial cell division gene *ftsZ* prevents cell division (but not growth) and produces filamentous cells (Ward and Lutkenhaus 1985; see Sect. 11.5). The FtsZ protein is a structural and functional analog of eukaryotic tubulins and crucial for formation of a ring structure (Z-ring) at the inner surface of the cytoplasmic membrane at the division site (Shih and Rothfield 2006). Promoter activities of *psbAI*, *kaiBC*, and endogenous *ftsZ* genes were monitored using luciferase reporters in the filamentous cyanobacterial cells (Fig. 11.5). Whether the cells stopped division by overexpression of *ftsZ* or not, expression of bioluminescence from any of these promoters (*psbAI*, *kaiBC*, *ftsZ*) maintained robust circadian rhythms for 4–5 days in LL (Mori and Johnson 2001; Fig. 11.5).

Fig. 11.3 (Continued) the cell number data of (**A**). The *diagonal line* indicates a doubling time (*DT*) of 11.8h. **C** The middle trace from (**A**) is replotted as an instantaneous rate of increase in cell number compensated for the rate of medium dilution: *Division/Hour* = ln(change of cell number h^{-1}). The points in (**C**) are the rate of change between each successive pair of points in the data of (**A**), and the *line* is a three-point moving average. On this graph, a value of zero means that the cell number did not increase at that time. From Mori et al. (1996); reprinted and adapted with permission of the National Academy of Sciences of the United States of America

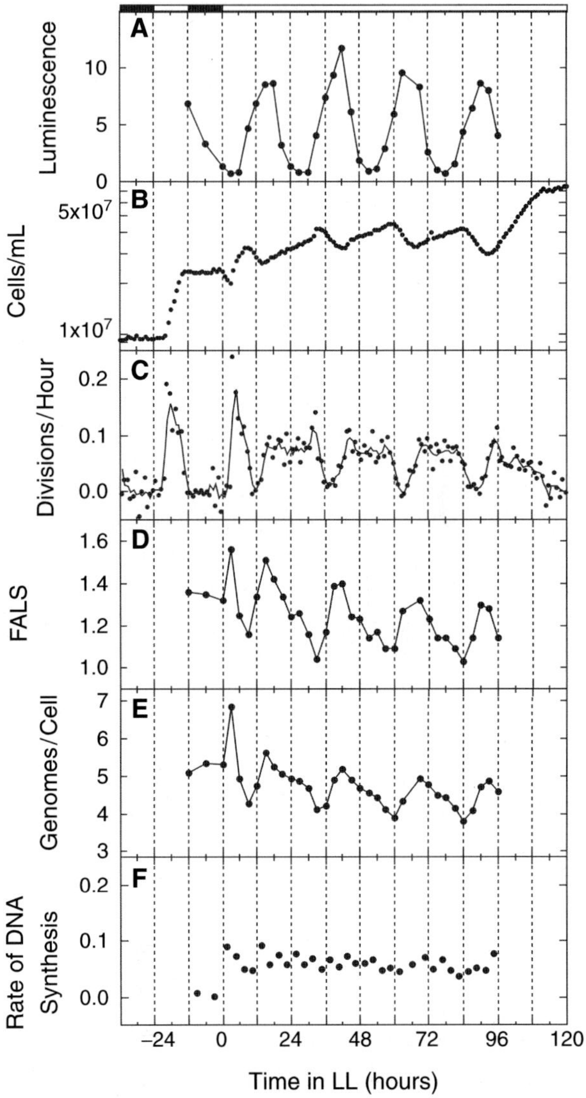

Fig. 11.4 Cell number, luminescence, cell size (FALS), and DNA content in continuously diluted cultures of AMC149. Cells were entrained to LD12:12 in batch cultures and then released into LL at time zero, at which time continuous dilution began and was maintained until 96h. **A** Luminescence expressed by luciferase reporter construct (P*psbAI*::*luxAB*). **B** Number of cells per millilter of culture (raw data). Estimation of period in LL by the maximum entropy method was 23.3h. **C** Instantaneous rate of increase in cell number (in divisions per hour) compensated for the rate of medium dilution. **D** Average FALS per cell. **E** Average number of genomes per cell (DNA content per cell). **F** Rate of DNA synthesis, expressed as the DNA-specific rate of increase in DNA in the culture (units = 1/time). For a given time point *t*, this rate was calculated as the sum of the dilution rate and the observed rate of change in total DNA (between times *t* and *t* + 1) divided by total DNA at time *t*; total DNA was calculated as the product of cells ml^{-1} and mean DNA $cell^{-1}$ (**B** and **E**, respectively). A slope of zero means that the specific rate of DNA synthesis is constant over time. From Mori et al. (1996); reprinted and adapted with permission of the National Academy of Sciences of the United States of America

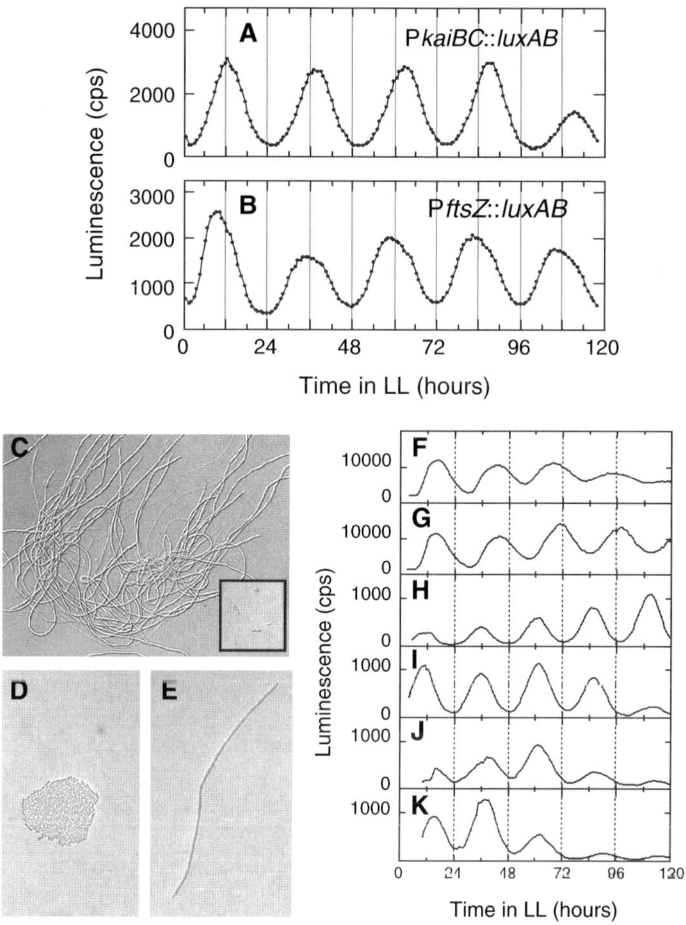

Fig. 11.5 Luminescence rhythms of luminescence in P*ftsZ*::*luxAB* reporter strain and in dividing and nondividing cyanobacteria. **A**, **B** Measurement of in vivo luminescence from 3-ml batch liquid cultures of P*kaiBC*::*luxAB* (**A**) and P*ftsZ*::*luxAB* (**B**) strains of *S. elongatus* PCC 7942. **C**, **D**, **E** Overexpression of FtsZ stopped the cell division of growing cyanobacteria, resulting in filamentous cells. **C** The P*trc*::*ftsZ* strain was grown for 101 h in liquid BG-11 medium supplemented with 0.5 mM IPTG. *Insert* P*trc*::null cells as a control under the same conditions. The P*trc*::*ftsZ* and P*trc*::null cells were also grown on solid (1.5% agar) BG-11 medium supplemented with 1 mM IPTG for 48 h. A colony of P*trc*::null cells (**D**) and a filamentous P*trc*::*ftsZ* cell (**E**) are shown. Presumably both the colony in (**D**) and the filament in (**E**) were derived from a single initial cell. **F–K** Luminescence rhythms in dividing and nondividing cyanobacteria in liquid cultures. In vivo luminescence was monitored in the following reporter strains: P*psbAI*::*luxAB*(**F**), FtsZ overexpression in the P*psbAI*::*luxAB* strain (**G**), P*kaiBC*::*luxAB* (**H**), FtsZ overexpression in the P*kaiBC*::*luxAB* strain (**I**), P*ftsZ*::*luxAB* (**J**), FtsZ overexpression in the P*ftsZ*::*luxAB* strain (**K**). FtsZ protein was overexpressed continuously with 0.5 mM IPTG in (**G**, **I**, **K**) and filamentous morphology in those cultures was confirmed microscopically. From Mori and Johnson (2001); reprinted and adapted with permission of the American Society for Microbiology

By using single, live-cell imaging of bioluminescence, Mihalcescu et al. (2004; see Chap. 13) tracked the bioluminescence rhythms of individual cells during continuous growth. Each cell exhibited a circadian rhythm of gene expression, and interestingly, after each cell division the daughter cells retained the inherited rhythm without altering the phase of the rhythm from the mother cell. These observations imply that the circadian timekeeping system that controls gene expression appears to be stable and there is no apparent feedback from the cell division cycle to the circadian oscillator.

Recently Nakajima et al. (2005) discovered that the oscillation of KaiC phosphorylation can be reconstituted in vitro by incubating KaiC with KaiA, KaiB, and ATP (see Chap. 5). Because cells will grow continuously in LL (Mori et al. 1996) and the Kai proteins are relatively abundant (Kitayama et al. 2003) in comparison with other regulatory proteins such as transcriptional factors, the concentration of Kai proteins in the cells would be approximately the same before and after cell division. Analogously, assuming each single cell is a "test tube" in which the KaiABC reaction takes place, cell division can simply be thought of as dispensing the reaction solution of a single test tube (mother cell) into two new test tubes (daughter cells). It would be reasonable to conclude that cell division does not affect the periodicity of the circadian oscillator, which is mainly governed by the post-translational KaiABC cycling reaction in each cyanobacterial cell.

11.5 Mechanisms of Circadian Control of Cyanobacterial Cell Division

As described in the previous section, rapidly growing continuous cultures of *S. elongatus* PCC 7942 display a circadian rhythm of cell division (cytokinesis). In synchronized cultures of *S. elongatus* PCC 6301, which is a closely-related species of *S. elongatus* PCC 7942, the period of the cell division cycle is characterized by the sequential and ordered synthesis and appearance of macromolecules, such as DNA, RNA, proteins, phospholipids, and peptidoglycan (Asato 2003). Cell division events such as DNA replication, chromosome segregation, and septum formation are thought to be linked to and/or coordinated with the synthesis of these macromolecular products (Asato 2003).

Interestingly, *S. elongatus* PCC 7942 possesses multiple copies of its chromosome (1–6 copies cell^{-1}). Microscopic and flow cytometric analyses indicated that the cell size (length by microscopic measurements, volume by flow cytometric light scattering measurements) and DNA content (chromosome copy number by flow cytometry measurements) of cells in the continuous culture were varied: a population of cells is heterogeneous and cell division cycles (or growth cycles) are not synchronized. The DNA content of the cells oscillated between an average of 4.0 and an average of 5.5 genomes per cell (Fig. 11.4E). The DNA content of the cells was strongly correlated to cell volume. Overall, the rate of DNA synthesis in a population tends to be constant, and periodic cell division results in the rhythm of

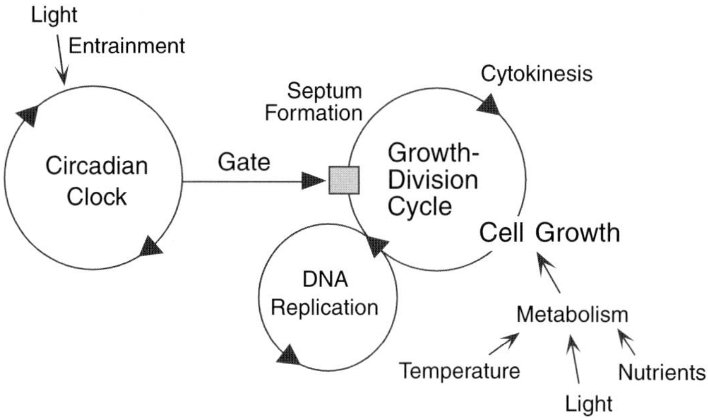

Fig. 11.6 Schematic model of how the circadian clock controls cell division cycle in cyanobacteria. The circadian pacemaker (*left circle*) is self-sustained with a period of about 24h, completely independent of cell division, and can be entrained by daily environmental cycles such as LD and temperature cycles. The progression of the cell cycle (*right circles*) is strongly influenced by light and other environmental factors. In *S. elongatus*, cell growth and DNA replication are weakly coupled. DNA replication is initiated synchronously or asynchronously with cell growth. The circadian pacemaker gates cell division (*arrow* from left to right), inhibiting septum formation or cytokinesis

DNA content per cell. Under exponential growth conditions in LL, the cell volume or mass grows continuously as it does with other bacteria, such as *E. coli*, and the rate of DNA synthesis is constant. However, the timing of cytokinesis is rhythmic. It has been hypothesized that *S. elongatus* always proceeds to cell growth depending on the availability of resources such as light, but the circadian clock prohibits the cells from dividing in specific phases of the circadian cycle (Fig. 11.6; Mori et al. 1996; Mori and Johnson 2000).

How does the circadian clock control cell division? Identification of genes involved in bacterial cell division as well as the mechanisms by which cell division is controlled have been extensively studied in systems including *E. coli*, *Bacillus subtilis*, and *Caulobacter crescentus* (Rothfield et al. 1999; Hiraga 2000; McAdams and Shapiro 2003; Romberg and Levin 2003; Löwe et al 2004; Lewis 2004; Angert 2005; Vicente et al 2006; Graumann 2007). Genetic regulatory networks – including gene expression, protein structure, biochemical properties, protein–protein and protein–DNA interactions, and cellular localization of the cell division proteins – have been studied to elucidate fundamental processes of the highly coordinated bacterial cell division cycle (Vicente et al. 2006; Graumann 2007).

In *E. coli*, many genes related to cell division are expressed at specific times during the cell division cycle. In *S. elongatus*, transcriptional activities from more than 90–95% of gene promoters exhibit a circadian rhythm (Liu et al. 1995). The rhythmic expression of genes related to cell division in the cyanobacteria might lead to the circadian control of the process of cell division. For example, the *ftsZ* gene is an essential cell division gene in bacteria. FtsZ protein is a structural and functional analog of eukaryotic tubulins and is crucial for the formation of a ring structure (Z-ring) at the inner surface of the cytoplasmic membrane at the division site (Shih and Rothfield 2006). The expression of *ftsZ* oscillates during the cell cycle in *E. coli*, with *ftsZ* mRNA levels increasing during the initiation of DNA replication (Garrido et al. 1993). Because the circadian clock controls the timing of cytokinesis in *S. elongatus* but not the growth cycle or DNA replication, *ftsZ* is a good candidate for the circadian regulation of cell division.

To determine whether the circadian clock of cyanobacteria controls the timing of cell division by regulating the expression of *ftsZ*, the *ftsZ* promoter region from *S. elongatus* PCC 7942 was used to drive expression of a *luxAB* reporter (Fig. 11.5B). The expression from the *ftsZ* promoter exhibits a circadian rhythm with peak activity in the early night and minimal activity in the early day (Fig. 11.5B; Mori and Johnson 2001). The circadian expression pattern of the *ftsZ* promoter does not support a direct correlation between *ftsZ* expression and the circadian control of cell division, which is forbidden in the early night when *ftsZ* expression is at its peak.

In *E. coli*, a low-level overproduction (five-fold increase) of FtsZ induces mini-cell formation and increases the frequency of cell divisions (Ward and Lutkenhaus 1985). In contrast, a high-level overproduction (12-fold or more) of FtsZ inhibits cytokinesis, which results in filamentous cells. The endogenous promoter activity of *ftsZ* exhibits the circadian rhythm even in filamentous cells that are continuously growing but cannot undergo cytokinesis due to the overexpression of *ftsZ* from a strong heterologous promoter (Fig. 11.5K). It is unclear why expression of *ftsZ* exhibits the circadian rhythm and peaks during the early night, when the cells stop dividing. Perhaps the FtsZ protein level in *S. elongatus* is maintained at high levels, and the increased transcriptional activity from the *ftsZ* promoter during the early night exceeds a threshold FtsZ protein level to allow for the inhibition of cell division. This hypothesis could be tested by observing cell division in cells expressing different levels of *ftsZ* under the control of an inducible promoter, as well as quantifying FtsZ protein levels in S. *elongatus* cells.

The rhythmic expression of *ftsZ* is in the phase of gene expression (Liu et al. 1996) in which most genes (~85%), including *psbAI*, *psbAIII* (encoding form II of D1 protein of photosystem II), *gnlA* (glutamine synthetase), and *rrnA* (rRNA), are rhythmically expressed (Liu et al. 1996). The circadian clock may be regulating the expression of many genes, including *ftsZ* in this phase, to coordinate energy and other cellular metabolisms in LD cycles under natural environmental conditions. Because the transcriptional activity of *ftsZ* is rhythmic, the level of FtsZ protein may be oscillating, whereas the change in protein level may not significantly affect cell division.

Comparative and mutational analyses by Miyagishima et al. (2005) identified many genes involved in cell division in *S. elongatus* PCC 7942. The genes that they identified, or genes which had previously been identified by others, include *ftsE*, *ftsI*, *ftsQ*, *ftsW*, *ftsZ*, *minC*, *minE*, *sulA*, *cdv1*, *cdv2* (*ylmF*), *cdv3* (*divIVA*-like), *ftn2*, *ftn6*, and *cikA* (Koksharova and Wolk 2002; Miyagishima et al. 2005). Some candidate cell division genes such as *ylmE*, *ylmG*, and *ylmH* have also been found (Miyagishima et al. 2005). Interestingly, no orthologs of the *E. coli* or *B. subtilis* cell division genes *ftsA*, *ftsL*, *ftsN*, *zipA*, *eztA*, *zapA*, and *ftsX* were found, and some genes such as *ftn2*, *ftn6*, and *cikA* are unique to cyanobacteria (Koksharova and Wolk 2002; Miyagishima et al. 2005). Many genes related to cell division and cell wall synthesis are found in a chromosomal region, the *dcw* cluster (*d*ivision and *c*ell *w*all) in bacterial organisms from distant groups (Viente et al. 2006). However, in *S. elongatus*, many of them are spread throughout the genome (Miyagishima et al. 2005). It was also demonstrated that, unlike in other bacteria, Z rings could be formed at sites occupied by nucleoids in *S. elongatus,* which has multiple copies of the chromosome (Miyagishima et al. 2005). These findings suggest that the molecular mechanism (regulatory networks of the expression of cell division genes, biochemical functions of the cell division proteins, etc.) in the regulation of septum formation in *S. elongatus* differs considerably from those in other bacteria. Any cell division-related gene, either unique to cyanobacteria or particularly common in bacteria, could be a candidate for the circadian regulation of cell division in cyanobacteria.

One mechanism that may be involved in the orchestration of rhythmic expression of genes related to cell division is topological changes in DNA structure. Circadian rhythms of chromosome compaction (Smith and Williams 2006) and superhelical status of DNA (Woelfle et al. 2007) have both been demonstrated in *S. elongatus*. Using single living-cell imaging, *kaiBC* promoter activity was monitored in cells in which the P*kaiBC*::*YFP-SsrA*(*LVA*) fusion gene [an SsrA-tagged (destabilized) yellow-shifted variant (YFP) of green fluorescent protein under the control of *kaiBC* promoter] was introduced into two separate locations of the cyanobacterial chromosome. Analysis of reporter gene expression from the two different strains allowed the magnitude of gaussian noise in *kaiBC* gene transcription to be separated into local and global contributions (Chabot et al. 2007). The global error appeared to scale linearly with the transcription rate (maximal at CT14), however, the local error was maximal at the circadian phase in which the transcription rate was minimal (CT0). The authors suggested that the calculated local noise, which could not simply be explained as only intrinsic noise, could be due to differences in local cellular environments between the two different chromosomal locations of the transgene reporter (e.g., chromosomal topology). These observations are consistent with a hypothesis that structural or topological changes in chromosomal DNA affect gene expression and could govern global circadian gene expression, including that of known clock-related genes (Woelfle et al. 2007). It could be possible that the expression of many genes (not necessarily any one specific gene) involved in cell division, which are influenced by chromosome compaction and DNA supercoiling, may contribute to the circadian control of cell division. For example, genes involved in the septal recruitment pathway (Harry et al. 2006) could be expressed under the control of the clock.

Alternatively, circadian control of cell division could be a result of post-translational modifications of cell division proteins mediated by two-component signal transduction systems linked to the core oscillator. Interestingly, a phytochrome-related histidine kinase CikA (circadian input kinase; Schmitz et al. 2000; see Chap. 8) plays a role in the circadian system as well as in that of cell division

(Miyagishima et al. 2005). Targeted disruption of *cikA* produced cells two to three times longer than those of the wild type (Miyagisawa et al. 2005; Zhang et al. 2006). *Zoanthus* sp. GFP (ZsGreen)-tagged Trx-CikA fusion protein overexpressed under a strong, inducible promoter shows a polar localization pattern with one or two foci per cell (Zhang et al. 2006), which is reminiscent of some protein kinases related to cell division, such as CckA, DivJ, and PleC, that localize at the poles in *Caulobacter crescentus* (Quardokus and Brun 2003). CikA may form a complex with proteins at the poles or septum site to regulate cell division (Zhang et al. 2006). Identification of proteins that interact with CikA may help to understand the interconnection between the circadian clock and cell division (Mackey et al. 2008).

11.6 Outlook and Future Perspectives

Circadian control of cell division in prokaryotes has been demonstrated in some unicellular (Sweeney and Borgese 1989; Mori et al. 1996) and filamentous (Lee and Rhee 1999) cyanobacterial species. Diurnal cell division and gene expression patterns in synchronized cultures of the marine cyanobacterium *Prochlorococcus* (Jacquet et al. 2001; Holtzendorff et al. 2001; Holtzendorff et al. 2002) raise the possibility that the cell division cycle is regulated by a circadian clock in this marine organism, whose small genomes have *kaiB* and *kaiC* genes but lack *kaiA* (Dvornyk et al. 2003). A recent study reported that the cell division and *psbA* gene expression rhythms damp very rapidly in *Prochlorococcus* under continuous light (Holtzendorff et al. 2008). This finding suggests that the genome reduction (i.e., deletion of *kaiA*) in *Prochlorococcus* has resulted in a loss of robustness in the endogenous oscillator but the basic *kaiBC* system still serves as a resettable hourglass timer in this marine cyanobacteria under daily environmental conditions (LD and/or temperature cycles). Do bacteria other than cyanobacteria have a circadian rhythm of the cell division cycle? Gene expression in the purple photosynthetic bacterium *Rhodobacter sphaeroides* has been investigated and appears to be rhythmic with a period of 20.5 h under aerobic conditions and a period of 10.6–12.7 h under anaerobic conditions (Min et al. 2005), which suggests the presence of a self-sustained oscillator. Cell division cycles in *Rhodobacter* could potentially be regulated by the oscillator that controls gene expression under aerobic and/or anaerobic conditions.

Why do cyanobacteria have circadian gating of cell division? In eukaryotic organisms, Pittendrigh suggested that an "escape from light" has played a significant role in the evolution of the circadian clock (Pittendrigh 1965, 1993; Paietta 1982). Solar irradiation contains ultraviolet (UV) light that may damage cellular components such as DNA, protein, and other organic molecules and may be harmful to cells. Specifically, UV radiation causes damage to DNA and leads to mutation. In some eukaryotic cells, it is known that the cells are most sensitive to UV irradiation in the G1/S phase of the cell division cycle (Cremer et al. 1981; Siede and Friedberg 1990). Selective pressure may have led to the evolution of the circadian clock to forbid the progression of UV-sensitive phases in the cell division cycle during the day (Nikaido and Johnson 2000; Mori and Johnson 2001). This hypothesis coincides with experimental observations in some eukaryotic organisms, such as *Chlamydomonas* (Goto and Johnson 1995) and *Euglena* (Edmunds 1989).

In contrast, the escape from light hypothesis is not a plausible explanation for the evolution of circadian control of cell division in the cyanobacterium *S. elongatus* PCC 7942 because it divides in the daytime (Figs. 11.2, 11.3, 11.4) and stops or slows cell division at the end of day or in the early night. It has been reported that *S. elongatus* PCC 6301 can grow and divide with a generation time of 2 h in optimum growth conditions (38°C, 5% CO_2; Herdman et al. 1970; Asato 2003). Additionally, it is known that many cyanobacteria synthesize UV-absorbing/screening compounds (photoprotectants) such as scytonemin and mycosporine-like amino acids (MAAs; Garcia-Pichel and Castenholz 1993; Sinha and Häder 2008). Further, in *S. elongatus*, each cell has multiple copies of the chromosome (Binder and Chisholm 1990; Mori et al. 1996), which may facilitate post-radiation DNA repair by homologous chromosome recombination. In fact, cyanobacteria are considerably more resistant to UV light than *E. coli* (Domain et al. 2004). Because of their relatively fast growth in light and resistance to UV, it could be hypothesized that the maximum reproductive fitness will presumably be achieved by dividing as fast as possible in light rather than waiting for darkness to accomplish DNA replication or cytokinesis. Since *S. elongatus* is an obligate photoautotroph and only has limited energy storage, cells may not proceed with the growth cycle in the dark. Therefore, the circadian clock may enable cells to anticipate darkness and stop dividing at the end of daytime or in the early subjective night.

In addition to unicellular protozoa, algae, and cyanobacteria, circadian control of cell division or cell proliferation has been reported in multicellular organisms, including mammals (Edmunds 1989; Scheving 1981). Recently, it was shown that expression of the cell cycle-related genes *cyclin B1*, *cdc2*, and *wee1* exhibit a circadian rhythm in regenerating liver cells from wild-type mice but not from *Cryptochrome*-deficient mice in vitro (Matsuo et al. 2003). Among these, expression of *wee1* is directly regulated by the CLOCK-BMAL1 heterodimeric transactivator, an essential component of the mammalian circadian clock, through its binding of E-box (CACGTG) elements in 5'-upstream regions of the *wee1* gene. Additionally, recent studies suggest an involvement of PERIOD1 and PERIOD2 in ATM-Chk1/Chk2 DNA damage response pathways (Fu et al. 2002).

The molecular mechanism of the regulation of cell division by the circadian clock system is poorly understood. Recent developments in genomics, proteomics, and interactomics utilizing high-throughput measurement techniques and bioinformatics (Laub et al. 2000; Bonneau et al. 2007; Ishii et al. 2007) allow systematic analyses of cellular events and overcome the limitations of gene-by-gene approaches in conventional molecular genetics. Systematic mutagenesis and high-throughput functional analyses (Holtman et al. 2005) will help to elucidate the mechanism of the circadian control of cell division. Implementing such approaches could facilitate the search for key regulatory networks in the circadian control of cell division.

Acknowledgements I thank Dr. Leland N. Edmunds Jr for allowing me to reprint his figure in this chapter, and I thank Brian Roberts and Dr. Carl H. Johnson for their editing and suggestions on an early draft of this chapter.

References

Angert ER (2005) Alternatives to binary fission in bacteria. Nat Rev Microbiol 3:214–224

Aronson BD, Johnson KA, Loros JJ, Dunlap JC (1994) Negative feedback defining a circadian clock: autoregulation of the clock gene *frequency*. Science 263:1578–1584

Asato Y (1984) Characterization of cell cycle events in synchronized cultures of *Anacystis nidulans*. J Gen Microbiol 130:2535–2542

Asato Y (2003) Toward an understanding of cell growth and the cell division cycle of unicellular photoautotrophic cyanobacteria. Cell Mol Life Sci 60:663–687

Barák I, Wilkinson AJ (2007) Division site recognition in *Escherichia coli* and *Bacillus subtilis*. FEMS Microbiol Rev 31:311–326

Barnett A (1969) Cell division: a second circadian clock system in *Paramecium multimicronucleatum*. Science 164:1417–1419

Binder BJ, Chisholm SW (1990) Relationship between DNA cycle and growth rate in *Synechococcus* sp. strain PCC 6301. J Bacteriol 172:2313–2319

Bonneau R, Facciotti MT, Reiss DJ, Schmid AK, Pan M, Kaur A, Thorsson V, Shannon P, Johnson MH, Bare JC, Longabaugh W, Vuthoori M, Whitehead K, Madar A, Suzuki L, Mori T, Chang DE, Diruggiero J, Johnson CH, Hood L, Baliga NS (2007) A predictive model for transcriptional control of physiology in a free living cell. Cell 131:1354–1365

Bouget FY, Moulager M, Corellou F (2008) Circadian regulation of cell division. In: Verma DP, Hong Z (eds) Cell division control in plants. Springer, Heidelberg, pp 3–12

Bré MH, Ferjani EE, Lefort-Tran M (1981) Sequential protein dependent steps in the cell cycle. Initiation and completion of division in vitamin B_{12} replenished *Euglena gracilis*. Protoplasma 108:301–318

Bruce VG, Bruce NC (1981) Circadian clock-controlled growth cycle in *Chlamydomonas reinhardi*. In: Schweiger HG (ed) International cell biology, 1980–1981. Proc 2nd Int Congr Cell Biol, Berlin, 1980. Springer, Berlin, pp 823–830

Chabot JR, Pedraza JM, Luitel P, van Oudenaarden A (2007) Stochastic gene expression out-of-steady-state in the cyanobacterial circadian clock. Nature 450:1249–1252

Chen T-H, Chen T-L, Hung L-M, Huang T-C (1991) Circadian rhythm in amino acid by *Synechococcus* RF-1. Plant Physiol 97:55–59

Chisholm SW, Morel FMM, Slocum SW (1980) The phasing and distribution of cell division in cycles in marine diatoms. In: Falkowski P (ed) Primary productivity and biogeochemical cycles in the sea. Plenum, New York, pp 281–300

Cremer C, Cremer T, Zorn C, Zimmer J (1981) Induction of chromosome shattering by ultraviolet irradiation and caffeine: comparison of whole-cell and partial-cell irradiation. Mutat Res 84:331–348

Domain F, Houot L, Chauvat F, Cassier-Chauvat C (2004) Function and regulation of the cyanobacterial genes *lexA*, *recA* and *ruvB*: LexA is critical to the survival of cells facing inorganic carbon starvation. Mol Microbiol 53:65–80

Dvornyk V, Vinogradova O, Nevo E (2003) Origin and evolution of circadian clock genes in prokaryotes. Proc Natl Acad Sci USA 100:2495–2500

Edmunds LN Jr (1966) Studies on synchronously dividing cultures of *Euglena gracilis* Klebs (strain Z). III. Circadian components of cell division. J Cell Physiol 67:35–43

Edmunds LN (1988) Cellular and molecular bases of biological clocks. Springer, New York

Edmunds LN Jr, Adams KJ (1981) Clocked cell cycle clocks. Science 211:1002–1013

Edmunds LN Jr, Funch RR (1969) Circadian rhythm of cell division in *Euglena*: effects of random illumination regimen. Science 165:500–503

Ehret CF (1980) On circadian cybernetics, and the innate and genetic nature of circadian rhythms. In: Scheving LE, Halberg F (eds) Chronobiology: principles and applications to shifts in schedules. Springer, Heidelberg, pp 109–126

Ehret CF, Wille JJ (1970) The photobiology of circadian rhythms in protozoa and other eukaryotic microorganisms. In: Halldal P (ed) Photobiology of microorganisms. Wiley, London, pp 369–416

Fu L, Lee CC (2003) The circadian clock: pacemaker and tumour suppressor. Nat Rev Cancer 3:350–361

Fu L, Pelicano H, Liu J, Huang P, Lee C (2002) The circadian gene *Period2* plays an important role in tumor suppression and DNA damage response in vivo. Cell 111:41–50

Garrido T, Sánchez M, Palacios P, Aldea M, Vicente M (1993) Transcription of *ftsZ* oscillates during the cell cycle of *Escherichia coli*. EMBO J 12:3957–3965

Goto K, Johnson CH (1995) Is the cell division cycle gated by a circadian clock? The case of *Chlamydomonas reinhardtii*. J Cell Biol 129:1061–1069

Graumann PL (2007) Cytoskeletal elements in bacteria. Annu Rev Microbiol 61:589–618

Grobbelaar N, Huang TC, Liu HY, Chow TJ (1986) Nitrogen-fixing endogenous rhythm in *Synechococcus* RF-1. FEMS Microbiol Lett 37:173–177

Halberg F, Conner RL (1961) Circadian organization and microbiology: variance spectra and a periodogram on behavior of *Escherichia coli* growing in fluid culture. Proc Minn Acad Sci 29:227–239

Halberg F, Cornélissen G (1991) The spectrum of rhythms in microorganisms revisited. Chronobiologia 18:114

Halberg F, Cornélissen G, Katinas G, Syutkina EV, Sothern RB, Zaslavskaya R, Halberg F, Watanabe Y, Schwartzkopff O, Otsuka K, Tarquini R, Frederico P, Siggelova J (2003) Transdisciplinary unifying implications of circadian findings in the 1950s. J Circadian Rhythms 1:2

Hardin PE, Hall JC, Rosbash M (1990) Feedback of the *Drosophila period* gene product on circadian cycling of its messenger RNA levels. Nature 343:536–540

Harmer SL, Panda S, Kay SA (2001) Molecular bases of circadian rhythms. Annu Rev Cell Dev Biol 17:215–253

Harry E, Monahan L, Thompson L (2006) Bacterial cell division: the mechanism and its precison. Int Rev Cytol 253:27–94

Herdman M, Faulkner BM, Carr NG (1970) Synchronous growth and genome replication in the blue green alga *Anacystis nidulans*. Arch Mikrobiol 73:238–249

Hiraga S (2000) Dynamic localization of bacterial and plasmid chromosomes. Annu Rev Genet 34:21–59

Holtman CK, Chen Y, Sandoval P, Gonzales A, Nalty MS, Thomas TL, Youderian P, Golden SS (2005) High-throughput functional analysis of the *Synechococcus elongatus* PCC 7942 genome. DNA Res 12:103–115

Holtzendorff J, Partensky F, Jacquet S, Bruyant F, Marie D, Garczarek L, Mary I, Vaulot D (2001) Diel expression of cell cycle-related genes in synchronized cultures of *Prochlorococcus* sp. strain PCC 9511. J Bacteriol 183:915–920

Holtzendorff J, Marie D, Post AF, Partensky F, Rivlin A, Hess WR (2002) Synchronized expression of *ftsZ* in natural *Prochlorococcus* populations of the Red Sea. Environ Microbiol 4:644–653

Holtzendorff J, Partensky F, Mella D, Lennon JF, Hess WR, Garczarek L (2008) Genome streamlining results in loss of robustness of the circadian clock in the marine cyanobacterium *Prochlorococcus marinus* PCC 9511. J Biol Rhythms 23:187–199

Homma K, Hastings JW (1989) Cell growth kinetics, division asymmetry, and volume control at division in the marine dinoflagellate *Gonyaulax polyedra*: a model of circadian clock control of the cell cycle. J Cell Science 92:303–318

Huang T-C, Tu J, Chow T-J, Chen T-H (1990) Circadian rhythm of the prokaryote *Synechococcus* sp. RF-1. Plant Physiol 92:531–533

Ishii N, Nakahigashi K, Baba T, Robert M, Soga T, Kanai A, Hirasawa T, Naba M, Hirai K, Hoque A, Ho PY, Kakazu Y, Sugawara K, Igarashi S, Harada S, Masuda T, Sugiyama N, Togashi T, Hasegawa M, Takai Y, Yugi K, Arakawa K, Iwata N, Toya Y, Nakayama Y, Nishioka T,

Shimizu K, Mori H, Tomita M (2007) Multiple high-throughput analyses monitor the response of *E. coli* to perturbations. Science 316:593–597
Ishiura M, Kutsuna S, Aoki S, Iwasaki H, Andersson CR, Tanabe A, Golden SS, Johnson CH, Kondo T (1998) Expression of a gene cluster *kaiABC* as a circadian feedback process in cyanobacteria. Science 281:1519–1523
Jacquet S, Partensky F, Marie D, Casotti R, Vaulot D (2001) Cell cycle regulation by light in *Prochlorococcus* strains. Appl Environ Microbiol 67:782–790
Jarrett RM, Edmunds LN Jr (1970) Persisting circadian rhythm of cell division in a photosynthetic mutant of *Euglena*. Science. 167:1730–1733
Johnson CH, Golden SS, Ishiura M, Kondo T (1996) Circadian clocks in prokaryotes. Mol Microbiol 21:5–11
Kitayama Y, Iwasaki H, Nishiwaki T, Kondo T (2003) KaiB functions as an attenuator of KaiC phosphorylation in the cyanobacterial circadian clock system. EMBO J 22:2127–2134
Koksharova OA, Wolk CP (2002) A novel gene that bears a DnaJ motif influences cyanobacterial cell division. J Bacteriol 184:5524–5528
Kondo T, Strayer CA, Kulkarni RD, Taylor W, Ishiura M, Golden SS, Johnson CH (1993) Circadian rhythms in prokaryotes: luciferase as a reporter of circadian gene expression in cyanobacteria. Proc Natl Acad Sci USA 90:5672–5676
Kondo T, Tsinoremas NF, Golden SS, Johnson CH, Kutsuna S, Ishiura M (1994) Circadian clock mutants of cyanobacteria. Science 266:1233–1236
Kondo T, Mori T, Lebedeva NV, Aoki S, Ishiura M, Golden SS (1997) Circadian rhythms in rapidly dividing cyanobacteria. Science 275:224–227
Laub MT, McAdams HH, Feldblyum T, Fraser CM, Shapiro L (2000) Global analysis of the genetic network controlling a bacterial cell cycle. Science 290:2144–2148
Lee D-Y, Rhee G-Y (1999) Circadian rhythm in growth and death of *Anabaena flosaquae* (cyanobacteria). J Phycol 35:694–699
Lewis PJ (2004) Bacterial subcellular architecture: recent advances and future prospects. Mol Microbiol 54:1135–1150
Liu Y, Golden SS, Kondo T, Ishiura M, Johnson CH (1995) Bacterial luciferase as a reporter of circadian gene expression in cyanobacteria. J Bacteriol 177:2080–2086
Liu Y, Tsinoremas NF, Johnson CH, Lebedeva NV, Golden SS, Ishiura M, Kondo T (1996) Circadian orchestration of gene expression in cyanobacteria. Genes Dev 9:1469–1478
Löwe J, van den Ent F, Amos LA (2004) Molecules of the bacterial cytoskeleton. Annu Rev Biophys Biomol Struct 33:177–198
Mackey SR, Choi JS, Kitayama Y, Iwasaki H, Dong G, Golden SS (2008) Proteins found in a CikA-interaction assay link the circadian clock, metabolism, and cell division in *Synechococcus elongatus*. J Bacteriol 190:3738–3746
McAdams HH, Shapiro L (2003) A bacterial cell-cycle regulatory network operating in time and space. Science 301:1874–1877
Matsuo T, Yamaguchi S, Mitsui S, Emi A, Shimoda F, Okamura H (2003) Control mechanism of the circadian clock for timing of cell division in vivo. Science 302:255–259
Mihalcescu I, Hsing W, Leibler S (2004) Resilient circadian oscillator revealed in individual cyanobacteria. Nature 430:81–85
Miller BH, McDearmon EL, Panda S, Hayes KR, Zhang J, Andrews JL, Antoch MP, Walker JR, Esser KA, Hogenesch JB, Takahashi JS (2007) Circadian and CLOCK-controlled regulation of the mouse transcriptome and cell proliferation. Proc Natl Acad Sci USA 104:3342–3347
Min H, Guo H, Xiong J (2005) Rhythmic gene expression in a purple photosynthetic bacterium, *Rhodobacter sphaeroides*. FEBS Lett 579:808–812
Mitsui A, Kumazawa S, Takahashi A, Ikemoto H, Cao S, Arai T (1986) Strategy by which nitrogen-fixing unicellular cyanobacteria grow photoautotrophically. Nature 323:720–722
Miyagishima SY, Wolk CP, Osteryoung KW (2005) Identification of cyanobacterial cell division genes by comparative and mutational analyses. Mol Microbiol 56:126–143
Mori T, Johnson CH (2000) Circadian control of cell division in unicellular organisms. Prog Cell Cycle Res 4:185–192

Mori T, Johnson CH (2001) Independence of circadian timing from cell division in cyanobacteria. J Bacteriol 183:2439–2444

Mori T, Binder B, Johnson CH (1996) Circadian gating of cell division in cyanobacteria growing with average doubling times of less than 24 hours. Proc Natl Acad Sci USA 93:10183–10188

Moulager M, Monnier A, Jesson B, Bouvet R, Mosser J, Schwartz C, Garnier L, Corellou F, Bouget FY (2007) Light-dependent regulation of cell division in *Ostreococcus*: evidence for a major transcriptional input. Plant Physiol 144:1360–1369

Nagoshi E, Saini C, Bauer C, Laroche T, Naef F, Schibler U (2004) Circadian gene expression in individual fibroblasts: cell-autonomous and self-sustained oscillators pass time to daughter cells. Cell 119:693–705

Nakajima M, Imai K, Ito H, Nishiwaki T, Murayama Y, Iwasaki H, Oyama T, Kondo T (2005) Reconstitution of circadian oscillation of cyanobacterial KaiC phosphorylation in vitro. Science 308:414–415

Nikaido SS, Johnson CH (2000) Daily and circadian variation in survival from ultraviolet radiation in Chlamydomonas reinhardtii. Photochem Photobiol 71:758–765

Paietta JV (1982) Photooxidation and the evolution of circadian rhythmicity. J Theor Biol 97:77–82

Pittendrigh CS (1965) Biological clocks, the functions, ancient and modern, of biological oscillations. In: Air Force Office of Scientific Research (ed) Science and the sixties, the proceedings of the cloudcroft symposium. Air Force Office of Scientific Research, Arlington, pp 96–111

Pittendrigh CS (1993) Temporal organization: reflections of a Darwinian clock-watcher. Annu Rev Physiol 55:17–54

Quardokus EM, Brun YV (2003) Cell cycle timing and developmental checkpoints in *Caulobacter crescentus*. Curr Opin Microbiol 6:541–549

Rogers LA, Greenbank GR (1930) The intermittent growth of bacterial cultures. J Bacteriol 19:181–190

Romberg L, Levin PA (2003) Assembly dynamics of the bacterial cell division protein FtsZ: poised at the edge of stability. Annu Rev Microbiol 57:125–154

Rothfield L, Justice S, García-Lara J (1999) Bacterial cell division. Annu Rev Genet 33:423–448

Scheving LE (1981) Circadian rhythms in cell proliferation: their importance when investigating the basic mechanism of normal versus abnormal growth. Prog Clin Biol Res 59C:39–79

Schmitz O, Katayama M, Williams SB, Kondo T, Golden SS (2000) CikA, a bacteriophytochrome that resets the cyanobacterial circadian clock. Science 289:765–768

Shih YL, Rothfield L (2006) The bacterial cytoskeleton. Microbiol Mol Biol Rev 70:729–754

Siede W, Friedberg EC (1990) Influence of DNA repair deficiencies on the UV sensitivity of yeast cells in different cell cycle stages. Mutat Res 245:287–292

Sinhaa RP, Häderb DP (2008) UV-protectants in cyanobacteria. Plant Sci 174:278–289

Smith RM, Williams SB (2006) Circadian rhythms in gene transcription imparted by chromosome compaction in the cyanobacterium *Synechococcus elongatus*. Proc Natl Acad Sci USA 103:8564–8569

Sturtevant R (1973a) Circadian patterns in linear growth of *Escherichia coli*. Anat Rec 175:453

Sturtevant R (1973b) Circadian variability in *Klebsiella* demonstrated by cosinor analysis. Int J Chronobiol 1:141–146

Sweeney BM, Borgese MB (1989) A circadian rhythm in cell division in a prokaryote, the cyanobacterium *Synechococcus* WH7803. J Phycol 25:183–186

Sweeney BM, Hastings JW (1958) Rhythmic cell division in populations of *Gonyaulax polyedra*. J Protozool 5:217–224

Vicente M, Rico AI, Martínez-Arteaga R, Mingorance J (2006) Septum enlightenment: assembly of bacterial division proteins. J Bacteriol 188:19–27

Ward JE Jr, Lutkenhaus J (1985) Overproduction of FtsZ induces minicell formation in *E. coli*. Cell 42:941–949

Wille JJ Jr, Ehret CF (1968) Light synchronization of an endogenous circadian rhythm of cell division in *Tetrahymena*. J Protozool 15:785–789

Woelfle MA, Xu Y, Qin X, Johnson CH (2007) Circadian rhythms of superhelical status of DNA in cyanobacteria. Proc Natl Acad Sci USA 104:18819–18824

Zhang X, Dong G, Golden SS (2006) The *pseudo*-receiver domain of CikA regulates the cyanobacterial circadian input pathway. Mol Microbiol 60:658–668

Chapter 12
The Adaptive Value of the Circadian Clock System in Cyanobacteria

Mark A. Woelfle and Carl Hirschie Johnson

Abstract Circadian clocks are thought to enhance the fitness of organisms by improving their ability to adapt to daily changes in the environment; however, there have been few rigorous tests of this proposal. Competition between cyanobacterial strains with different circadian periods showed that strains compete most effectively in a rhythmic environment when the frequency of their internal biological oscillator and that of the environmental cycle are similar. These observations demonstrate the adaptive value of the circadian system in cyanobacteria, but this adaptive value is only fulfilled in cyclic environments. Many questions still remain. What cellular mechanism(s) mediated by the circadian clock confers this adaptive value in cyanobacteria and what selective pressures lead to the evolution of circadian systems?

12.1 Why do Organisms have Circadian Clocks?

Circadian clocks are found in a wide range of organisms from bacteria to mammals. Due to the diversity of organisms possessing a circadian clock, the idea that circadian systems enhance fitness is a basic tenet of circadian biology. There is a great deal of literature on the circadian regulation of behaviors and metabolic events that are interpreted to enhance fitness, e.g., the hypothesis that an internal clock allows the anticipation of regular daily events such as dawn or dusk (Daan 1981; Horton 2001; Sharma 2003; Dunlap et al. 2004). Nonetheless, the value of circadian clocks as an evolutionary adaptation that enhances the fitness of species possessing them has been speculation based more on plausibility than on rigorous testing (DeCoursey 2004). But what do we mean by fitness and adaptation? The fitness of a given genotype is the average lifetime contribution on a per capita basis of individuals of that genotype to the population after one or more generations (Futuyma 1998). Thus, in terms of evolution, fitness is a measure of reproductive

M.A. Woelfle and C.H. Johnson(✉)
Department of Biological Sciences, Vanderbilt University, Nashville, TN 37235, USA,
e-mails: carl.h.johnson@vanderbilt.edu, mark.woelfle@vanderbilt.edu

J.L. Ditty et al. (eds.), *Bacterial Circadian Programs.*

success and this success can be influenced by secondary factors such as longevity, survival, growth, and development, etc. However, the measurement of these secondary factors is not a direct measure of reproductive success. To date, there have been very few studies that have directly tested reproductive success as a measure of enhanced fitness conferred by a circadian clock system.

An adaptation is a feature or characteristic of an organism that is the product of evolution by natural selection. Because a particular feature was selected in a specific environment, it represents a solution to some challenge presented by that environment. Thus, an adaptation is a feature of an organism that enhances its reproductive success relative to other possible features (Futuyma 1998). Of course, there is also the process of adaptation, which is the evolutionary change in phenotype/genotype driven by natural selection in a given environment. In the strictest sense, a new feature that is produced by natural selection can only be assumed to be adaptive when it first appears. Over time, the feature may persist and continue to be adaptive because the selective pressure remains. In contrast, a feature is no longer adaptive if the selective pressure has relaxed, but there is no selection against the feature and it continues to persist passively. In addition, a feature may persist and no longer be adaptive for the original reason; other features may become linked to the original feature such that in the absence of the original selective pressure the feature persists because so many other processes depend upon it. Rigorous use of the term adaptation requires that a feature remain under some selective pressure.

12.2 Extrinsic Versus Intrinsic Adaptive Value of Circadian Clocks

In the eyes of many biologists, the selective force that drove the evolution of circadian systems was the daily cycle of light, temperature and humidity present in the natural environment. In this view, circadian clocks are an adaptation that would be expected to enhance the fitness of an organism by entraining behavioral and physiological processes so that they occur at optimal phases in the day/night cycle, yielding an "extrinsic" adaptive value (Sharma 2003). Several studies do support the idea that an intact circadian clock enhances fitness in a variety of organisms in cyclic environments (DeCoursey et al. 2000; Beaver et al. 2002, 2003; Michael et al. 2003; Sharma 2003; DeCoursey 2004). For example, free-living chipmunks with lesions in the supra-chiasmatic nucleus that inactivated their circadian system were more susceptible to predation than were animals with an intact circadian clock (DeCoursey et al. 2000). The lesioned animals displayed nighttime restlessness that presumably led to the increase in their susceptibility to predators. In *Drosophila melanogaster*, clock mutations that disrupt circadian rhythmicity showed reduced sperm production in males (Beaver et al. 2002). Clock mutations also affected oogenesis in *Drosophila* females, but this effect appears to be pleiotropic and probably does not directly involve the circadian clock (Beaver et al. 2003). Michael et al. (2003) have shown that there is a positive correlation between the circadian period and the latitude from which samples of the plant *Arabidopsis* have been isolated suggesting that day length and temperature are relevant to circadian clocks. Despite all of these intriguing observations that suggest circadian systems provide extrinsic adaptive value, none of these studies measured reproductive fitness directly.

In contrast to extrinsic value, others have proposed that circadian clocks may have evolved to provide an "intrinsic" adaptive value (Pittendrigh 1993; Paranjpe et al. 2003). Over the course of evolution, circadian pacemakers have become an integral part of internal temporal organization and as such, may have become intertwined with other traits that influence reproductive fitness in addition to their original role for adaptation to environmental cycles. If circadian clocks retained extrinsic value and over time accrued an intrinsic value, they would still be considered an adaptation. In such a case, circadian clocks would be expected to be of adaptive value to an organism in constant conditions as well as in cyclic environments. In support of this hypothesis, populations of *D. melanogaster* raised for hundreds of generations in constant conditions retain the ability to entrain to various light/dark (LD) cycles indicating that even in the absence of environmental selection the components of the circadian system are maintained (Paranjpe et al. 2003). In contrast, a counter example is that of cave animals that frequently lose robust behavioral rhythmicity in the constant environment of caverns, which suggests that there is no intrinsic value of having a clock for these cave creatures (Blume et al. 1962).

12.3 Does the Circadian Clock System in Cyanobacteria Provide Adaptive Value?

To fully address this subject, two questions must be addressed. Does the circadian clock increase the reproductive fitness of an organism? And, is the clock of extrinsic or intrinsic adaptive value to that organism? Although these questions seem to be straightforward, most studies to date address only one or the other question and often only indirectly. Furthermore, studies designed to address these questions have often produced results that contradict previous conclusions. So what type of study might be able to fully address these questions and what model organism could be used?

Cyanobacteria are an ideal model system for the questions posed above. These prokaryotes are evolutionarily ancient microorganisms and several species are known to have circadian systems. In *Synechococcus elongatus* PCC 7942, the *kaiABC* gene cluster encodes the cyanobacterial clock proteins that regulate circadian oscillations (see Chap. 5). A number of *kaiA*, *kaiB* and *kaiC* mutants have been isolated that result in a variety of circadian phenotypes including short and long period mutants as well as mutants that are arhythmic (Kondo et al. 1994). Growth in competition for many generations among cyanobacterial strains with differing clock properties can be used to test directly whether circadian clocks enhance reproductive fitness. For bacteria and other asexually reproducing microbes, differential growth of one strain in competition with another is a direct measure of reproductive fitness (Lenski and Travisano 1994; Ouyang et al. 1998; Woelfle et al. 2004). In addition,

S. elongatus can be grown in both LD cycles and in constant light (LL). Thus the *S. elongatus* system allows both the opportunity to directly measure reproductive fitness and the ability to address whether the circadian clock system of this species provides the organism extrinsic versus intrinsic adaptive value.

12.4 Growth in Competition Shows that Reproductive Fitness is Enhanced by the Cyanobacterial Clock

The adaptive significance of the circadian clock in cyanobacteria was tested by competing strains of *S. elongatus* that differed in their circadian phenotype against each other in controlled environments (Ouyang et al. 1998; Woelfle et al. 2004). In pure culture, cyanobacterial strains that displayed a wild-type, short- or long-period, or arhythmic phenotype grew at the same rate in both LL and LD cycles; therefore, there was no apparent advantage or disadvantage in having a particular circadian period when the strains were grown in pure cultures. As a test of reproductive fitness, two different cyanobacterial strains were mixed together and grown in competition to determine whether the composition of the population changes as a function of time (Fig. 12.1). Competition between different strains was conducted in LL or in LD cycles that had equal intervals of light versus darkness, but the light intervals varied in their total length of periodicity (i.e., LD11:11, LD12:12, or LD15:15). The mixed cultures were diluted at intervals to allow growth to continue for ~30–45 generations and were sampled at regular time intervals to determine the composition of the population.

To address the question of whether fitness is enhanced by having a circadian clock that was in synchrony with the environmental cycle, mutant strains that varied in the periods of their circadian rhythms were used (Fig. 12.2). Two strains, one with a shorter than normal free-running period (FRP; 22 h) resulting from a mutation in *kaiB* (B22a) and one with a longer than normal FRP (30 h) due to a mutation in *kaiA* (A30a) were competed against a wild-type strain with a period of about 25 h. In a series of experiments, when these strains were mixed and grown together in competition, a clear pattern emerged that depended on the frequency of the LD cycle and the inherent circadian period of the strain. In a 22-h cycle (LD11:11), the strain with a 22-h period could out-compete the wild-type strain in mixed cultures (Ouyang et al. 1998). Similarly, the 30-h period mutant could defeat the wild type in a 30-h cycle (LD15:15). When competed in a "normal" 24-h cycle (12 h of light followed by 12 h of darkness; LD12:12), the wild-type strain could overtake either period mutant (Ouyang et al. 1998; Woelfle et al. 2004). Competitions using mutant *kaiC* strains with similar short- and long-period phenotypes yielded similar results, indicating that these observations are not dependent upon which of the clock genes is mutated (Ouyang et al. 1998; Woelfle et al. 2004). These results clearly show that the strain whose period most closely matched that of the LD cycle outgrew the competitor. Each of the LD conditions used has equal amounts of light and dark

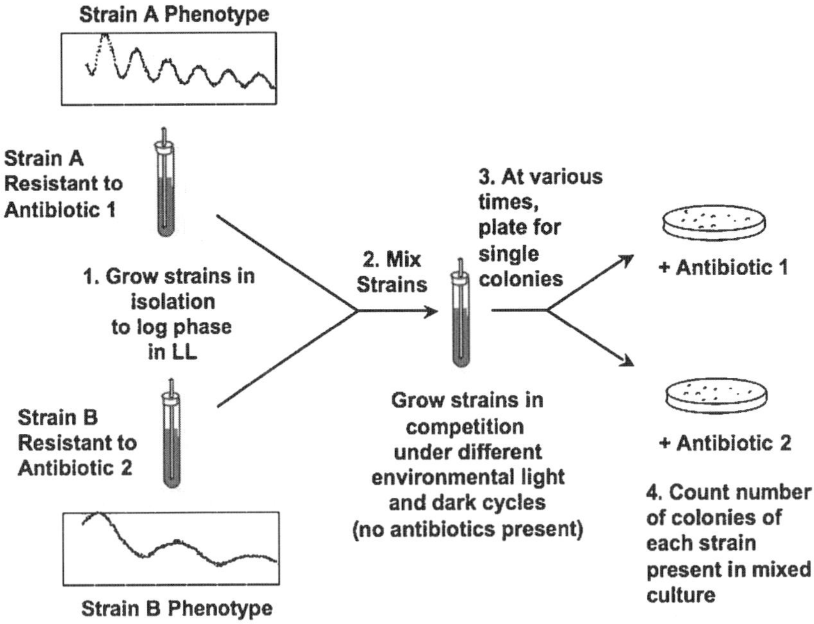

Fig. 12.1 Illustration of the competition experiment between cyanobacterial strains with different circadian phenotypes. The circadian phenotypes of two strains, A and B, are shown as luminescence produced from the promoter activity of the *Synechococcus elongatus psbA1* gene fused to *luxAB*. Strains also differ in genes that confer resistance to antibiotics such that A is resistant to one antibiotic while B is sensitive and vice versa. Strains A and B are grown separately in pure culture to log phase, diluted to the same optical density and equal volumes of the two strains are mixed. An initial sample of this mixed culture is taken, plated on each of the two selective media for growth of single colonies of each strain, and the mixed culture is placed in an LD cycle for ~8 generations. After ~8 generations, a sample of the mixed culture is taken, plated on each of the two selective media, and the remainder of the culture is diluted into fresh medium. The mixed culture is returned to the same LD cycle for an additional ~8 generations. This schedule of sampling and dilution of the mixed culture is continued until the strains have grown in competition with each other for a total of ~30–45 generations, at which time a final sample is taken and plated on selective media. The fraction of each strain present in the mixed culture at any sampling time is the number of colonies of that strain detected divided by the total number of colonies of both strains. Colonies of each strain at different sampling times are also used to monitor the luminescence rhythms to confirm circadian phenotypes

exposure over the total course of the experiment (a critical point for a photosynthetic organism, such as a cyanobacterium); therefore, it is only the frequency of light versus dark that differs among these LD cycles. Furthermore, the fact that strains with anomalous circadian phenotypes could defeat the wild-type strain when the period of the LD cycle is similar to their endogenous period strongly suggests that the differential effects observed are a result of the differences in the circadian clock and not to some pleiotropic genetic effect. Thus, a circadian clock in tune with the environmental cycle appears to enhance the reproductive fitness of cyanobacteria.

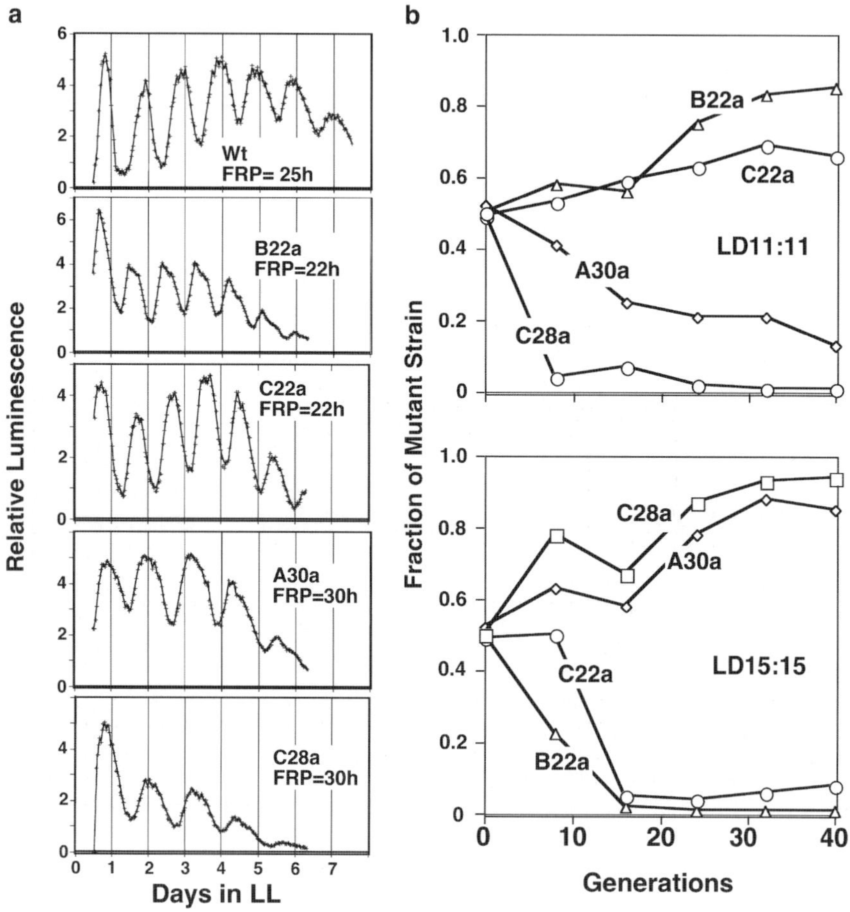

Fig. 12.2 Competition of mixed cultures in LD11:11 and LD15:15 cycles. (**a**) The circadian phenotypes of the wild type [*Wt*, AMC343, free-running period (FRP) ~25 h] and mutants [mutations in *kaiB* (*B22a*, FRP ~22 h), *kaiA* (*A30a*, FRP ~30 h), and *kaiC* (*C22a*, FRP ~22 h; *C28a*, FRP ~30 h)]. All strains have a luciferase construct that reports the promoter activity of the *psbA1* gene assayed with a CCD camera/turntable device. (**b**) Kinetics of competition in mixed cultures between wild-type and the mutant strains in LD11:11 (*upper panel*) and LD15:15 (*lower panel*). Data are plotted as the fraction of the mutant strain in the mixed culture (*y*-axis) versus the estimated number of generations (*x*-axis). Figure used with permission from Ouyang et al. (1998)

If the circadian clock system of cyanobacteria provides an intrinsic adaptive value, this biological timekeeper would be expected to be of adaptive value to cells in both constant conditions as well as in cyclic environments. To test this hypothesis, strains with a "normal" circadian clock were competed against an arhythmic strain without a functioning clock (CLAb) and against a strain with a rapidly damping oscillator (CLAc). The CLAb arhythmic strain was rapidly defeated (within ~20 generations) by the wild type, as was the strain with the rapidly damping oscillator

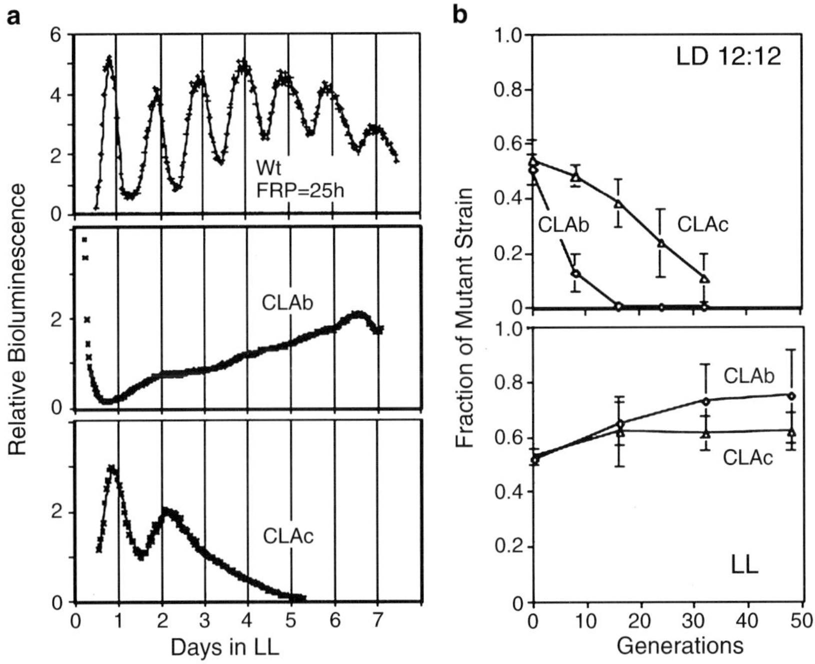

Fig. 12.3 Competition of arhythmic strains with the wild type. (**a**) The circadian phenotypes from: wild type (*Wt*, AMC343) and two *kaiC* mutants (*CLAb*, *CLAc*). The wild type has a FRP of ~25 h; the *kaiC* mutant CLAb appears to be arhythmic and the *kaiC* mutant CLAc displays a rapidly damped oscillation. (**b**) Competition between wild-type and mutant strains in LD12:12 (*upper panel*) and LL (*lower panel*) plotted as the fraction of mutant in the mixed culture (*y*-axis) versus the estimated number of generations (*x*-axis). Figure used with permission from Woelfle et al. (2004)

(CLAc), when the competition was conducted in LD12:12 cycles (Fig. 12.3B; Woelfle et al. 2004). This result demonstrates that a functioning circadian clock enhances the reproductive fitness of cyanobacteria in a periodic environment. However, when the competition was conducted in LL, the arhythmic strain not only was maintained in mixed cultures with the wild-type strain, it grew slightly better than the wild-type strain. This somewhat surprising result suggests that the clock system does not appear to be of any intrinsic value to cyanobacteria in constant conditions and, in fact, a functioning circadian clock may be a slight disadvantage to cells in such an environment. In anticipation of the daily onset of darkness, the circadian clock in wild-type cells may temporarily interrupt photosynthesis or some other clock-mediated metabolic process, which may be one reason that the wild-type strain is at a competitive disadvantage in constant conditions relative to clock-disrupted strains.

As noted previously, the growth rates of the cyanobacterial strains used in each of these competition experiments were not significantly different from one another when the strains were grown in single-strain pure cultures. Thus, these results are quite likely an example of "soft selection" (Futuyma 1998); i.e., the reduced fitness of one genotype compared to another is seen only in a competitive environment. Together, these two studies clearly demonstrate that an intact circadian clock whose FRP is consonant with the environment significantly enhances the reproductive fitness of cyanobacteria in rhythmic environments. However, this same clock system seems to provide no adaptive advantage to cyanobacteria in constant environments and may even be slightly detrimental.

12.5 What is the Mechanism of Clock-Mediated Fitness Enhancement?

What is the mechanistic basis of the adaptive advantage conferred by the cyanobacterial circadian clock system? A number of hypotheses have been proposed, but currently there is little evidence that clearly supports or refutes any of these proposals. The first possibility is that the circadian clock allows the optimal utilization of some limiting resource such as light, nutrients, or carbon dioxide (Fig. 12.4). Optimal utilization of a limiting resource (the Limiting Resource Model; see Sect. 12.5.1) by cells with a functioning circadian clock that harmonizes with the cycling environment would provide those cells an opportunity to synchronize their metabolism and reproduction to their specific environmental conditions that would allow those cells to out-compete others that are out of synchrony. A second possibility is that the cyanobacterial circadian clock regulates the rhythmic secretion of and/or sensitivity to some diffusible factor(s) (the Diffusible Factor Model; see Sect. 12.5.2) capable of inhibiting the growth of other cyanobacterial strains (Fig. 12.5). In this case, cells out of tune with their fellow competitors and the environment would be subject to growth inhibition by those who are in tune. A third possibility is that the circadian system regulates some form of cell-to-cell communication system operating in populations of cyanobacteria (the Cell-to-Cell Communication Model; see Sect. 12.5.3); this communication system results in the enhanced fitness of the population as a whole (Fig. 12.6).

12.5.1 Limiting Resource Model

It is known that the phasing of *psbAI* gene expression is disrupted among different cyanobacterial strains maintained in non-optimal LD cycles (Ouyang et al. 1998). Since the circadian clock in *S. elongatus* regulates the rhythmic expression of virtually all genes studied to date, it is reasonable to propose that the tight temporal coordination of cellular processes such as photosynthesis, carbon fixation and possibly others would produce an overall benefit in fitness, and thus an adaptive advantage to cells that have a functioning circadian clock that is in tune with

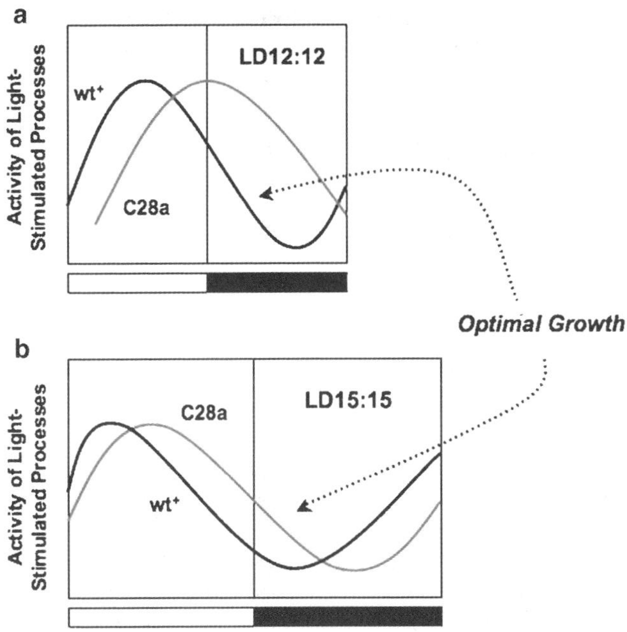

Fig. 12.4 Limiting resource model. The rhythmic expression of some light-stimulated process allows utilization of a limiting resource such as a nutrient or CO_2; the activity of this light-stimulated process is plotted (**a**, **b**) for a wild-type (*wt*+) strain (*black*) with a FRP of ~25 h and for a long-period mutant (*C28a*, *gray*) with a FRP of ~30 h. Periods of light (*white rectangles*) and darkness (*black rectangles*) are indicated beneath each plot. The activity of the light-stimulated process in the wild type and in the long-period mutant is plotted in LD12:12 (**a**) and in LD15:15 (**b**). In (**a**) the wild-type strain is in resonance with the LD12:12 cycle resulting in optimal growth, while in (**b**) the long-period mutant is better in tune with the LD15:15 cycle resulting in a fitness advantage over the wild-type strain

the cycling environment (Fig. 12.4). Those cells whose clock is dissonant with the environment would be at a comparative disadvantage in a mixed population because certain cellular processes would either be early or late in turning on or off. Furthermore, cells without a functioning circadian clock would be reduced to simply responding to the onset of light or darkness and would be unable to anticipate an incipient change. By this line of reasoning, one might propose that arhythmic strains would suffer the greatest reduction in fitness in rhythmic environments because they respond only to changes in light and darkness. Strains with a "damped", but a residually functioning oscillator would be somewhat more fit than clock-deficient strains, but less fit when compared to strains with a functional circadian clock. Furthermore, cyanobacterial strains whose clock is out of tune with the environment would be expected to have a reduction in fitness that is, in some degree, reflective of the difference between its FRP and the period of the environmental cycle. To a large extent, competition experiments support this line of reasoning. In LD12:12, the clock-deficient strain (CLAb) is more rapidly

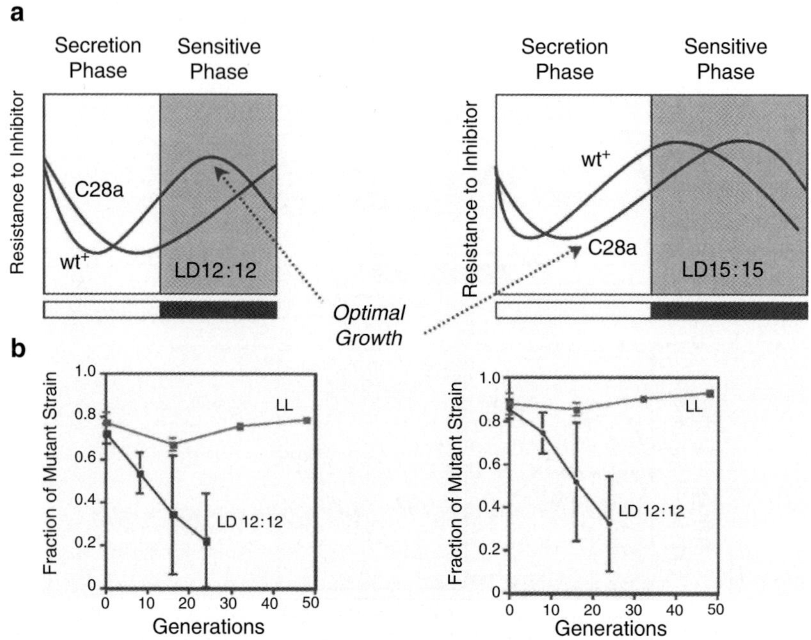

Fig. 12.5 Diffusible factor model. (**a**) Rhythms of resistance to a secreted inhibitor for the wild type (*wt*+) and a long-period mutant (*C28a*) are modeled in LD12:12 and LD15:15. Over the daily cycle, the cells alternate between: (i) a phase of secreting an inhibitor while being relatively resistant to the inhibitor and (ii) a phase of sensitivity to the inhibitor. The secretion phase is indicated in *white* and the sensitive (night) phase is shown in *gray*. Periods of light (*white rectangles*) and darkness (*black rectangles*) are indicated below. In LD12:12, the wild-type strain shows high resistance to the inhibitor during the time that coincides with high levels of inhibitor secretion, while peak resistance is delayed in the long-period mutant as compared to the wild type and therefore not optimal. In LD15:15, the resistance of the long-period mutant coincides with the greatest level of inhibitor so that it is now optimally adapted. (**b**) The kinetics of competition between the wild type (AMC343) and the arrhythmic strain, CLAb, with the CLAb strain at ~75% (*left*) and at ~90% of the starting population (*right*). The fraction of CLAb is plotted versus the estimated number of generations for competition in LL and LD12:12

defeated by the wild-type strain in mixed cultures than is a strain with a damped oscillator (CLAc; Fig. 12.3; Woelfle et al. 2004). Moreover, in LD12:12, a strain (C28a) with a FRP that is 4 h longer than the 24-h period of the environmental cycle is defeated more rapidly by a wild-type strain than is a strain (C22a) with a FRP that is 2 h shorter than the 24-h period of the environmental cycle (Ouyang et al. 1998).

One experimental attempt to address the potential differences in the optimal utilization of some limiting factor was to examine whether there are small differences in the initial growth rates between wild-type strains and clock mutant strains when stationary phase cells are transferred to fresh culture medium. Small

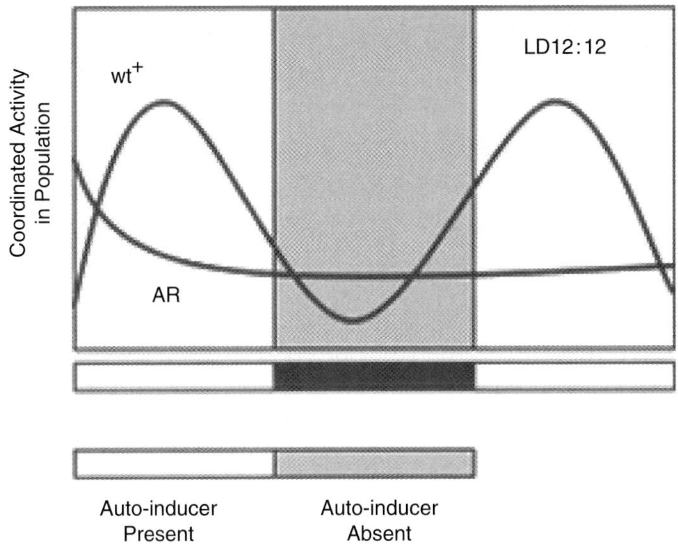

Fig. 12.6 Cell-to-cell communication model. Coordinated activity in a population in response to rhythmic production of an auto-inducer is plotted for both a wild-type strain (*wt*+) and an arhythmic strain (*AR*). Periods of light (*white rectangles*) and darkness (*black rectangles*) are indicated below. Periods of auto-inducer synthesis and release are shown in *white*; periods in the absence of auto-inducer are shown in *gray*. The response to the rhythmic synthesis and release of an auto-inducer produces a coordinated behavior that leads to a growth advantage in the population of wild-type cells; either non-rhythmic synthesis, release, or perception of auto-inducer by the arhythmic strain results in the lack of coordinated response in the population, placing these cells at a competitive disadvantage

differences in the adaptation to new medium repeated over several cycles of growth could result in large differences in the composition of a mixed population after many generations in competition. Multiple pure cultures of wild-type and arhythmic CLAb cells were grown to stationary phase in LD12:12, then a small fraction of these cells was diluted into fresh medium. The growth of these new cultures in LD12:12 was monitored until stationary phase was again reached. This process was repeated several times and the growth rates, especially the initial growth rates, were compared between wild-type and arhythmic cultures. No significant difference in the growth rates was found between the wild-type and arhythmic strains suggesting that both are equally capable of adapting to new medium when grown in pure culture and not in competition (Mori and Maini, unpublished observations). Although there appears to be a possible correlation between circadian phenotype and the rate at which the trend in competition becomes apparent, there is no compelling evidence to date that supports the hypothesis that the reduction in fitness suffered by clock mutant strains is due to an inability to compete for some limiting resource. In addition, there is no experimental evidence to suggest which, or if any, cellular processes might be adversely affected in clock mutant strains. Future experiments might address the differences between wild-type and clock mutant strains in their ability to determine whether there is clock control in the uptake of specific nutrients, whether there are differences in the rates of photosynthesis or carbon fixation, or whether there are differences in other physiological processes that do not affect the overall growth rate of cells, but do reduce reproductive rates.

12.5.2 Diffusible Factor Model

Assumptions of the Diffusible Factor Model (Fig. 12.5) are that cyanobacteria rhythmically secrete an auto-inhibitory molecule to which they are sensitive only in antiphase to the secretory phase. For example, cells might secrete this inhibitor during the day (light-dependent) or subjective day (clock-dependent) such that inhibitor concentrations peak near dusk. Resistance to this inhibitor molecule is also clock-controlled and resistance is greatest during the night phase when inhibitor concentrations are high. Mathematical modeling of cyanobacterial competition experiments favored the "rhythmic inhibitor" alternative over the limiting resources hypothesis discussed previously (Roussel et al. 2000; Gonze et al. 2002). These mathematical models made the prediction that less fit strains could compete effectively with more fit strains if the less fit strain is in a large enough excess in the starting populations. This prediction was tested by competition between an arhythmic strain (CLAb) and a wild-type strain; the arhythmic strain used in this test was chosen because it displayed a reduction in fitness equal to or greater than period mutant strains. This suggests that the arhythmic strain may have an elevated sensitivity to an inhibitor during all phases of the circadian cycle and therefore would be a good indicator of the validity of the models predictions. The mixed populations were composed of ~75% or ~90% arhythmic cells at the start of the competition (Fig. 12.5B). These proportions were selected because they were significantly larger than those predicted by the mathematical model to be at a bifurcating fraction (~60% to ~70%). The arhythmic strain was maintained at or slightly above these high starting levels in mixed cultures in LL; however, the fraction of the arhythmic strain in the mixed populations dropped dramatically in LD12:12 cycles (Fig. 12.2; Woelfle et al. 2004). This observation suggests that the reduction in fitness suffered by clock inactivation cannot be completely overcome simply by starting with more individuals in the population in contrast to the prediction of the models (Roussel et al. 2000; Gonze et al. 2002). Furthermore, there was no significant association in the rate at which arhythmic cells declined in the population and the initial starting proportion of the less fit strain which suggests that the relative fitness of the weaker strain does not change in a frequency-dependent manner (Woelfle et al. 2004). However, this observation does not exclude the possibility that cyanobacteria rhythmically secrete a factor that inhibits the growth of other cyanobacterial strains.

12.5.3 Cell-to-Cell Communication Model

A third, and as yet untested hypothesis, is a hybrid of the two hypotheses discussed above (Sects. 12.5.1, 12.5.2). It proposes that some form of cell-to-cell communication is regulated by the cyanobacterial circadian clock and benefits overall growth in the population (Fig. 12.6). Viewed in this way, communication between cells promotes a coordinated response in the population allowing the population as a whole to take advantage of some limiting resource such as light, nutrients, or carbon dioxide. Cells that are unable to communicate effectively with other cells in the population due to a non-functional clock or a clock out of tune with the environment would be at a competitive disadvantage relative to cells in tune with other cells and the environment in their ability to utilize some resource or in executing some necessary group behavior. Thus, a reduction in the reproductive fitness of these out of tune cells would be expected. Quorum sensing is a process of bacterial cell-to-cell communication involving the production and detection of extracellular signaling molecules called autoinducers (Xavier and Bassler 2003). This mode of communication allows populations of bacteria to coordinately control gene expression and synchronize group behaviors – specifically, behaviors that are not productive unless many individual cells participate. Most autoinducers enable intraspecies communication; however, autoinducer-2 (AI-2; derived from the recycling of *S* adenoyl-homocysteine, SAH, to homocysteine) has been proposed to serve as a universal signal for interspecies communication (Xavier and Bassler 2003). The functions of AI-2 that have been reported include production of bioluminescence in *Vibrio harveyii* and regulation of virulence factors in a number of bacterial species (Xavier and Bassler 2003).

Examination of the genome sequences of a number of cyanobacterial species including *Synechocystis* sp. PCC 6803 and *Thermosynechococcus elongatus* BP-1 revealed the presence of homologs of *sahH* (a gene encoding SAH hydrolase, which is a key component in the recycling of SAH; Sun et al. 2004). In addition, the genome sequences of both of these cyanobacterial species contain potential homologs of *luxQ* and *luxU*, components of the AI-2 signaling pathway; however, neither species contains a convincing match to *luxP*, which encodes the receptor of AI-2 in *V. harveyii* and a number of other bacterial species (Sun et al. 2004). The presence of such genes in the genome of cyanobacterial species such as *Synechocystis* sp. PCC 6803 suggests that other cyanobacteria may also be capable of quorum sensing. This observation encourages the possibility that an autoinducer system might also be operating in *S. elongatus*. Intraspecies communication that is regulated by the circadian clock and occurs among cells within a population may be beneficial for coordinately responding to environmental signals such as light or some limiting resource. The ability to sense the relative numbers of other microbes in the environment through AI-2 could be involved in regulating the secretion of some inhibitory molecule that gives cyanobacteria a reproductive advantage (see Chap. 13).

12.6 Additional Roles for the Circadian Clock in Cyanobacteria: the "Escape from Light" Hypothesis

The question of whether circadian clocks are adaptive is linked with identifying the forces of natural selection that originally encouraged the evolution of these systems. The ancestors of modern cyanobacteria appear in the fossil record approximately 3.5×10^9 years ago, suggesting that circadian clocks were an ancient invention of evolution (see Chap. 2); however, it is also possible that circadian clocks may have been absent in ancient cyanobacteria and evolved relatively recently (see Chap. 14). A driving force for the evolution of circadian clocks in cyanobacteria, and perhaps other organisms as well, could have been the advantage of phasing cellular events that are damaged by sunlight to occur only at night. This idea has been called the "escape from light" hypothesis (Fig. 12.7A; Pittendrigh 1965, 1993). For obligate phototrophs such as cyanobacteria, light is the sole energy source for cellular processes, but in addition to providing energy, sunlight can also cause damage. DNA can be mutated by exposure to ultraviolet

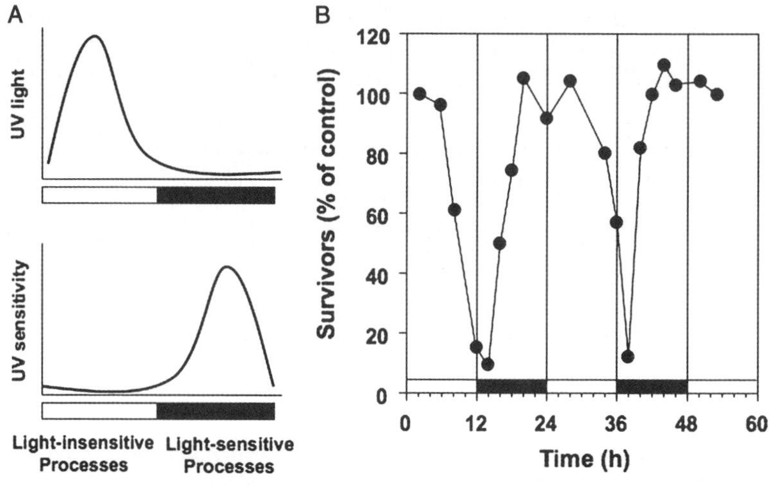

Fig. 12.7 The "escape from light" hypothesis. **A** Predictions of the "escape from light" hypothesis. In the *upper panel*, the amount of ultraviolet (*UV*) light is plotted as a function of time in an LD cycle; the *lower panel* depicts the prediction from the "escape from light" hypothesis, namely that UV sensitive processes will be phased to the night to minimize light-induced damage. The model predicts that cells would be most sensitive to exposures to UV light during the night. Periods of light (*white rectangles*) and darkness (*black rectangles*) are indicated below each panel. **B** Survival of *Chlamydomonas* cells after irradiation by UV light as a function of the time in an LD cycle. *Chlamydomonas* cultures were plated onto agar medium and treated with equal amounts of UV light at different phases of an LD12:12 cycle. Survival was measured as the colony-forming ability of cells following treatment as compared to that of cells that were not irradiated with UV light (modified from Nikaido and Johnson 2000). Note that the lower panel of (**A**) is complementary to the data in (**B**); in (**A**) "sensitivity" is plotted whereas the inverse function of survivability is plotted in (**B**)

(UV) light, and the genome may be at greater risk to UV irradiation at some phases of the cell division cycle, namely when the DNA is being replicated. In a number of microorganisms, DNA replication and cell division are restricted to the night (Edmunds 1984).

If the "escape from light" hypothesis about the early evolution of circadian clocks is correct, then organisms today might retain a restriction of light-sensitive processes to the night. The eukaryotic alga, *Chlamydomonas reinhardtii* displays rhythmic sensitivity to UV light; cells are more sensitive near sunset and into the early night than at other times during the daily cycle (Fig. 12.7B; Nikaido and Johnson 2000). This rhythm of UV sensitivity persists in constant conditions, although with a reduced amplitude. The circadian clock in *Chlamydomonas* also regulates the timing of the cell division cycle (Goto and Johnson 1995) and the phases of the cell cycle that show the greatest sensitivity to UV irradiation corresponded with S/G2 phases. These observations suggest that the daily cycle of UV radiation may have been a strong selective pressure favoring the evolution of circadian clocks in *Chlamydomonas* and are consistent with the "escape from light" hypothesis (Pittendrigh 1993).

The role of cryptochromes in circadian systems provides additional support for this hypothesis. Cryptochromes are pigmented photoreceptors involved in blue-light-mediated entrainment and photoperiodism in a wide variety of plant and animal species; these proteins share sequence homology to another blue-light-activated protein, DNA photolyase, which uses blue-light energy to repair UV-induced damage of DNA. Based on the "escape from light" hypothesis, a clock-related role for a DNA photolyase-type enzyme may have evolved from an ancestral photolyase that repaired DNA damage caused by the daily cycle of UV light. This ancestral DNA repair protein may have, over time, become an integral part in biological timing mechanisms and evolved into cryptochromes (Nikaido and Johnson 2000; Gehring and Rosbash 2003). Perhaps the daily cycle of exposure to UV radiation experienced by *S. elongatus* provided similarly strong selective pressures for the evolution of a circadian clock system that allows this photosynthetic microbe to restrict light-sensitive processes to occur only at night.

12.7 Future Directions

Cyanobacteria have proved to be an ideal model system to address questions about the adaptive significance of circadian clock systems; and we now know that the clock in cyanobacteria provides a means to enhance reproductive fitness of cells whose clock is in tune with the environmental cycle. In the absence of an environmental cycle, the circadian clock appears to provide little or no fitness advantage and might even be detrimental to cyanobacterial cells, which suggests that the clock system does not confer intrinsic adaptive value to this organism.

However, many important questions still remain unsolved and with the sequencing of the *Synechococcus elongatus* genome now complete (Joint Genome Institute; http://genome.jgi-psf.org/finished_microbes/synel/synel.home.html), new approaches to addressing these questions may be at hand. First and foremost are questions surrounding the mechanism of how reproductive fitness is enhanced by the clock system in cyanobacteria. Candidate genes of cell communication pathways, such as the homologs of *sahH*, *luxO* and *luxU* mentioned previously, could potentially be identified and inactivated. The effect of inactivating components of this communication pathway could then be assessed in competition experiments. In addition, the systematic construction of knockout mutations of each of the genes in the genome, which is currently underway (*Synechococcus elongatus* PCC 7942 Functional Genomics Project, http://www.bio.tamu.edu/synecho/index.html), and an examination of the resulting phenotypes may lead to the identification of unexpected candidate genes that may play a role in reproductive fitness in cyanobacteria. The circadian phenotype of these potential candidate mutant strains would need to be characterized and the fitness of these strains could then be examined in competition experiments.

Another potentially interesting avenue of research that might identify the mechanism of clock-controlled fitness enhancement might be an evolution experiment. In all of the competition experiments performed thus far, the number of cells of the "defeated" strain in the population is greatly reduced, but is not completely eliminated. Some of these survivors that remain after 30–45 generations of competition could be isolated and then competed again against the "winning" strain for another 30–45 generations. It would be interesting to determine whether the kinetics of the competition experiment remain the same, or whether the survivors show the ability to better compete with the more fit strain. If the kinetics of the competition between survivors and the originally more fit strain were different, that result might suggest the presence of compensatory mutation(s) in the survivors. These compensatory mutations would be expected to be in genes involved in the clock-controlled fitness enhancement pathway.

Finally, cyanobacteria also represent a good model system to further examine the "escape from light" hypothesis using experiments similar to those performed in *Chlamydomonas*. These types of experiments might reveal evidence of a light-dependent DNA repair pathway under control of the circadian clock system.

Great progress has been made in addressing the adaptive significance of circadian clocks using cyanobacteria, but many questions remain. We have not yet identified conclusively the selective pressure(s) that led to the evolution of these timekeeping systems. Moreover, the adaptive significance of biological rhythms has not been rigorously demonstrated in most other organisms. Perhaps the strides made in addressing these questions in cyanobacteria will provide clues that allow these mysteries to be unraveled in a host of other organisms.

References

Beaver LM, Gvakharia BO, Vollintine TS, Hege DM, Stanewsky R, Giebultowicz JM (2002) Loss of circadian clock function decreases reproductive fitness in males of *Drosophila melanogaster*. Proc Natl Acad Sci USA 99:2134–2139

Beaver LM, Rush BL, Gvakharia BO, Giebultowicz JM (2003) Noncircadian regulation and function of clock genes period and timeless in oogenesis of *Drosophila melanogaster*. J Biol Rhythms 18:463–472

Blume J, Bünning E, Gunzler E (1962) Zur Aktivitätsperiodik bei Höhlentieren. Naturwissenschaften 49:525

Daan S (1981) Adaptive daily strategies in behavior. In: J Aschoff (ed) Handbook of behavioral neurobiology; biological rhythms, vol 4. Plenum, New York, pp 275–298

DeCoursey PJ (2004) The behavioral ecology and evolution of biological timing systems. In: Dunlap JC, Loros JJ, DeCoursey PJ (eds) Chronobiology; biological timekeeping. Sinauer, Sunderland, Mass., pp 48–58

DeCoursey PJ, Walker JK, Smith SA (2000) A circadian pacemaker in free-living chipmunks: essential for survival? J Comp Physiol A 186:169–180

Dunlap JC, Loros JJ, DeCoursey PJ (2004) Chronobiology: biological timekeeping. Sinauer, Sunderland, Mass

Edmunds LN (1984) Circadian oscillators and cell cycle controls in algae. In: Nurse P, Streiblova' E (eds) The microbial cell cycle. CRC, Boca Raton, pp 209–230

Futuyma DJ (1998) Evolutionary biology, 3rd edn, Sinauer, Sunderland, Mass

Gehring W, Rosbash M (2003) The coevolution of blue-light photoreception and circadian rhythms. J Mol Evol 57:S286–S289

Gonze D, Roussel MR, Goldbetter A (2002) A model for the enhancement of fitness in cyanobacteria based on resonance of a circadian oscillator with the external light–dark cycle. J Theor Biol 214:577–597

Goto K, Johnson CH (1995) Is the cell division cycle gated by a circadian clock? The case of *Chlamydomonas reinhardtii*. J Cell Biol 129:1061–1069

Horton TH (2001) Conceptual issues in the ecology and evolution of circadian rhythms. In: Takahashi JS, Turek FW, Moore RY (eds) Handbook of behavioral neurobiology; circadian clocks, vol 12. Plenum, New York, pp 45–57

Kondo T, Tsinoremas NF, Golden SS, Johnson CH, Kutsuna S, Ishiura M (1994) Circadian clock mutants of cyanobacteria. Science 266:1233–1236

Lenski RE, Travisano M (1994) Dynamics of adaptation and diversification: a 10,000 generation experiment with bacterial populations. Proc Natl Acad Sci USA 91:6808–6814

Michael TP, Salome PA, Yu HJ, Spencer TR, Sharp EL, McPeek MA, Alonso JM, Ecker JR, McClung CR (2003) Enhanced fitness conferred by naturally occurring variation in the circadian clock. Science 302:1049–1053

Ouyang Y, Andersson CR, Kondo T, Golden SS, Johnson CH (1998) Resonating circadian clocks enhance fitness in cyanobacteria. Proc Natl Acad Sci USA 95:8660–8664

Paranjpe DA, Anitha D, Kumar S, Kumar D, Verkhedkar K, Chandrashekaran MK, Joshi A, Sharma VK (2003) Entrainment of eclosion rhythm in *Drosophila melanogaster* populations reared for more than 700 generations in constant light environment. Chronobiol Int 20:977–987

Pittendrigh CS (1965) Biological clocks: the functions, ancient and modern, of circadian oscillations. Air Force office of scientific research, science and the sixties. Proc Cloudcraft Symp 1965:96–111

Pittendrigh CS (1993) Temporal organization: reflections of a Darwinian clock-watcher. Annu Rev Physiol 55:16–54

Roussel MR, Gonze D, Goldbetter A (2000) Modeling the differential fitness of cyanobacterial strains whose circadian oscillators have different free-running periods:comparing the mutual inhibition and substrate depletion hypotheses. J Theor Biol 205:321–340

Sharma VK (2003) Adaptive significance of circadian clocks. Chronobiol Int 20:901–919

Sun J, Daniel R, Wagner-Döbler I, Zeng A-P (2004) Is autoinducer-2 a universal signal for interspecies communication: a comparative genomic and phylogenetic analysis of the synthesis and signal transduction pathways. BMC Evol Biol 4:36–47

Woelfle MA, Ouyang Y, Phanvijhitsiri K, Johnson CH (2004) The adaptive value of circadian clocks: an experimental assessment in cyanobacteria. Curr Biol 14:1481–1486

Xavier KB, Bassler BL (2003) LuxS quorum sensing: more than just a numbers game. Curr Opin Microbiol 6:191–197

Chapter 13
Stability and Noise in the Cyanobacterial Circadian Clock

Irina Mihalcescu

Abstract By monitoring single cyanobacterial cells in vivo we show that individual cells generate impressively stable circadian rhythms. In multicellular organisms, the circadian clock accuracy is achieved via intercellular coupling of the individual noisy oscillators. Here we demonstrate that cyanobacterial clock stability is a built-in property. We first theoretically design our experiment to be able to distinguish coupling, even weak, from phase diffusion (noise). As the precision of our evaluation increases with the length of the experiments, we continuously monitor, for a couple of weeks, mixtures of cell populations with different initial phases. The inherent experimental noise contribution, initially dominant, is reduced by enhanced statistics. We report a value of the coupling constant that is small compared to the diffusion constant of the phase. It appears therefore that the clock stability a built-in property for each bacterium.

13.1 Introduction

The scientific way of thinking during the nineteenth century was essentially deterministic such that, for a given set of initial conditions, an understanding of the laws of interaction among molecules allowed for the ulterior state of the system to be perfectly determined. The twentieth century brought the notion of stochasticity, which described the random nature of physical interactions at the chemical and biochemical levels: a molecule has a given probability to remain or not in the same state, to interact or not with another molecule, etc. Consequently, even for a chemical reaction at equilibrium, the number of molecules of a given compound fluctuates near the constant value of the deterministic solution. The relative amplitude of these fluctuations (noise) is higher when there is a lower number of molecules involved in the reaction. The biochemical environment within a cell implies interactions among

I. Mihalcescu
Laboratoire de Spectrométrie Physique, Université de Grenoble–CNRS UMR5588, 38402 Saint Martin d'Hères, France, e-mail: Irina.Mihalcescu@ujf-grenoble.fr

J.L. Ditty et al. (eds.), *Bacterial Circadian Programs.*

a small number of molecules, beginning with the DNA molecule (which is in one or a few copies) and all the proteins associated with DNA-associated processes (replication, activation, transcription) may each suffer stochastic fluctuations. The collection of these fluctuations then propagates downhill to their enslaved biochemical reactions, to give rise to an overall noisy intracellular environment. Despite these random fluctuations, accuracy in cellular functions has to be achieved, and organisms deploye diverse strategies to attain this goal (Raser and O'Shea 2005).

The circadian clock as a limit cycle oscillator responds differently to stochastic noise with respect to its two defining characteristics: amplitude and phase. Following a perturbation, the amplitude has a stable behaviour, such that the perturbation of the amplitude decays rapidly to the stable value. The phase has a neutral behaviour; the perturbation of the phase neither grows nor decays, such that any phase perturbation results in a lag that is kept until the system is perturbed again. Consequently, in the presence of stochastic noise, the amplitude of the oscillation fluctuates within a limited range while the phase accumulates errors, like a random walk[1]. Therefore, the temporal stability of the clock is crucial and two strategies are to be considered: (i) a stable clock is built in each cell, meaning that the internal mechanism producing the oscillation assures its stability (Barkai and Leibler 2000), or (ii) each cell may have a sloppy oscillator but increases its stability by communication with the nearby cells (Pikovsky et al. 2001). The circadian clock in multicellular organisms is an example of the latter strategy. In vivo monitoring revealed that individual cells generate autonomous circadian rhythms in protein abundance (Nagoshi et al. 2004; Welsh et al. 2004; Carr et al. 2005) but these rhythms appear to be noisy with drifting phases and frequencies. However, the whole organism is significantly more accurate through the temporal precision that results from intercellular coupling of the individual noisy oscillators (Liu et al. 1997; Herzog et al. 2004).

In this chapter we review the characteristics of single cell oscillators in the cyanobacterium *Synechococcus elongatus* sp. PCC 7942 (Mihalcescu et al. 2004) and present investigations related to the origin of its apparent strong temporal stability (Amdaoud et al. 2007a). It is shown in the end that the cyanobacterium has adopted the first strategy: for this unicellular organism, temporal stability is a built-in property.

13.2 Single Cell Oscillator in Cyanobacteria

We were able to observe slow growth of *S. elongatus* microcolonies from single "progenitor" cells over long periods of time. In order to measure the circadian rhythm of gene expression in a single cell, we used a bacterial luciferase reporter

[1] A random walk describes the path of a particle which takes successive steps each in a random direction. This describes the small particle diffusion in gas or liquids and has also been intuitively imaged as the "drunkard's walk". In a non-biased diffusive process, the average distance travelled by the particle is zero, while the variance of this distance increases linearly with time.

system (Katayama et al. 1999), which consists of two neutral site chromosomal insertions P*psbAI::luxAB* and P*psbAI::luxCDE*. This autobioluminescence system, which does not necessitate exogenously added aldehyde substrate, had been successfully used to monitor output generated by the circadian oscillator in populations of cyanobacteria. Most of the genes in this strain are under circadian clock control (Liu et al. 1995) and the *psbAI* promoter used here is a strong promoter, which provides higher overall levels of bioluminescence as compared to other cyanobacterial reporter strains. However, detecting and imaging individual 2.3–6 μm bacteria requires at least 50× higher sensitivity than current protocols, a factor roughly equal to the ratio between the number of cells in the smallest population yet monitored (Kondo et al. 1997) and a single individual. To achieve this sensitivity, we implemented an experimental set-up based on a back-illuminated cooled CCD detection camera with high quantum efficiency coupled to high numerical aperture lens to capture most of the emitted photons (Fig. 13.1c). Because the light levels obtained from a single cyanobacterium were typically on the order of 10–20 photons min^{-1} $cell^{-1}$, we used relatively long integration times (30 min) in order to maximize the signal-to-noise ratio without significantly affecting the circadian rhythm. A computerized control of internal light and temperature as well as an entirely automated data acquisition system allowed measurements over prolonged periods (upto 2 weeks) in constant conditions. The growth chambers for the microscope were made using Petri dishes (50 × 9 mm), with a coverslip bottom. The movement of the cells deposited on the coverslip was highly restrained by a thin layer of low melting agarose gel and separated from the growth medium by a paraffin sealed membrane. Using this method, each cell had a homogeneous access to growth medium and light, while a sufficiently transparent light pathway allowed for phase contrast microscopy.

Figure 13.1a, b shows an example of microcolony growth during the first 5.5 days (1 day = 24 h) monitored by both phase-contrast and bioluminescence microscopy. The inoculated bacterium (marked F in Fig. 13.1a) was slowly growing without undergoing cell division for the first ≅ 1.2 days. Its density of bioluminescence (defined as the total luminescence divided by the cell size) was clearly oscillating (Fig. 13.1d). When cell F divided, the siblings of the progenitor cell produced oscillations in bioluminescence with a striking synchronicity to one another (Fig. 13.1d). Each of the cells had different amplitude of oscillations, but each was characterized by similar period and phase. The average oscillation of all the progeny is remarkably well described by a simple periodic function[2], $\langle d(t) \rangle = B + A\cos(2\pi t / T_0 + \mu)$, not only with a constant period and phase, but also with constant amplitude (Fig. 13.1e). In the same manner, we analysed a few other neighbouring microcolonies that were derived from individual cells. Each progeny oscillated similarly and synchronously with its progenitor cell, maintaining a closely similar period and an average phase characteristic to each microcolony (s.d./mean <0.5%) for all the cells

[2] The ensemble average of a random variable $x(t)$ is denoted here by $\langle x(t) \rangle$ and the time average by $\bar{x}$.

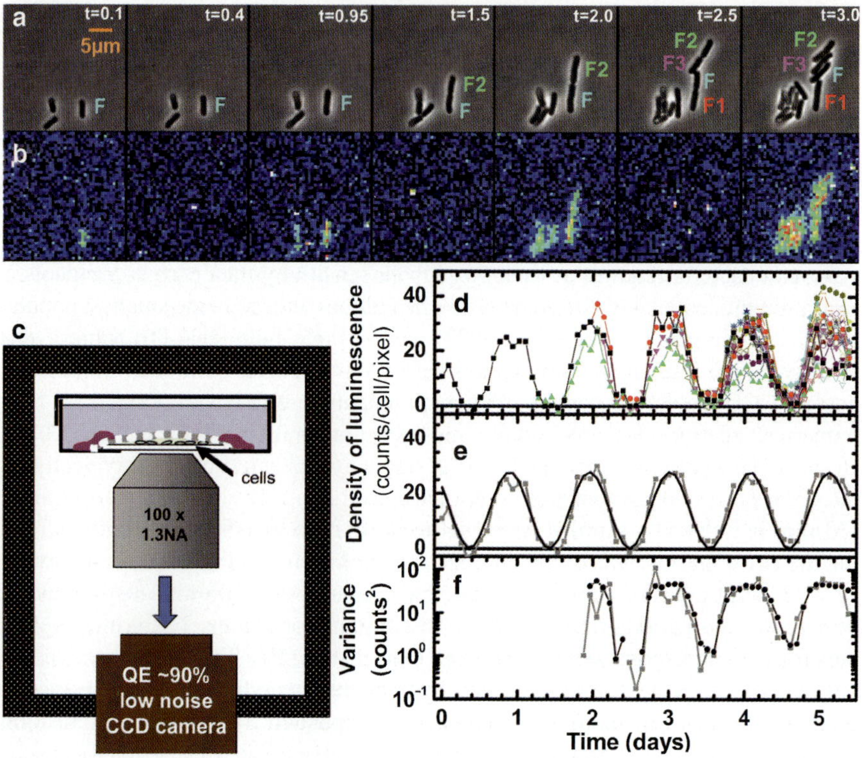

Fig. 13.1 Circadian oscillation of bioluminescence in individual bacteria. Snapshots of phase contrast image (**a**) and related bioluminescence image (**b**) at different times *t* (given in days, a 24-h period of time) from the beginning of the measurement. For the bioluminescence we used pseudo-colour, where red is high signal intensity and blue is low signal intensity. (**c**) Schematic representation of the experimental set-up. The bacteria grown in situ, in a Petri dish with a glass slide, are monitored trough a high magnification (100×) and high numerical aperture lens (NA = 1.3). Their bioluminescence is detected by a CCD camera with a high quantum efficiency (QE ≃ 90%) and very low electronic noise. (**d**) Density of bioluminescence for the progenitor cell F and all its progeny as a function of time. The density of bioluminescence is defined as the total bioluminescence detected from a cell divided by its size. (**e**) The average density of bioluminescence versus time (*grey line*) and its fit with a sine-like function (*black line*): $\langle d(t)\rangle = B + A \cos(2\pi t / T_0 + \mu)$. The resulting period is $T_0 = 25.4 \pm 0.12$ h, the offset $B = 14.8 \pm 0.3$ counts cell^{-1} pixel^{-1}, the amplitude $A = 12.9 \pm 0.3$ counts cell^{-1} pixel^{-1} and the mean phase $\mu = 52 \pm 2.8$. (**f**) Fit of the experimental variance ■$\sigma_{osc}^2(t) = \sigma_d^2(t) - \sigma_{detection}^2(t)$ with the theoretical variance ●$\sigma_d^2(t) = \langle (d_j(t) - \langle d(t)\rangle)^2$. The fit parameters are: the amplitude relative error $\eta_g = \sigma_g / \langle g \rangle = 0.25 \pm 0.01$ and the phase diffusion constant $D = 0.012 \pm 0.007$ days^{-1}. Adapted from Mihalcescu et al. (2004a)

studied in the experiment. This was noticeably different from the individual cell measurements in multicellular organisms where period distributions can be up to 10% (s.d./mean; Nagoshi et al. 2004).

As each individual cell is a self-sustained biochemical oscillator, which is defined roughly by a noisy amplitude and phase (Pikovsky et al. 2001), we express the bioluminescence of the clock (cell) j as $d_j(t) = g_j(t)[1 + b \cos(\omega_0 t + \varphi_j(t))]$. Fluctuations in time and from one cell to another are described by the two random functions, the amplitude $g_j(t)$ and the phase $\varphi_j(t)$ of the oscillator j; ω_0, is the free running frequency common to all cells and b the relative amplitude of oscillation/cell as determined by the specific promoter used as reporter[3] (Amdaoud et al. 2007b). We next take the simplest stochastic models for the amplitude and phase fluctuations: (i) for the amplitude $g_j(t)$ a stationary Gaussian process, with constant average $<g(t)> = g$ and standard deviation σ_g, (ii) for the phase $\varphi_j(t)$ a Wiener (a random walk) process, a Gaussian process with constant mean $<\varphi(t)> = \mu$ but a variance growing linearly with time $\sigma_\varphi^2 = Dt$, with D the diffusion constant. The theoretical variance of the oscillators $\sigma_d^2(t) = <(d_i(t) - <d(t)>)^2>$ quantifies the deviation of the density of bioluminescence $d_j(t)$ of each cell j from the mean and can be expressed as a function of the mean $<d(t)>$. Its fit to the experimental variance[4] $\sigma_{osc}^2(t)$ (Fig. 13.1f) gives the amplitude noise, $\eta_g = \sigma_g / <g> = 0.25$ and the phase diffusion constant $D = 0.012 \pm 0.007$ day^{-1}. This result confirms what we first mentioned: the oscillators have high amplitude fluctuations, $\eta_g = 0.25$, but remain strongly in synchronicity with a correlation time, $\tau = 2/D = 166 \pm 100$ days.

Stochastic effects in gene expression fluctuations, such as molecular noise, may be the origin of these fluctuations; and a noise level of gene expression of $\eta = 0.25$ is not unusual (Elowitz et al. 2002). But what is surprising is small temporal noisiness. One hypothesis is that the genetic network of each individual cell reduces temporal fluctuations. Another hypothesis is that the individual oscillators suppress temporal noise by coupling with other oscillators. Indeed, it has been shown that increasing the coupling between periodic and even chaotic chemical oscillators results in an onset of synchronization at a critical coupling level (Kiss et al. 2002).

In a first attempt to evaluate this possibility, we followed four microcolonies growing in close vicinity, but originating from individual cells having different initial phases of circadian oscillations. The progenitor cells, initially separated, progressively formed microcolonies which, after 10–12 successive division cycles, came into close contact with one another. Figure 13.2 presents the temporal evolution of the phase for some of the cell lines superimposed over three snapshots of the colonies. The phase of the circadian oscillator, in degrees and quantified as described in Fig. 13.2, is represented at each time interval by the colour of the overlaid temporal track. Here again, the cells from each microcolony oscillate with essentially the same period, while their phase is specific to each colony. It is easy to see that the difference in phases of different cell lines does not decrease in any significant way when they are driven closer to each other (Fig. 13.2, arrows).

[3] Both ω_0 and b depend slightly on overall metabolic conditions, like lighting or the pH of the medium.

[4] As the instrumental contribution $\sigma^2_{detection}(t)$ is six times smaller than the rough experimental variance, the net variance $\sigma^2_{osc}(t) = \sigma^2_d(t) - \sigma^2_{detection}(t)$ is the experimental measure of the fluctuation between the internal oscillators of the cells.

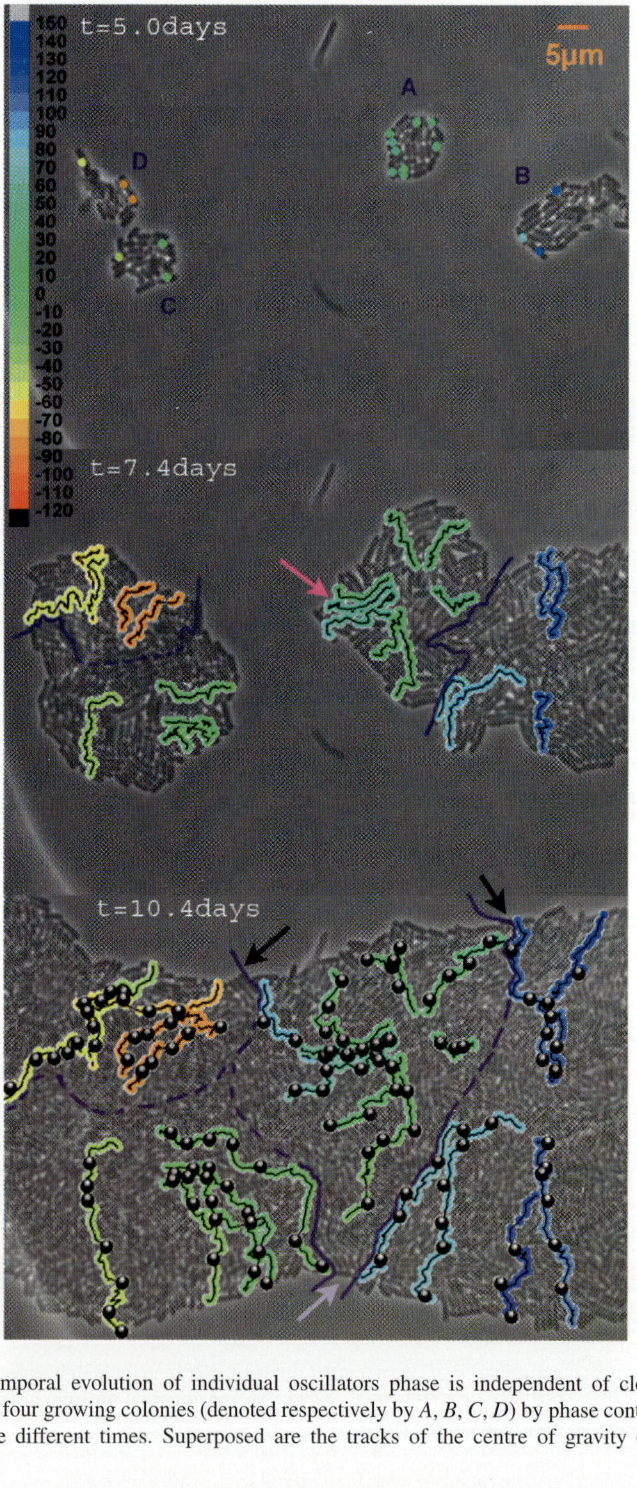

Fig. 13.2 Temporal evolution of individual oscillators phase is independent of close vicinity. Snapshots of four growing colonies (denoted respectively by *A*, *B*, *C*, *D*) by phase contrast microscopy at three different times. Superposed are the tracks of the centre of gravity of each cell

It seems that any interactions between closely packed cells have negligible effect on their relative phase of oscillations; however, as the precision of the phase measurement is $\sigma_{measure} \approx 6°$ (Mihalcescu et al. 2004b), one cannot exclude a weak coupling that would result in differences of phases that are not enough large to be detected. The next section presents a simplified model of a phase oscillator, which allows us to evaluate the detection limit of the coupling strength between oscillators.

13.3 Phase Oscillator

As single cell experiments have ruled out a strong coupling between circadian clocks in cyanobacteria, we consider only a weak coupling limit. In this case, we approximate the circadian oscillators to be phase oscillators, i.e. the interaction between oscillators is portrayed mainly by their phase dynamics while their amplitude is constant. Phase models can capture important synchronization properties of populations with weak interactions, as confirmed by theoretical, numerical (Strogatz 2000; Pikovsky et al. 2001) and experimental methods (Kiss et al. 2005). In the simplest approximation, the interaction between two oscillators is described by the first Fourier term, i.e. the sine of the difference of the two oscillator phases (Pikovsky et al. 2001). The phase dynamics of the oscillator j interacting with a second oscillator k is then given by:

$$\frac{d\phi_j}{dt} = \omega_j + \varepsilon \sin(\phi_k - \phi_j) \tag{13.1}$$

where $\phi_j(t)$, $\phi_k(t)$ are the instantaneous phases of the oscillators, ω_j is the free-running frequency of oscillator j and ε is the coupling constant between the two oscillators. For N mutually coupled oscillators, the influences of each oscillator are added. If the oscillators have the same free-running frequency (ω_0), Eq. 13.1 becomes the Kuramoto equation (Strogatz 2000), represented in a ω_0-rotating frame with $\varphi_j = \phi_j - \omega_0 t$:

Fig. 13.2. (continued) followed in time. The colour of each track is given by the phase (measured in degrees) of the circadian oscillation of the cell quantified by a fit over three intervals of time: the first two days (days 5–7), the entire time (days 5–10.5) and the last two days of the measurement (days 8.5–10.5). The colour of the interval 7–8.5 days of the time-track represents the average phase of the given individual oscillator. The fit function is $d(t) = B + A \cos(2\pi t / T_0 + \varphi)$, with $T_0 = 24.78$ h. Each *black dot* represents a cell-division event. The *indigo lines* show the precise (*continuous line*) or estimated (*dashed line*) boundaries between the merging colonies. The *arrows* point to examples of spatially close cells that are oscillating with different phases. Adapted from Mihalcescu et al. (2004a)

$$\frac{d\varphi_j}{dt} = \frac{\varepsilon}{N}\sum_{k=1}^{N}\sin(\varphi_k - \varphi_j) + \xi_j(t) \qquad (13.2)$$

Here $\xi_j(t)$ is a stochastic noise term having the characteristics of the previously defined Wiener process with the diffusion constant D. The time evolution of phases is then obtained by numerical simulation of the equations written for each of the interacting oscillators. In some particular cases the equations can be solved analytically.

One trivial case that has an exact solution is the one of uncoupled oscillators ($\varepsilon = 0$). When all oscillators have the same initial phase, the probability of observing a given bacterium with phase φ at time t is a normal distribution widening in time:

$P(\varphi,t) = \frac{1}{\sqrt{2\pi Dt}}\exp\left[\frac{-(\varphi-\mu)^2}{2Dt}\right]$ with μ their mean phase. However as the phase is 2π periodic, the random walkers diffuse on a circle and the solution becomes the wrapped normal distribution. Within the same periodicity condition the statistics also become circular (Mardia and Jupp 2000). In a population of oscillators, the mean phase and a measure of the synchronization are obtained from the mean field:

$$Z = \frac{1}{N}\sum_{j=1}^{N} e^{i\varphi_j} = \frac{1}{2\pi}\int_{\pi}^{\pi} P(\varphi,t)e^{i\varphi}d\varphi = \rho e^{i\mu} \qquad (13.3)$$

The mean field argument $\mu \in [(-\pi, \pi)$ is the mean phase while the amplitude $\rho \in [0, 1])$ of the mean field is the order parameter. ρ is a direct measure of the synchronization: for oscillators uniformly distributed $\rho = 0$, whereas $\rho = 1$ when all oscillators have exactly the same phase (Fig. 13.3).

For an arbitrary value of the coupling ε, Eq. 13.2 has an analytic solution only at steady-state (Amdaoud et al. 2007b), which relies on the comparative values of the phase diffusion constant D and the coupling constant ε. If $\varepsilon > D$, the coupling is stronger and the phases can be driven toward a distribution of limited width (Fig. 13.3, upper panels) given by the von Mises distribution (von Mises 1918): $P(\varphi) = \frac{1}{2\pi I_0(k)} e^{k\cos(\varphi-\mu)}$ where $\mu \in [-\pi, \pi]$ is the mean phase, k a parameter bi-univocally related to the order parameter $\rho = \frac{I_1(k)}{I_0(k)}$ and $I_n(k)$ is the modified Bessel function of the first kind of order n. In contrast, if $D > \varepsilon$, the noise is stronger and the phases ultimately are uniformly distributed between $[(-\pi, \pi)$ (Fig. 13.3, lower panels). As a result, the outcome of apparent precision of oscillators will also depend on the ranking of D with respect to ε.

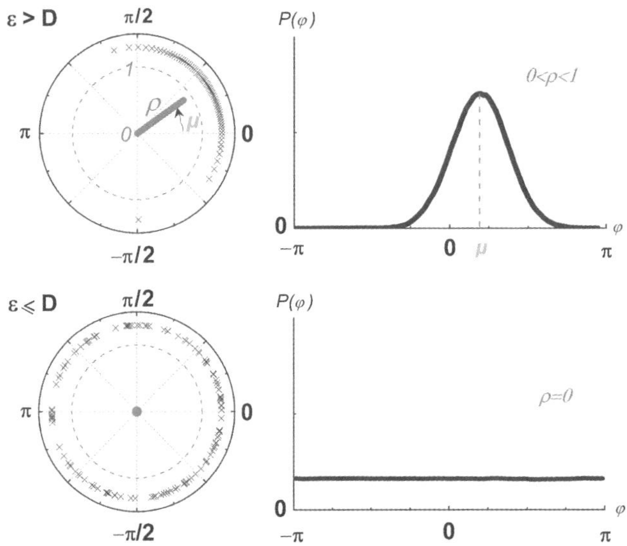

Fig. 13.3. The fate of the steady-state synchronization versus disorder for a population of coupled phase oscillators is determined by how the phase diffusion constant D relates to the coupling constant ε. *Upper panels* Synchronization, $\varepsilon > D$. *Lower panels* Disorder $\varepsilon \leq D$. *Left panels* The phases φ_j of individual oscillators (×) in the population are represented in radians on the trigonometrical circle. Their corresponding order parameter ρ is in *grey* and the mean phase μ in *light grey*. Both ρ and μ are obtained from Eq. 13.3. As one can see, if $\varepsilon > D$ (*upper panels*) the oscillators reach a partial synchronization, the order parameter $\rho > 0$ and one can define the mean phase μ. In contrast, if $\varepsilon \leq D$ (*lower panels*), the steady-state is complete disorder, the order parameter $\rho = 0$ and μ is undefined. *Right panels* The resulting distribution $P(\varphi)$ at steady state is described by a von Mises distribution for the synchronous case (*upper panels*) or by an uniform distribution for the complete disorder (*lower panels*)

In order to define an upper bound for the coupling constant and tentatively compare it to D, we performed (Mihalcescu et al. 2004b) numerical simulations following Eq. 13.2 for two distinct configurations designated in Fig. 13.2 by a pink arrow and a violet arrow. In the first configuration two neighbouring cells originating from the same initial single cell show a phase difference of $\Delta\varphi = 30°$ after 6 days of close contact. The coupling must therefore be smaller than the upper bound in order for the oscillators to drift away. A numerical simulation with two coupled oscillators, starting with the same initial phase, allows us to map the variance of their phase difference as a function of the coupling constant. Consequently, in order for $\Delta\varphi$ to have a value of 30° within a 95% confidence interval, the coupling constant has to be $\varepsilon \leq 0.13$ day^{-1}. In the second configuration (Fig. 13.2, violet arrow), the phase of the cell evolving along the boundary between the two colonies is roughly constant over the 5 days of measurement. From the other side of the boundary, the closest cell accessible for tracking has its phase slightly drifting away (initial 5°, final −11°). As these two cells were separated by at approximately three others, in a second simulation, we mimicked this particular geometry by creating a row of 11 interacting cells with the two borders between fixed phases. Simulations corresponding to approximately 2.5 days of experiments led us to an upper limit of $\varepsilon \leq 0.05$ day^{-1}. These estimations for the upper bound of ε confirm our first conclusion that the coupling is weak if not inexistent. However, the resolution of this experiment is insufficient for a direct comparison of ε with D: the lowest value of the upper limits for $\varepsilon \leq 0.05$ day^{-1} exceeds the determined interval of values $D = 0.012 \pm 0.007$ day^{-1}.

13.4 Minority Against Majority

13.4.1 Theoretical Considerations

We designed a new experiment which gave us a 30× improvement in the experimental resolution for the coupling constant evaluation. Two different populations of circadian oscillators, grown in liquid cultures, are mixed. The first, denoted as "majority" (M) is introduced preponderantly in a ratio 20:1 against a second population denoted as "minority" (m). Both populations have identical phase distributions, albeit centered around different averages, $\mu_M^{\ 0}$ and $\mu_m^{\ 0}$, respectively.

We first show theoretically that, when two populations with widely different abundances are mixed, the time evolution of the mean phase of the minority population allows the coupling between bacteria to be measured directly. This result is highlighted by Eq. 13.5. For that, we first separate the contribution of each population in the expression for the mean field: $Z = \frac{1}{N_M + N_m}\left(\sum_{k=1}^{N_M} e^{i\varphi_k} + \sum_{k=1}^{N_m} e^{i\varphi_k}\right) = \frac{N_M}{N_M + N_m} Z_M + \frac{N_m}{N_M + N_m} Z_m \approx \frac{20}{21} Z_M$ where $Z_M = \frac{1}{N_M}\sum_{k=1}^{N_M} e^{i\varphi_k} = \rho_M e^{i\mu}$ and $Z_m = \frac{1}{N_m}\sum_{k=1}^{N_m} e^{i\varphi_k} = \rho_m e^{i\mu}$ are the mean field of the majority and minority population, respectively, each one taken alone. Being entrained in the same way, both populations have initially the same amplitude of the mean field $\rho_M^{\ 0} = \rho_m^{\ 0}$. We then suppose that, through the progression in time, the distributions have similar widths and therefore comparable values of the mean field. In these conditions, the contribution of the minority population to the overall mean field can be neglected: $Z \cong \rho_M e^{i\mu_M} = Z_M$. The Fokker–Planck equations (Gardiner 1985) for the probability densities, $P_M(\varphi,t)$ and $P_m(\varphi,t)$, of the majority and minority populations are respectively:

$$\frac{\partial P_M(\varphi,t)}{\partial t} = \varepsilon \frac{\partial}{\partial \varphi}\left[\rho_M \sin(\varphi - \mu_M) P_M(\varphi,t)\right] + \frac{D}{2}\frac{\partial^2 P_M(\varphi,t)}{\partial \varphi^2}$$

$$\frac{\partial P_m(\varphi,t)}{\partial t} = \varepsilon \frac{\partial}{\partial \varphi}\left[\rho_M \sin(\varphi - \mu_M) P_m(\varphi,t)\right] + \frac{D}{2}\frac{\partial^2 P_m(\varphi,t)}{\partial \varphi^2}$$

Initial ($t = 0$) and final (steady state) probability densities, $P_M(\varphi,t)$ and $P_m(\varphi,t)$, are von Mises distributions:

1. $P_M(\varphi, t = 0)$ and $P_m(\varphi, t = 0)$ have been roughly approximated (Amdaoud et al. 2007b) by normal distributions with standard deviation $\sigma_0 = 0.9$ radian. As the densities of wrapped normal and von Mises distributions are very similar for any value of variance (Mardia and Jupp 2000; Amdaoud et al. 2007b), it is justifiable here to consider them as von Mises, with the parameter $k_0 = 1.8$.
2. For $t \rightarrow \infty$, we have seen previously that the steady-state solution is the von Mises distribution, which is common to both populations $P_M(\varphi) = P_m(\varphi)$.

We subsequently interpolate between these two limits and approximate $P_M(\varphi,t)$ and $P_m(\varphi,t)$, at any point in time, with a von Mises distribution. $P_M(\varphi,t)$ and $P_m(\varphi,t)$ are defined respectively by the time-dependent parameters $\mu_M(t)$, $k_M(t)$ and $\mu_m(t)$, $k_m(t)$. Their time progression can be easily calculated as follows: we multiply both equations by $e^{i\varphi}$ and integrate $\int_{-\pi}^{\pi} \ldots d\varphi$. This gives the set of equations:

$$\begin{aligned}
\frac{\partial \rho_M}{\partial(\varepsilon t)} &= \frac{\rho_M}{2} \cdot \left(\frac{2\rho_M}{k_M} - \frac{D}{\varepsilon} \right) \\
\frac{\partial \rho_m}{\partial(\varepsilon t)} &= \frac{\rho_m}{2} \cdot \left[\frac{2\rho_M}{k_m} \cdot \cos(\mu_M - \mu_m) - \frac{D}{\varepsilon} \right] \\
\frac{\partial \mu_m}{\partial(\varepsilon t)} &= \frac{\rho_M}{\rho_m} \cdot \left(1 - \frac{\rho_m}{k_m} \right) \cdot \sin(\mu_M - \mu_m)
\end{aligned} \tag{13.4}$$

and $\mu_M(t) = \mu_M^0$ (the majority mean phase remains constant). In Eq. 13.4, εt appears as the natural reduced time. By excluding the initial conditions, the ratio D/ε remains the only parameter affecting the evolution of the minority mean phase $\mu_m(\varepsilon t)$ and both order parameters $\rho_M(\varepsilon t)$ and $\rho_m(\varepsilon t)$. Figure 13.4 compares the numerical simulation of the Eq. 13.2, for different values of ε and D/ε, with the solution of Eq. 13.4. The good agreement validates our previous approximations, in the parameter range of our experimental circumstances. It is remarkable that the influence of D/ε on $\mu_m(\varepsilon t)$ is weak: the insert of Fig. 13.4a shows similar variation of $\mu_m(\varepsilon t)$ for D/ε values from 0.01 to 10. For $\varepsilon t << 1$, $\mu_m(t)$ is independent of the diffusion coefficient value D:

$$\mu_m(t) - \mu_m^0 \cong (1 - \rho_0 / k_0) \cdot \sin(\mu_M^0 - \mu_m^0) \cdot \varepsilon t \tag{13.5}$$

The coupling constant ε can be therefore evaluated independently of D. The appropriate physical measure is the slope of the time progression of the minority mean phase, $\mu_m(t)$.

13.4.2 Experimental Results: Minority Impassive to Majority

To follow the circadian oscillations of a population of cells we used a 96-well plate luminometer. Each well was filled with the same total number of cells, $\approx 3\times10^7$ cells well^{-1}. A custom-made external chamber kept the plates in constant external

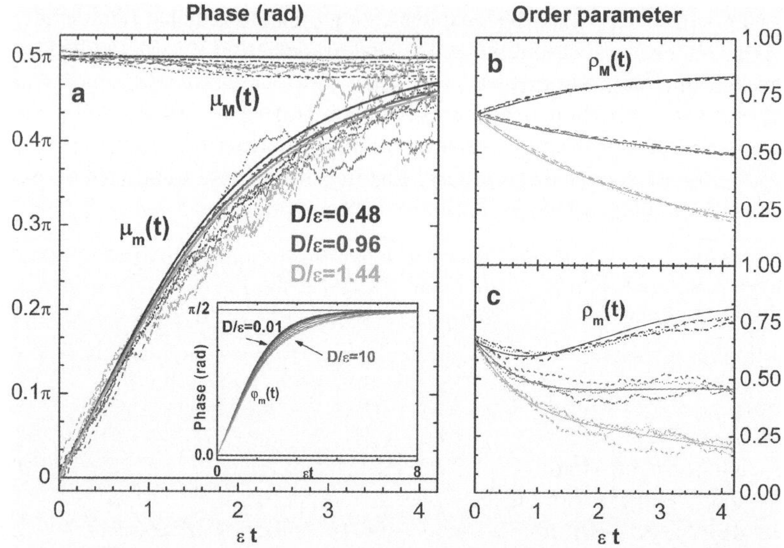

Fig. 13.4. Comparison between the solution of Eq. 13.4 (*solid line*) and numerical simulations of Eq. 13.2 for 10,500 interacting phase oscillators: (**a**) the mean phases $\mu_M(\varepsilon t)$ and $\mu_m(\varepsilon t)$, (**b**) order parameter of the majority population $\rho_M(\varepsilon t)$ and (**c**) order parameter of the minority population $\rho_m(\varepsilon t)$ are represented for three values of D/ε, 0.48 (*black*), 0.96 (*grey*) and 1.44 (*light grey*). Three simulations are superimposed for each value of D/ε, with different values of the coupling constant: $\varepsilon = 0.2$ (*dashed line*), $\varepsilon = 0.1$ (*dash-dot line*), $\varepsilon = 0.05$ (*dotted line*). The initial phase of the oscillators, independently distributed, has one majority group of $N_M = 10{,}000$, normally distributed around $\mu_M^0 = \pi/2$ and one minority group of $N_m = 500$ normally distributed around $\mu_m^0 = 0$. Initially, both groups have the same standard deviation $\sigma_0 = 0.9$ radian. For the numerical simulation we used a forward Euler scheme with time step $\Delta t = 2.6 \times 10^{-3}$. The noise ξ, a Wiener process with a diffusion constant D, is simulated by a phase jump that is equally probable forward or backward, at each time step. Insert: similar variation of $\mu_m(\varepsilon t)$, obtained from Eq. 13.4 for D/ε varying from 0.01 to 10.0. Adapted from Amdaoud et al. (2007a)

conditions: 900 lux (1 lux = 1 cd. sr. m^{-2}) white-light illumination and 30°C. Each experimental condition was represented by 8–12 independent wells. To avoid any loss of the minority signal by a 20-fold more abundant majority, we used for the majority population, the wild-type strain (*S. elongatus* PCC 7942) with no reporter, i.e. non-bioluminescent. For the minority population we used AMC462 (Katayama et al. 1999), a bioluminescent strain which has two neutral site chromosomal insertions, P*kaiBC::luxAB* and P*psbAI::luxCDE*. Note that any feedback influence of the bioluminescent light on the bacterial clock is excluded as the white-light illumination provides each cell with at least 10^7 more photons than the bioluminescent reporter (Amdaoud et al. 2007b).

To obtain samples with the desired mean phase and the same distribution of individual cell phases around this mean, the cultures were first entrained by the same 12 h light/12 h dark (LD12:12) cycle, then simultaneously frozen and finally thawed at different time intervals (Amdaoud 2007). Freezing cyanobacteria stops the circadian clock ticking, while thawing them restarts it. This leaves the phase difference directly related to the time interval between each thawing[5]. Figure 13.5a shows an example of the circadian oscillation from the luminescent strain alone, here shown for two initially opposite mean phases followed approximately 40 days. This example reveals a remarkable well-to-well reproducibility of the oscillations. Moreover, the initial opposition of phases between the two conditions is maintained until the end of the experiment.

The light detected from a well, $i(t)$, sums up the luminescence of all emitting cyanobacteria inside the well: $i(t) = \langle d(t)\rangle \cdot N_E$, where N_E is the number of emitting cells and $\langle d(t)\rangle = g \cdot [1 + b\rho\cos(\omega_0 t + \mu)]$ is the average luminescence of the clock cells. Here ρ is the order parameter and μ is the mean phase of the detected population of clocks (Eq. 13.2). During the experiment, the number of cells in each well continues to grow[6]. This growth rapidly limits the detection of the luminescence only to the top layers of cells, as the chloropyll-containing cyanobacteria re-absorb the bioluminescence photons emitted by the bottom layers (Amdaoud et al. 2007b). In addition, the nutrient resources shared among increasingly numerous members will reduce the gain of the biochemical luminescent reaction/cell depicted here by the average amplitude g. The sensitivity of the bioluminescent reporter to the metabolic conditions is circumvented by defining the oscillatory signal: $s(t) = i(t)/\overline{i(t)} - 1$, with $i(t)$ the luminescence recorded from a well and $\overline{i(t)}$ the temporal mean calculated by smoothing the bioluminescence curve. The baseline $\overline{i(t)}$ appears to be proportional to the relative concentration of the emitters/total number of cells in a well (Amdaoud et al. 2007b). We used this property to continuously monitor the ratio between the minority and the majority population. In the experiment showed in Fig. 13.5, this ratio slowly decreased from 1:20 to 1:30, which should increase even further the influence of the majority on the minority.

The oscillatory signal written as $s(t) = b\rho\cos(\omega_0 t + \mu)$ is proportional to the real part of the mean field of the detected oscillators in the well[7]. The time variation of the amplitude of oscillations $b\rho$ describes the apparent damping of $s(t)$ oscillations in Fig. 13.5a, b. As b is time-dependent[8], variable from one experiment to another and accounts for nearly all of the amplitude damping, it is not possible to get a reliable experimental determination of $\rho(t)$ for a quantitative comparison with the theoretical expectations (Eq. 13.4). By contrast, the phase of $s(t)$ is precisely the mean phase μ; and for this reason we used this quantity to monitor the minority mean phase.

[5] The standard deviation of the phase distribution (Amdaoud 2007) obtained by this method was approximately $\sigma_0 \cong 0.9$ rad, which corresponds to an order parameter $\rho_0 \cong 0.67$ and to a concentration parameter $k_0 = 1.8$

[6] Colour photographs of the plates taken every week illustrate the wells which become greener due to the increasing chlorophyll concentration (Amdaoud 2007; Amdaoud et al. 2007b)

[7] In the mixture case, Z_m the mean field is the minority population

[8] The relative amplitude of oscillation/cell b, for the promoter monitored here P*kaiBC* is initially $b \sim 1$, then decreases during the experiment (Amdaoud 2007; Amdaoud et al. 2007b).

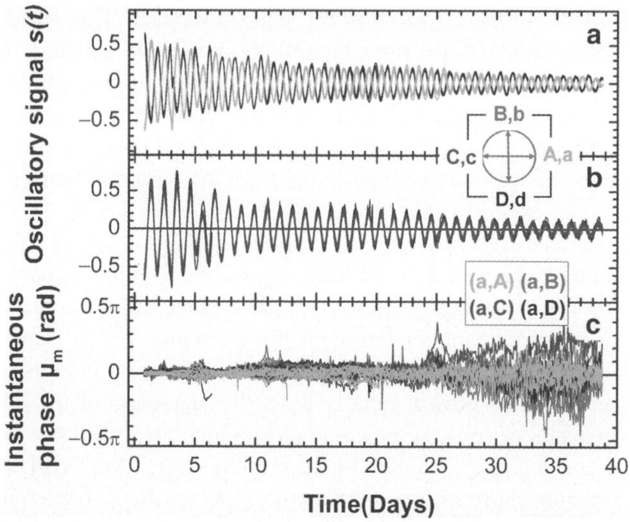

Fig. 13.5. Communication experiments: the oscillation of the minority population is unperturbed by the majority presence. **a** The circadian oscillation $s(t)$ for the bioluminescent strain (AMC462) alone previously entrained at opposite phases (*light grey* and *black lines*, **A** and **C** respectively) and (**b**) for the mixtures. *Light grey* to *black*: mixtures of a luminescent minority with a 20× larger population of wild-type cells, as follows: (a, A), (a, B), (a, C) and (a, D). *Lower case* denotes minority and *upper case* majority mean phase. **c** Instantaneous mean phase of the minorities μ_μ (t) extracted from $s(t)$, represented with the same colours as in (**a**). *Insert* The four phases of entrainment are separated by ~ —$\pi/2$ radians, with A leading B, in opposition to C and lagging D. Adapted from Amdaoud et al. (2007a)

We chose to work with four initial phases, denoted A, B, C, D, separated by $\cong\pi/2$ radians (Fig. 13.5, insert). Mixtures were made in 96-well plates using different pairs (μ_m^0, μ_M^0): from (a, A), (a, B) … to (d, C) and (d, D). Lower case denotes mean phases of the minority population, which contains the autobioluminescent reporter; and upper case depicts those of the majority population, which does not harbor a reporter. Figure 13.5b presents the oscillatory signal $s(t)$ from individual wells containing mixtures of the same minority (a) with four different majorities: A, B, C, D. The oscillation of the minority population appears to be unperturbed by the majority presence: throughout the experiment, their oscillations overlap regardless of the majority mean phase.

For a more detailed analysis, we used the Hilbert Transform to extract the instantaneous phase from the oscillatory signal. The Hilbert Transform reconstructs the analytical signal $z(t)$, starting from its real part $s(t)$: $z(t) = s(t) + is_{HT}(t) = A(t)\,e^{i\varphi(t)}$, which gives the instantaneous amplitude $A(t)$ and phase $\varphi(t)$. Here $s_{HT}(t)$ is the Hilbert Transform of $s(t)$: $s_{HT}(t) = \frac{1}{\pi} P.V. \int_{-\infty}^{+\infty} \frac{s(\tau)}{t-\tau} d\tau$ (where P.V means the principal Cauchy value). The instantaneous phase $\varphi(t)$ is thus given by $\tan^{-1}\left(\frac{s_{HT}(t)}{s(t)}\right)$ if $s_{HT}(t) > 0$, and by $\tan^{-1}\left(\frac{s_{HT}(t)}{s(t)}\right) \pm \pi$ if $s_{HT}(t) < 0$. Figure 13.5c follows the temporal progression of the minority phase $\mu_m(t)$ for individual wells containing mixtures of the same minority with four different majorities: (a, A), (a, B), (a, C) and (a, D), shown previously in Fig. 13.5b. The instantaneous phase was portrayed relative to the reference $<\mu_{(a,A)}(t)>$, the average of wells containing a mixture with the same initial phase $\mu_m^{\,0} = \mu_M^{\,0} = A$. Theoretically, in the other mixtures where $\mu_m^{\,0} \neq \mu_M^{\,0}$, we would have expected the minority to drift gradually towards the phase of the majority (Fig. 13.4a), if the coupling was strong enough. Figure 13.5c, however, shows no apparent phase deviation. The instantaneous phases spread out around zero similar to reference mixture (a,A) wells, the minority population appeared unaffected by the presence of a majority population with a different phase. Either there is no intercellular interaction between oscillators or, more probably, the experimental noise causing this dispersion masks a possible weak coupling.

A way of extracting a possible minority phase variation from the experimental noise is by improving the statistics. We repeated the same experiments and then averaged individual wells from all the measurements. The average was taken over ≅80 individual wells of equivalent experiments, i.e., wells with the same initial phase difference $\mu_M^{\,0} - \mu_m^{\,0}$ between majority and minority. As previously, the reference for each well was its corresponding average of same-phase mixtures $\mu_m^{\,0} = \mu_M^{\,0}$. Again, the minority phase variation remained buried in the noise and exhibited a similar variation for all conditions. The precision of the experiment, however, enabled us to set an upper limit for the coupling constant ε. As the condition $\varepsilon t << 1$ is obviously valid, we used Eq. 13.5. The expected variation of the minority phase is then linear, with the slope directly related to ε: $(1 - \rho_0 / k_0) \bullet \sin(\mu_M^{\,0} - \mu_m^{\,0})$. ρ_0 and k_0 are known[5] and we considered the extreme scenario $\mu_M^{\,0} - \mu_m^{\,0} = \pm \pi/2$ where we would have expected the strongest variation of the minority mean phase. A coupling constant, if existent, should be confined within a 95% confidence interval to $|\varepsilon| < 1.5 \cdot 10^{-3}$ day^{-1}. This time one can compare the diffusion constant D evaluated from single cell experiments, with the upper limit of the coupling constant. It states that $D > \varepsilon$, therefore the measured stability of the clock at the single cell level cannot be the result of intercellular communication. The precision of the circadian clock in cyanobacteria is built-in and their genetic and metabolic network must be responsible for this stability.

13.5 Built-In Stability

The cyanobacterial clock has a central post-translational pacemaker based on the repeated interaction of the three clock proteins KaiA, KaiB and KaiC. In vitro experiments (Nakajima et al. 2005) have shown that the level of KaiC phosphorylation oscillates with a circadian period when only KaiC, KaiA, KaiB and ATP are present. This phosphorylation rhythm persists for at least 10 days without damping (Ito et al. 2007). During the oscillation, the master clock protein KaiC goes cyclically through four different phosphoforms (Nishiwaki et al. 2007). The key point is that the product of each step in the phosphorylation cycle regulates the reaction in the next step. The mixture of six in vitro samples, in different phosphorylation phases synchronizes this time rapidly within one circadian cycle (Ito et al. 2007). It seems that the strong coupling is not between cells but between the molecular KaiC-based oscillators. The coupling is linked to KaiA and KaiB proteins (Rust et al. 2007), where (at least) one of the KaiC phosphoforms inhibits the activity of KaiA through interaction with KaiB. The post-translational oscillator that emerges from this description is potentially less noisy than a transcription/translation oscillator (Raser and O'Shea 2005). However, the amount of each of the proteins in a test tube is infinitely larger that their abundance within a single cell. In a cell the amount of Kai proteins has been estimated to be 5,000–15,000 KaiC molecules, 7,000–30,000 KaiB and only to 200–500 KaiA molecules (Kitayama et al. 2003). It remains an open question whether the in vitro mechanism of the pacemaker is sufficient to explain the observed stability in vivo or if supplementary feedback loops are needed.

References

Amdaoud Malika (2007) Stabilité du rythme circadien des cyanobactéries: Investigation d'un couplage entre oscillateurs. PhD, University of Grenoble. http://www-lsp.ujf-grenoble.fr/pdf/theses/adma.pdf

Amdaoud M, Vallade M, Weiss-Schaber C, Mihalcescu I (2007a) Proc Natl Acad Sci USA 104(17):7051–7056

Amdaoud M, Vallade M, Weiss-Schaber C, Mihalcescu I (2007b) Supporting information text: cyanobacterial clock, a stable phase oscillator with negligible intercellular coupling. http://www.pnas.org/cgi/data/0609315104/DC1/9. Accessed 16 Apr 2007

Barkai N, Leibler S (2000) Circadian clocks limited by noise. Nature 403:267–268

Carr AJ, Whitmore D (2005) Imaging of single light-responsive clock cells reveals fluctuating free-running periods. Nat Cell Biol 7:319–321

Elowitz MB, Levine AJ, Siggia ED, Swain PS (2002) Stochastic gene expression in a single cell. Science 297:1183–1186

Gardiner CW (1985) Handbook of stochastic methods for physics, chemistry and the natural sciences. Springer, Heidelberg

Herzog ED, Aton SJ, Numano R, Sakaki Y, Tei H (2004) Temporal precision in the mammalian circadian system: a reliable clock from less reliable neurons. J Biol Rhythms 19:35–46

Ito H, Kageyama H, Mutsuda M, Nakajima M, Oyama T, Kondo T (2007) Autonomous synchronization of the circadian KaiC phosphorylation rhythm. Nat Struct Mol Biol 14:1084–1088

Katayama M, Tsinoremas NF, Kondo T, Golden SS (1999) *cpmA*, a gene involved in an output pathway of the cyanobacterial circadian system. J Bacteriol 181:3516–3524

Kiss IZ, Zhai Y, Hudson JL (2002) Emerging coherence in a population of chemical oscillators. Science 296:1676–1678

Kiss IZ, Zhai Y, Hudson JL (2005) Predicting mutual entrainment of oscillators with experiment-based phase models. Phys Rev Lett 94:248–301

Kitayama Y, Iwasaki H, Nishiwaki T, Kondo T (2003) KaiB functions as an attenuator of KaiC phosphorylation in the cyanobacterial circadian clock system. EMBO J 22:2127–2134

Kondo T, Mori T, Lebedeva NV, Aoki S, Ishiura M, Golden SS (1997) Circadian rhythms in rapidly dividing cyanobacteria. Science 275:224–227

Liu C, Weaver DR, Strogatz SH, Reppert SM (1997) Cellular construction of a circadian clock: period determination in the suprachiasmatic nuclei. Cell 91:855–860

Liu Y, Tsinoremas NF, Johnson CH, Lebedeva NV, Golden SS, Ishiura M, Kondo T (1995) Circadian orchestration of gene expression in cyanobacteria. Genes Dev 9:1469–1478

Mardia KV, Jupp PE (2000) Directional statistics, 2nd edn. Wiley series in probability and statistics. Wiley, Chichester

Mihalcescu I, Hsing W, Leibler S (2004a) Resilient circadian oscillator revealed in individual cyanobacteria. Nature 430:81–85

Mihalcescu I, Hsing W, Leibler S (2004b) Supplementary discussion: resilient circadian oscillator revealed in individual cyanobacteria. http://www.nature.com/nature/journal/v430/n6995/extref/nature02533-s1.pdf. Accessed 4 Jul 2004

Nagoshi E, Saini C, Bauer C, Laroche T, Naef F, Schibler U (2004) Circadian gene expression in individual fibroblasts: cell-autonomous and self-sustained oscillators pass time to daughter cells. Cell 119:693–705

Nakajima M, Imai K, Ito H, Nishiwaki T, Murayama Y, Iwasaki H, Oyama T, Kondo T. (2005) Reconstitution of circadian oscillation of cyanobacterial KaiC phosphorylation in vitro. Science 308:414–415

Nishiwaki T, Satomi Y, Kitayama Y, Terauchi K, Kiyohara R, Takao T, Kondo T (2007) A sequential program of dual phosphorylation of KaiC as a basis for circadian rhythm in cyanobacteria EMBO J 26:4029–4037

Pikovsky A, Rosenblum M, Kurts J (2001) Synchronization. A universal concept in nonlinear sciences. Cambridge University Press, Cambridge

Raser JM, O'Shea EK (2005) Noise in gene expression: origins, consequences, and control. Science 309:2010–2013

Rust MJ, Markson JS, Lane WS, Fisher DS, O'Shea EK (2007) Ordered phosphorylation governs oscillation of a three-protein circadian clock. Science 318:809–812

Strogatz SH (2000) From Kuramoto to Crawford: exploring the onset of synchronization in populations of coupled oscillators. Physica D 143:1–20

von Mises R (1918) Über die 'Ganzzahligkeit' der Atomgewichte und verwandte Fragen. Phys Z 19:490–500

Welsh DK, Yoo SH, Liu AC, Takahashi JS, Kay SA (2004) Bioluminescence imaging of individual fibroblasts reveals persistent, independently phased circadian rhythms of clock gene expression. Curr Biol 14:2289–2295

Chapter 14
The Circadian Clock Gear in Cyanobacteria: Assembled by Evolution

Volodymyr Dvornyk

Abstract The circadian system of cyanobacteria has a long and complex evolutionary history. Some of its genetic elements are probably as old as cyanobacteria themselves. Currently available data from evolutionary studies suggest that, in the course of evolution, the whole system as well as its elements experienced a number of major structural modifications, which resulted in diversification of the circadian system. There are probably at least three main types of the circadian system in cyanobacteria, which differ by their set of elements. Whether these differences result in any functional modifications or malfunction is yet to be determined. Some evidence exists that major steps in macroevolution of the cyanobacterial circadian system were adaptive and associated with large-scale changes in global environment. Further studies will help to fully reconstruct a scenario by which the circadian system of cyanobacteria evolved into a finely tuned regulatory mechanism.

14.1 Introduction

The rotation of the Earth about its axis and revolution around the sun result in orderly fluctuations of micro- and macro-environments related to the respective periodic changes in light, temperature, and other conditions. The vast majority of living things have developed endogenous mechanisms to adapt to these changes by controlling a variety of biological rhythms of different periodicities, such as circadian, infradian, annual, and others. Among those, the mechanism controlling physiological rhythms with approximately daily periodicity is termed circadian. It has been comprehensively studied in eukaryotes (for a review, see Dunlap et al.

V. Dvornyk
School of Biological Sciences, The University of Hong Kong, Pokfulam Rd, Hong Kong SAR, P.R. China, e-mail: dvornyk@hku.hk

J.L. Ditty et al. (eds.), *Bacterial Circadian Programs*.

2004). Among prokaryotes, circadian rhythmicity was first reported in cyanobacteria (Huang et al. 1990; Kondo et al. 1994). Recently, some evidence was obtained that suggests such rhythms exist in purple photosynthesizing bacteria (Min et al. 2005). However, cyanobacteria remain a principal subject of studies in prokaryotic chronobiology.

Cyanobacteria are thought to have appeared on Earth about 3.5×10^9 years ago (Bya; Schopf and Packer 1987; see Chap. 2). Throughout the course of their evolution, these prokaryotes have endured enormous changes in a wide range of environmental conditions on the Earth and, yet more importantly, were able to develop highly efficient adaptive mechanisms (Whitton 1987). Because of their ability to control many key cellular processes, circadian clock components are thought to be a cornerstone of the remarkable adaptiveness of cyanobacteria (Johnson 2005). Estimates yield a proportion of genes in the cyanobacterial genome that are expressed rhythmically to vary from 2% in *Synechocystis* sp. PCC 6803 as determined by microarray analyses (Kucho et al. 2005), to up to 30% in *Synechococcus elongatus* PCC 7942 using promoter-trap experiments (Liu et al. 1995). Apparently, the circadian clock itself is a result of adaptive evolution; however, it remains largely unclear how and when various circadian clock genes acquired circadian function. The unicellular cyanobacterium *S. elongatus* PCC 7942 is the model species for the studies of the circadian system in prokaryotes. During the past decade, researchers have been able to identify several genes with circadian function in this organism (Table 14.1). These data, in addition to the growing volume of available genomic data, have made it possible to reconstruct the evolution of some of these genes and formulate a hypothesis about the origin and evolution of the circadian system.

Table 14.1 A list of the currently known circadian genes from the model species *Synechococcus elongatus* PCC 7942

Division of the circadian system	Gene	References
Input	*cikA*	Schmitz et al. (2000)
	ldpA	Katayama et al. (2003)
	pex	Kutsuna et al. (1998)
Central oscillator	*kaiA*	Ishiura et al. (1998)
	kaiB	Ishiura et al. (1998)
	kaiC	Ishiura et al. (1998)
Output	*sasA*	Iwasaki et al. (2000)
	labA	Taniguchi et al. (2007)
	rpaA	Takai et al. (2006)
	cpmA	Katayama et al. (1999)
	Group 2 sigma factors	Tsinoremas et al. (1996), Nair et al. (2002)

14.2 Circadian Genes in Prokaryotes: Structure and Occurrence

14.2.1 Genes of the Central Oscillator

The central oscillator of the circadian system in cyanobacteria, as it is described in *S. elongatus* PCC 7942, is composed of three genes, *kaiA*, *kaiB*, and *kaiC* (Ishiura et al. 1998). The *kaiC* gene consists of two tandemly arrayed homologous domains (Ishiura et al. 1998). This gene is characteristic to cyanobacteria but also occurs in other prokaryotes, including Proteobacteria and some Euryarchaeota. KaiC does have single-domain homologs that do not occur in cyanobacteria but are ubiquitous in other bacteria and archaea (Dvornyk et al. 2003). Evolutionarily, the *kaiC* gene is likely the oldest among all circadian genes. It belongs to the RecA superfamily of ATP-dependent recombinases (Leipe et al. 2000). These proteins are essential for DNA repair (Roca and Cox 1990) and are thought to stem from the last universal common ancestor (DiRuggiero et al. 1999; Lin et al. 2006). The duplication and subsequent fusion of the ancestral single-domain *recA* resulted in the formation of the double-domain *kaiC* gene (Dvornyk et al. 2003). Interestingly, while double-domain *kaiC* homologs are common in Proteobacteria and Cyanobacteria, a single cyanobacterial species, *Gloeobacter violaceus* PCC 7421, has no *kai* genes (Nakamura et al. 2003). *Gloeobacter* is a quite particular species among cyanobacteria as it lacks thylakoid membranes (Jurgens and Schneider 1991).

The single-domain *kaiC* homologs, which presumably maintain their original function related to DNA metabolism and repair, have higher similarity to the N-terminal domain of the double-domain *kaiC* genes than the C-terminus (Dvornyk et al. 2003). This suggests that the C-terminal domain of *kaiC* diverged more significantly than the N-terminus towards its current circadian function. Indeed, recent data show that KaiA binds exclusively to the C-terminal domain of KaiC. Deletion of this entire domain, or only its 25 C-terminal amino acid residues, abolishes KaiA binding (Pattanayek et al. 2006). The high variability of the C-terminal region in KaiC of bacteria that lack KaiA (Dvornyk and Knudsen 2005) provides further support that the C-terminal domain evolved for circadian function.

The *kaiB* gene homologs can be divided by length into two major groups. The shorter genes (approximately 300–400 bp) are found in Archaea, Chloroflexi, Proteobacteria, and all Cyanobacteria (except the aforementioned *G. violaceus* PCC 7421), whereas the longer genes (up to approx. 900 bp) only appear in some Cyanobacteria. The *kaiA* gene homologs vary from 300 bp to 900 bp and demonstrate the highest degree of polymorphism among all the *kai* genes. The *kaiB* and double-domain *kaiC* genes usually form an operon, which is characteristic of the Cyanobacteria and also occurs in some Archaea, Proteobacteria, and Chloroflexi. In addition, some Proteobacteria and Cyanobacteria have additional copies of *kaiB* and *kaiC* scattered in their genomes. In contrast to *kaiB* and *kaiC*, the *kaiA* gene is found only in cyanobacterial species, always in a single copy. When all three *kai* genes occur in a genome, they always form a single *kaiABC* cluster (Dvornyk et al. 2003). In some cyanobacterial species, *kaiA* is absent (e.g., *Prochlorococcus* sp.).

14.2.2 Input to the Clock

Three genes (*cikA, ldpA, pex*) have been identified thus far as being involved in the input signal to the central oscillator (Table 14.1). The *cikA* gene has a three-domain architecture consisting of GAF, histidine protein kinase (HPK), and pseudo-receiver (PsR), and is a key element of circadian input (Schmitz et al. 2000; see Chap. 8). CikA belongs to a superfamily of two-component histidine kinases whose members are common in prokaryotes, such that many prokaryotes bear genes that display apparent homology to *cikA* (Baca et al., unpublished data); however, in most cases this homology is limited only to the histidine kinase domain. The GAF domain, which has been shown to be essential for circadian function (Mutsuda et al. 2003), is found in relatively few *cikA* homologs. Moreover, the *cikA* homologs that encode the three-domain architecture (GAF–HPK–PsR), as was described for the bona fide *cikA*, are present only in the Cyanobacteria and in a single copy. A comparative analysis of polymorphism was able to identify several conserved regions of probable functional importance for the gene. In particular, one of these motifs lies immediately upstream the GAF domain. The N-terminal region of ~180 amino acid residues in CikA was previously shown to enhance phosphorylation of the HPK domain (Ivleva et al. 2006); however, no specific fragment of this region was identified as a major contributor to this function. The analysis of polymorphism showed that a fragment corresponding to amino acid residues 168–183 in CikA of *S. elongatus* PCC 7942 is highly conserved in the bona fide CikA proteins, while being variable in their apparently non-circadian homologs, and thus is likely the enhancer of the HPK phosphorylation (Baca et al., unpublished data). The GAF–HPK–PsR gene architecture is not common for all Cyanobacteria. In the filamentous species *Nostoc* and *Anabaena*, the *cikA*-like gene lacks the PsR domain. This domain was shown to be essential for *cikA* function in *S. elongatus* PCC 7942 as a negative regulator of the HPK domain activity (Mutsuda et al. 2003). In such a case, the absence of a PsR domain in a *cikA*-like gene product suggests that: (i) the mechanism of regulating CikA phosphorylation, and hence its function, in these filamentous cyanobacteria is different from that in *S. elongatus* PCC 7942 and (ii) a separate gene attenuator of the HPK activity probably exists. Neither *cikA* nor any apparent GAF domain-containing homologs are found in some unicellular cyanobacteria (*Prochlorococcus* sp., *Synechococcus* sp. WH 8102) and in other bacteria (Baca et al., unpublished data). This fact limits the origin of the *cikA* gene to the Cyanobacteria and suggests that it was then lost in some cyanobacterial taxa.

Another component of circadian input, *ldpA*, belongs to the superfamily of 4Fe-4S ferredoxins (Katayama et al. 2003) and has many homologs in other bacteria (Dvornyk 2005). In the majority of *ldpA* homologs, the homology is fairly weak and limited only to the HycB domain. This domain is a part of the formate-hydrogenlyase system in *Escherichia coli* (*hyc* operon) and encodes a small subunit of hydrogenase-3 (Rossmann et al. 1991; Sauter et al. 1992). An interesting fact is that transcription of the *hyc* operon is controlled by FhlA (Rossmann et al. 1991; Sauter et al. 1992), which belongs to the same superfamily as the GAF domain of *cikA* (Aravind and Ponting 1997). This adds some evolutionary insights into the CikA-LdpA interaction (Ivleva et al. 2005). Only cyanobacteria possess genes homologous to the full-length *ldpA* of *S. elongatus* PCC 7942. The *ldpA* genes in the Cyanobacteria have several unique highly conserved regions and motifs absent in the homologs of other bacteria and return no matches in the Conserved Domain Database (Marchler-Bauer et al. 2007). Therefore, these regions may be potentially important to the circadian function of *ldpA* (Dvornyk 2005).

The *pex* gene belongs to the PadR family of transcriptional regulators (Kutsuna et al. 1998), which can be found in many bacteria. The *pex* genes from cyanobacteria are more conserved than their homologs in other bacteria. The *pex* gene does not occur in cyanobacterial strains that lack the *kaiA* gene (*Prochlorococcus* sp.); however, some cyanobacterial species that contain a *kaiA* gene (e.g., *Crocosphaera watsonii*, *Synechocystis* PCC 6803) may lack *pex* (Dvornyk, unpublished data).

14.2.3 Genes of the Circadian Output

Several genes have been identified to control an output signal from the central oscillator (Table 14.1; see Chap. 9). Among those, *sasA* and *cpmA* were subjected to an evolutionary analysis (Dvornyk et al. 2004; Dvornyk 2006b). Similar to *cikA*, *sasA* is a member of the two-component histidine kinase superfamily and has numerous homologs in prokaryotes (Dvornyk et al. 2004). However, all the homologs outside of the Cyanobacteria (save *G. violaceus* PCC 7421 which lacks all *kai* genes; Nakamura et al. 2003) lack the KaiB-like sensory domain. The absence of the KaiB-like domain in the non-cyanobacterial *sasA* homologs is further evidence that *kaiB* originated in cyanobacteria and was laterally transferred to other prokaryotes. In all studied cyanobacterial genomes, *sasA* occurs in a single copy.

Another gene of circadian output, *cpmA*, is also ubiquitous in the Cyanobacteria and other prokaryotes. Unlike *cikA*, *sasA*, and *ldpA*, the *cpmA* genes and their homologs share the same domain architecture. However, the *cpmA* genes of the Cyanobacteria are more conserved, especially in their C-terminal half (Dvornyk 2006b). This region in the coded protein contains two hydrophobic motifs (Katayama et al. 1999) and shows some sequence similarity to the PurE related proteins, AIR carboxylase and NCAIR mutase. which are involved in metabolism of purines (Watanabe et al. 1989; Meyer et al. 1992). Interestingly, neither *sasA* nor *cpmA* homologs were found in the photosynthetic α-proteobacteria or *Chloroflexus*, both of which have *kai* gene homologs (Dvornyk et al. 2003).

The remaining known circadian output genes, *rpaA* and *labA*, and their homologs show quite different patterns of occurrence among prokaryotic taxa. The *rpaA* gene and its homologs occur in all Cyanobacteria, are common in Firmicutes (while being relatively rare in Chloroflexi, Proteobacteria, Actinobacteria), and are absent in Archaea. In contrast, *labA* and its homologs are found only in the Cyanobacteria that contain *kaiA*, are ubiquitous in Proteobacteria (particularly those of the α- and γ-subdivisions), and occur in some Archaea, but are very rare in Firmicutes and missing in Chloroflexi (Dvornyk, unpublished data). So, the occurrence of *rpaA* and *labA* is quite predictable in Cyanobacteria and may reflect their circadian functions. In addition to these genes, sigma factors have been shown to play a role in circadian output (Nair et al. 2002). The group 2 sigma factors are ubiquitous in cyanobacteria, and a BLAST search of 40 fully sequenced cyanobacterial genomes in GenBank returned over 200 homologous sigma factors (Dvornyk, unpublished data). All these genes have the same basic structure of four sigma-70 domains. No data about the role of each domain of the sigma factors in the circadian function is available.

As the available data show, many circadian genes are members of large gene superfamilies that are widely distributed in prokaryotes. For example, *kaiC* belongs to RecA-like recombinases, *sasA* and *cikA* to two-component sensory transduction histidine kinases, and *ldpA* to Fe-S-cluster-containing ferredoxins. The question is: what makes these genes function as a circadian mechanism in cyanobacteria? One of the apparent factors is their distinctive domain architecture. Indeed, *kaiC* differs from its homologs by the presence of the second recombinase C-terminal domain, *sasA* and *cikA* have the KaiB-like and GAF domains, respectively, and *ldpA* possesses the unique C-terminal domain; these unique domains seem to have each evolved the circadian function in these genes.

14.3 Evolutionary Constraints and Altered Substitution Rates of the Cyanobacterial Circadian Genes

Circadian genes of cyanobacteria control a large proportion of genes in the genome (Liu et al. 1995; Kucho et al. 2005) and constitute the basis of adaptive reactions to diurnal changes in these organisms (Woelfle et al. 2004; see Chap. 12). Basic evolutionary theory predicts higher selective constraints for genes of fundamental functional importance (Kimura and Ohta 1974). Evolutionary studies of the circadian genes in cyanobacteria corroborate these basic predictions. Indeed, the circadian homologs have a significantly lower level of polymorphism as compared to their non-circadian homologs. For example, among the four subfamilies formed by the *cpmA* gene and its homologs, the subfamily that contains the bona fide *cpmA* from *Synechococcus elongatus* PCC 7942 is the least polymorphic (Dvornyk 2006b). The same patterns were recently determined for the *cikA* homologs as well (Baca et al., unpublished data). Importantly, even when the circadian genes are evolutionarily younger than their homologs from other bacteria (as in the case with *cpmA*), their low polymorphism is not due to their younger evolutionary age but is due to their significantly lower mutation rates (Dvornyk 2006b).

In the course of their evolution, the circadian genes experienced many duplications (Dvornyk and Nevo 2003; Dvornyk et al. 2004). Duplication, which leads to an acquisition of a new function by one of the duplicates, results in an evolutionary rate shift known as functional divergence. If duplicate genes differ by their evolutionary rates, the respective proteins are thought to have type I divergence (Gu 1999). When gene duplication does not cause changes in functional constraints, but instead results in a radical change in amino acid properties between the encoded proteins (e.g., hydrophobicity, charge, etc.), type II divergence is assumed (Gu 2001). The divergence is thought to occur at amino acid sites that are important for the new function of the gene.

Circadian genes manifest both types of the functional divergence. For example, according to the results of the likelihood-ratio tests of the rate shift (Knudsen and Miyamoto 2001; Knudsen et al. 2003), 92 amino acid sites, out of the ~600 that were analyzed, experienced either type of divergence in the KaiB and KaiC proteins of cyanobacteria (Dvornyk and Knudsen 2005). Among these sites, 67 residues manifested a shift towards a significantly lower rate of mutation than the overall average for the whole sequence. The KaiB and KaiC proteins from different cyanobacteria that either contain or lack KaiA also manifested significant functional divergence: about 5% of the sites were determined to have altered functional constraints. These sites may be related to the interaction with KaiA (Dvornyk and Knudsen 2005). As type II divergence is associated with radical amino acid changes (Gu 2001), it seems to have a larger effect on the protein structure and properties. For example, polar and uncharged residue T572 in KaiC of *S. elongatus* PCC 7942 is important for bonding the protein to the C-terminal domain of KaiA (Vakonakis and LiWang 2004); replacing the threonine with a non-polar alanine weakens the interaction and disrupts circadian rhythmicity (Ishiura et al. 1998). In non-circadian KaiC homologs from other prokaryotes, this residue is replaced by either polar and positively charged histidine or other radically different residues (Dvornyk and Knudsen 2005). Likewise, many sites critical for the divergence were identified either within or close to the highly conserved motifs and regions of known or putative functional importance in the other circadian proteins (Dvornyk et al. 2004; Dvornyk 2005; Dvornyk and Knudsen 2005).

14.4 Evolutionary Evidence for Diversification of the Circadian System in Cyanobacteria

From very early in the evolutionary analysis of circadian systems in cyanobacteria, when genomic data was limited, evidence suggested more than one type of circadian system might exist in cyanobacteria. The most obvious piece of evidence was that not all cyanobacteria appeared to possess the *kaiA* gene (Dvornyk et al. 2003), which is one of the three critical genes for clock function in *S. elongatus* PCC 7942 (Ishiura et al. 1998). Due to this difference, the existence of two types of circadian systems was hypothesized. The central gear of one type (the *kaiABC* system) is a cluster of all three *kai* genes, while the other (the *kaiBC* system) lacks the *kaiA* gene and is built solely upon the *kaiBC* operon (Dvornyk and Nevo 2003). The *kaiABC* system has been the subject of research since the discovery of the *kai* genes in the model species *S. elongatus* PCC 7942 and, based on the available genomic data, is probably the most common in cyanobacteria. The *kaiBC* system has been identified thus far only in the unicellular *Prochlorococcus*, where circadian activity has not yet been documented (see Chap. 15). Some ambiguity to this two-tiered classification was introduced by the identification of another potential circadian system in *Synechococcus* sp. WH 8102, as this system appeared to share features of both types. This system features *kaiA*; however the phylogenetic analyses of the *kaiBC* operon (Dvornyk and Knudsen 2005), *sasA* (Dvornyk et al. 2004), and *ldpA* (Dvornyk 2005) of *Synechococcus* sp. WH 8102 positions it more closely to the *kaiBC* system. The growing volume of publicly available genomic data provides further support for the diversification of circadian systems in the Cyanobacteria to three main types and makes it possible to determine the differences between the system types. Currently the three main types are classified according to their composition; the system of *Synechococcus* sp. WH 8102 is referred to as *kaiABC*$^{\Delta}$ (Table 14.2). The most recent results suggest that this system type is characteristic for unicellular cyanobacteria that are closely related to *Prochlorococcus* (Baca et al., unpublished data). The only difference between the *kaiABC*$^{\Delta}$ and the *kaiBC* systems is the presence of the *kaiA* gene.

While the three-tier system for potential circadian systems is the current hypothesis, some data suggest that the diversification of cyanobacterial circadian systems may be even larger. For example, while the *pex* gene is missing in cyanobacteria with the *kaiBC* system, it is also absent in some species with the *kaiABC* system (Table 14.2). Likewise, *cikA*, while being characteristic to the species with the *kaiABC* system, may in some systems either lack the PsR domain, as in heterocystous

Table 14.2 Composition of the three main types of circadian systems in Cyanobacteria. + Present, – missing, +/– may or may not be present, ? distantly related homologs are present, but their circadian function needs to be confirmed

Functional division and gene	System		
	kaiABC (*S. elongatus* PCC 7942)	*kaiABC*$^{\Delta}$ (*Synechococcus* sp. WH 8102)	*kaiBC* (*Prochlorococcus*)
Input			
cikA	+	–	–
ldpA	+	+	+
pex	+/–	+	–
Central oscillator			
kaiA	+	+	–
kaiB	+	+	+
kaiC	+	+	+
Output			
sasA	+	+	+
labA	+	–	–
rpaA	+	+	+
cpmA	+	–/?	–/?

Table 14.3 System-specific polymorphism of some circadian genes (non-synonymous substitutions; reproduced with permission from Dvornyk 2006a)

Gene	System type		Reference
	kaiABC	*kaiBC*	
kaiB	0.080 ± 0.016	0.105 ± 0.020	Dvornyk, unpublished data
kaiC	0.192 ± 0.012	0.097 ± 0.008	Dvornyk, unpublished data
sasA	0.541 ± 0.030	0.249 ± 0.019	Dvornyk et al. (2004)
ldpA	0.474 ± 0.026	0.519 ± 0.033	Dvornyk and Knudsen (2005)
cpmA	0.305 ± 0.024	0.442 ± 0.037	Dvornyk (2006b)

Nostocaceae, or feature several additional domains, as in the thermophilic Yellowstone strains *Synechococcus* sp. JA-2-3B′a(2–13) and *Synechococcus* sp. JA-3-3Ab (Baca et al., unpublished data).

The genetic elements of the different types of circadian systems demonstrate system-specific patterns of their polymorphism. For example, *sasA* is about twice more conserved in the *kaiBC* system than in *kaiABC*, and so is *kaiC* (Table 14.3). This fact further emphasizes different functional constraints of the genes between the systems and, respectively, functional modifications.

Results of these evolutionary studies provide the growing body of evidence that the bona fide *kaiABC* circadian system of *S. elongatus* PCC 7942 may represent only one of many possible types of circadian systems. Based on the current evolutionary data, there are at least two more types of circadian systems with a reduced set of elements (Table 14.2). In addition, each of the main types may have modifications, as described above. These compositional changes of the circadian system likely result in respective alterations of the clock mechanism. The question is whether the systems missing any of the components are still able to function in a circadian manner. The answer is: potentially yes. Numerous studies of the system in *S. elongatus* PCC 7942 have shown that inactivation of some circadian genes that are not part of the central oscillator itself impairs the clock, but does not destroy it completely (Katayama et al. 1999; Schmitz et al. 2000; Katayama et al. 2003). Moreover, a recent study by Tomita et al. (2005) demonstrated that having only the three Kai proteins and ATP is sufficient to reconstitute a temperature-compensated, circadian oscillation of KaiC phosphorylation in vitro. This fact further supports the assumption that lacking some auxiliary components, such as *cikA*, *pex*, or *cpmA* for example, does not cause the circadian system to malfunction.

The ability of the *kaiBC* circadian system to function is more complex. Inactivation of *kaiA* completely abolishes the clock in *S. elongatus* PCC 7942 (Ishiura et al. 1998). However, several studies reported diel patterns of cell cycle and associated gene expression in *Prochlorococcus* and suggested an endogenous clock to control these processes (Shalapyonok et al. 1998; Garczarek et al. 2001; Holtzendorff et al. 2001; Jacquet et al. 2001). Therefore, the functionality of the *kaiBC* system is still under investigation.

14.5 From the Very Beginning Until Now: How the Circadian System Was Built

The study of circadian biology in prokaryotes hardly totals 20 years. Evolutionary studies of the cyanobacterial clock system are even younger. Nevertheless, they have produced results that allow for the construction of a basic framework for evolution of the circadian system in cyanobacteria (Fig. 14.1). The evolutionary history of some circadian genes can be traced back to their predecessors of nearly 3.5 Bya, probably even before the split between Archaea and Bacteria occurred. These genes include *kaiC*, the cornerstone of the circadian system (Dvornyk et al. 2003), *sasA*, *cikA*, and *cpmA* (Dvornyk et al. 2004; Dvornyk 2006b; Baca et al., unpublished data). The other two *kai* genes seem to be younger: *kaiB* is estimated to have originated between 3,500 and 2,320 million years ago (Mya; Dvornyk et al. 2003). The *kaiA* gene initially was thought to emerge about 1,000 Mya (Dvornyk et al. 2003). However, the most recent data from the available completed cyanobacterial genomes show that the unicellular thermophilic strains *Synechococcus* sp. JA-2-3B′a and JA-3-3Ab and *Thermosynechococcus elongatus* BP-1, which is phylogenetically positioned between *G. violaceus* PCC 7421 (no *kai* genes) and *S. elongatus* PCC 7942 (Baca et al., unpublished data), possess *kaiA* and some other elements of the *kaiABC* system, thus suggesting the much earlier time of *kaiA* origin (Dvornyk 2006a).

At its beginning, the ancestral circadian system probably consisted of a minimal set of elements, as compared to the modern ones. The central oscillator was then composed of *kaiB* and *kaiC* not joined in an operon. Disjoined copies of these genes still exist in genomes of some extant cyanobacteria (e.g., *Synechocystis* sp. PCC 6803) and are significantly diverged from their bona fide counterparts (Dvornyk et al. 2003). The minimum time of the *kaiBC* operon origin in cyanobacteria was estimated at about 2,300 Mya. The *kaiBC* operon was then laterally transferred to the other prokaryotes sometime after (Dvornyk et al. 2003). Interestingly, not even a single occurrence of *kaiA* in the prokaryotes outside of cyanobacteria has been documented. This fact may have two possible explanations: the lateral transfer of the *kaiBC* operon occurred either before *kaiA* originated in cyanobacteria or before all three *kai* genes joined in the cluster.

It still remains unclear when the system acquired the ability to oscillate and drive cellular activity. Theoretically, this may have become possible upon the emergence of the *kaiA* gene that made available the minimum set of circadian components (Tomita et al. 2005). Alternatively, given the currently existing variability of the system's composition, there is a possibility that the circadian protosystem was able to function even without *kaiA*. However, a definite answer to this question is yet to be found.

Macroevolution of the circadian system in cyanobacteria was governed by a wide variety of factors. Multiple cases of lateral transfer were determined for the *kaiB* and *kaiC* genes and the *kaiBC* operon (Dvornyk et al. 2003; Dvornyk and Nevo 2004). Two kinases, *sasA* and *cikA*, were formed through domain accretion and gene fusion (Dvornyk et al. 2004; Baca et al., unpublished data). Duplication events were common for *kaiB* and *kaiC* (Dvornyk et al. 2003). The analysis of selection for *sasA* and *ldpA* suggested that, in contrast to purifying selection, positive selection likely played a minor role in their evolution (Dvornyk et al. 2004; Dvornyk 2005). Along with the recruitment of genes to the system, their loss in some cyanobacterial taxa has also taken place, as it happened to *kaiA* and *cikA* in the species with the *kaiBC* and *kaiABC*$^{\Delta}$ systems (Dvornyk 2006a).

14.6 Clock Around the Rock: What Can the Circadian System Tell Us About the Geological History of Earth?

Cyanobacteria are among the most ancient organisms on our planet, once being called the "pioneers of the early Earth" (Schopf 1996). Since nearly the very beginning of life they have evolved through dramatic large-scale changes in the geosphere, climate, and other conditions on the Earth. Importantly, cyanobacteria were able to adapt to these changes and survive. The circadian clock, which may be regarded equally as a mechanism and a product of the adaptation (Woelfle et al. 2004), had to acquire respective evolutionary modifications to withstand these enormous environmental changes. Some of the clock genes in cyanobacteria are evolutionarily old (Dvornyk et al. 2003) and, therefore, each of them (and the circadian system as a whole) may potentially carry information about major events that occurred in the geological and life history of our planet. In their recent study, Battistuzzi et al. (2004) provided time estimates for key steps in the evolution of prokaryotes, such as the origins of methanogenesis, anaerobic methanotrophy, and phototrophy. When compared against this timeline, the emergence of *kaiB* and double-domain *kaiC* about 3,000 Mya (Dvornyk et al. 2003) seems to be congruent with the appearance of anoxygenic photosynthesis. The upper time limit of approximately 2,320–2,500 Mya for the origin of the *sasA* gene (Dvornyk et al. 2004) and the *kaiBC* operon (Dvornyk et al. 2003) corresponds to the time when oxygenic photosynthesis evolved. In turn, this event is associated with the rapid increase of atmospheric oxygen known as the Great Oxidation Event (Rye and Holland 1998; Bekker et al. 2004).

The above key events in evolution of the circadian system seem to be related to yet another factor, the level of solar UV radiation in atmosphere. As proposed by Garcia-Pichel (1998), three main stages existed in the evolution of UV levels in atmosphere and its impact on cyanobacteria. The first stage was characterized by high levels of the most harmful UVC and UVB and lasted until about 2,500 Mya. During the second stage, the UVR level further increased due to the influx of the previously insignificant UVA portion of the spectrum. The first oxygenic cyanobacteria are thought to appear at this time. The effect of UVR began to reduce during the third period, as oxygen produced by cyanobacteria began forming the ozone shield. Relative to the Garcia-Pichel hypothesis, the origin of *kaiB*, double-domain *kaiC*, the *kaiBC* operon, and *sasA* may correlate to the first and second stages of UV levels in the atmosphere when, in addition to high fluxes of the most harmful UVC and UVB, the UVA began to rise.

The above correlations are in support of the hypothesis that the keystone advances in the evolution of the prokaryotic circadian system were induced by global environmental and geological changes. Most likely, these evolutionary changes were adaptive. Further studies may reveal an intrinsic link between the changes at molecular and global geological levels.

The primary function of the circadian system is to control expression of downstream genes, and therefore behavior, in anticipation of the daily light/dark changes. The current day length is about 24 h, but day length has changed drastically during the Earth's history. The causes of these changes are in part due to the tidal friction between the ocean–atmosphere system and the surface of Earth and the torque within the Earth–Moon system (Zahnle and Walker 1987; Krasinsky 1999). As a result, during the geological history of our planet, the rate of the Earth's rotation gradually decreased, while the day length increased correspondingly. A recent theoretical study suggested that 1.9 Bya, the Earth's rotation period was approximately only 4 h (Krasinsky 2002). Other studies of the fossil record and data from more recent geological epochs estimated an Earth-day to be about 20 h at 400 Mya (Wells 1963), approximately 21 h at 600 Mya (Zahnle and Walker 1987) and approximately 18 h at 900 Mya (Sonett et al. 1996). Serving as a main mechanism for adaptation, the circadian system increases Darwinian fitness through matching the endogenous clock with environmental light/dark cycle (Ouyang et al. 1998). Therefore, the circadian system in cyanobacteria should have acquired respective evolutionary modifications to maintain its adaptive potential in accordance with the day length increase. These modifications might include functionally important nucleotide substitutions in the existing clock genes, as well as the emergence of new circadian genes through the various evolutionary mechanisms or co-option of other, non-circadian, genes to perform circadian function. Hypothetically, if Krasinsky's estimate of the Proterozoic day length and modeling of its evolution over the Earth's history (Krasinsky 2002) are accurate, then during "the age of cyanobacteria" about 2.5 Bya (Schopf 1996), the primary evolutionary mechanism by which these organisms might have developed a circadian system was adaptation to the diurnal pattern of about 4 h. This may also implicate a common origin and evolutionary history of the systems controlling the circadian clock and the cell division cycle. Indeed, a recent study identified the *cikA* gene as an element shared by both circadian and cell division systems (Miyagishima et al. 2005). Circadian control of cell division is also a well established fact (Sweeney and Borgese 1989; Mori et al. 1996; Mori and Johnson 2000). Further support of this assumption comes from data on the association of cell cycle phases with the circadian period of various eukaryotes (Bjarnason et al. 2001; Bolige et al. 2005; Hirayama et al. 2005). However, the independence of the circadian clock from cell division cycle in cyanobacteria (Mori and Johnson 2001) suggests that at some point evolutionary tracks of these processes became diverged.

Cyanobacteria and their various circadian systems provide a unique possibility to link day length changes in the Earth history to evolution of the circadian clock. These organisms have their existence traced from the earliest Earth, dating back to at least 2.8 Bya and possibly earlier (Schopf 2000). Fossils from Precambrian life are scarce and can hardly provide direct data about changes of day length over geological time, as shelled animals that can fossilize and serve as a source of such data had not yet evolved during that era. Precambrian cyanobacteria did not display evident circadian-like growth changes in their morphology and thus cannot be utilized in this capacity to reconstruct evolution of day length. However, cyanobacteria are the only survivors of Earth's earliest biota and the only prokaryotes that existed during this time in which circadian rhythmicity has been experimentally confirmed in extant organisms. During their evolution, cyanobacteria successfully passed through a huge range of large-scale environmental changes and day lengths. Genomes of these prokaryotes likely keep records about these changes. Can they be read? Future evolutionary studies of the cyanobacterial circadian system will examine this and many other questions.

14.7 Forward in the Past: What Next?

The evolutionary history of the cyanobacterial circadian system has just begun to be uncovered and is far from its completion. It will be updated and corrected as new elements of the clock are identified. The rapidly growing volume of genomic data is a valuable source of material for a comparative evolutionary analysis of the bona fide circadian genes and their apparent homologs, especially as more cyanobacterial genomes are sequenced. At present, the current data are insufficient to conduct a comprehensive analysis because the databases contain fully annotated sequences of cyanobacteria from only nine genera. While 13 more sequencing efforts are currently in progress, the most recent taxonomical surveys of the Cyanobacteria list from 57 (Castenholz 2001) to 246 (Hoffmann et al. 2004) different genera, and this list itself is likely far from complete. Given that three putative types of circadian systems have been hypothesized from this limited available data set, it is possible that several more divergent circadian systems may exist in cyanobacteria.

Another issue is a controversy in the estimates for the lower time limit for the origin of cyanobacteria. The earliest estimates, based mainly on the fossil record, date their appearance back to as far as 3.3–3.5 Bya (Schopf and Packer 1987; Schopf 1993). The analysis of genomic data provides the value of ~2.6 Bya, with upper and lower bounds of about 2.3 Bya and 3.0 Bya, respectively (Battistuzzi et al. 2004). The most recent estimates from the combined paleobiological, paleogeochemical, and molecular data suggest the time for the origin of heterocystous cyanobacteria is between 2.45 Bya and 2.10 Bya (Tomitani et al. 2006), which assumes an even earlier time for the emergence of unicellular cyanobacteria. These issues make it difficult to reconstruct both phylogenetic patterns and timeline of the circadian system evolution with confidence. Therefore, the proposed scenario and dating of the events (Fig. 14.1) may be biased and should be interpreted with a reasonable caution.

Despite the existing problems, evolutionary studies of the circadian system in cyanobacteria provide important data about the factors that played a primary role

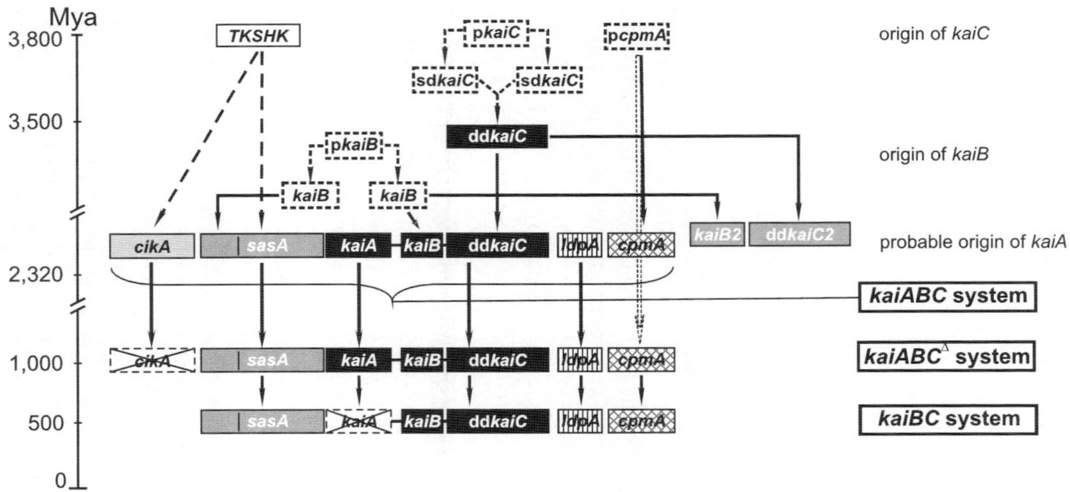

Fig. 14.1 The evolutionary reconstruction of the circadian clock system in Cyanobacteria, based on data currently available. The timescale is not proportional. Extinct genes are shown as *dashed boxes*. Lost genes are *crossed out*. *kaiB2* and *kaiC2* represent the genes not joined in an operon. p*kaiC*, predecessor of the *kaiC* gene; sd*kaiC*, single-domain *kaiC*; dd*kaiC*, double-domain *kaiC*; p*kaiB*, predecessor of the *kaiB* gene; p*cpmA*, predecessor of the *cpmA* gene; TCSHK, two-component sensory transduction histidine kinase (modified with permission from Dvornyk 2006a)

in assembling the circadian system and adjusting it to the global environmental changes. From this perspective, studying evolution of the circadian clock on a microscale has immense potential and may shed light on its immediate adaptive value under natural conditions (Dvornyk et al. 2002). For example, how do the various components of the systems respond to the complex stress? What is relative adaptive significance of each component under specific conditions? How do the scattered *kai* genes behave under stress? What is the level of polymorphism in the genes controlling input and output pathways of each system? Comparative analysis of the micro- and macroevolutionary patterns of the circadian clock will help to portrait the intriguing and exciting history of one of the finest control devices ever assembled by nature.

References

Aravind L, Ponting CP (1997) The GAF domain: an evolutionary link between diverse phototransducing proteins. Trends Biochem Sci 22:458–459

Battistuzzi FU, Feijao A, Hedges SB (2004) A genomic timescale of prokaryote evolution: insights into the origin of methanogenesis, phototrophy, and the colonization of land. BMC Evol Biol 4:44

Bekker A, Holland HD, Wang PL, Rumble D III, Stein HJ, Hannah JL, Coetzee LL, Beukes NJ (2004) Dating the rise of atmospheric oxygen. Nature 427:117–120

Bjarnason GA, Jordan RC, Wood PA, Li Q, Lincoln DW, Sothern RB, Hrushesky WJ, Ben-David Y (2001) Circadian expression of clock genes in human oral mucosa and skin: association with specific cell-cycle phases. Am J Pathol 158:1793–1801

Bolige A, Hagiwara SY, Zhang Y, Goto K (2005) Circadian G2 arrest as related to circadian gating of cell population growth in *Euglena*. Plant Cell Physiol 46:931–936

Castenholz RW (2001) Phylum BX. Cyanobacteria. Oxygenic photosynthetic bacteria. In: Garrity G, Boone DR, Castenholz RW (eds) Bergey's manual of systematic bacteriology, 2nd edn. Springer, New York

DiRuggiero J, Brown JR, Bogert AP, Robb FT (1999) DNA repair systems in archaea: mementos from the last universal common ancestor? J Mol Evol 49:474–484

Dunlap JC, Loros JJ, DeCoursey PJ (2004) Chronobiology: biological timekeeping. Sinauer, Sunderland, Mass

Dvornyk V (2005) Molecular evolution of *ldpA*, a gene mediating circadian input signal in cyanobacteria. J Mol Evol 60:105–112

Dvornyk V (2006a) Evolution of the circadian clock mechanism in prokaryotes. Isr J Ecol Evol 52:343–357

Dvornyk V (2006b) Subfamilies of *cpmA*, a gene involved in circadian output, have different evolutionary histories in cyanobacteria. Microbiology 152:75–84

Dvornyk V, Knudsen B (2005) Functional divergence of circadian clock proteins in prokaryotes. Genetica 124:247–254

Dvornyk V, Nevo E (2003) Genetic polymorphism of cyanobacteria under permanent natural stress: a lesson from the "Evolution Canyons". Res Microbiol 154:79–84

Dvornyk V, Nevo E (2004) Evidence for multiple lateral transfers of the circadian clock cluster in filamentous heterocystic cyanobacteria *Nostocaceae*. J Mol Evol 58:341–347

Dvornyk V, Vinogradova ON, Nevo E (2002) Long-term microclimatic stress causes rapid adaptive radiation of *kaiABC* clock gene family in a cyanobacterium, *Nostoc linckia*, from the "Evolution Canyons" I and II, Israel. Proc Natl Acad Sci USA 99:2082–2087

Dvornyk V, Vinogradova ON, Nevo E (2003) Origin and evolution of circadian clock genes in prokaryotes. Proc Natl Acad Sci USA 100:2495–2500

Dvornyk V, Deng HW, Nevo E (2004) Structure and molecular phylogeny of *sasA* genes in cyanobacteria: insights into evolution of the prokaryotic circadian system. Mol Biol Evol 21:1468–1476

Garcia-Pichel F (1998) Solar ultraviolet and the evolutionary history of cyanobacteria. Orig Life Evol Biosph 28:321–347

Garczarek L, Partensky F, Irlbacher H, Holtzendorff J, Babin M, Mary I, Thomas JC, Hess WR (2001) Differential expression of antenna and core genes in *Prochlorococcus* PCC 9511 (*Oxyphotobacteria*) grown under a modulated light-dark cycle. Environ Microbiol 3: 168–175

Gu X (1999) Statistical methods for testing functional divergence after gene duplication. Mol Biol Evol 16:1664–1674

Gu X (2001) Maximum-likelihood approach for gene family evolution under functional divergence. Mol Biol Evol 18:453–464

Hirayama J, Cardone L, Doi M, Sassone-Corsi P (2005) Common pathways in circadian and cell cycle clocks: light-dependent activation of Fos/AP-1 in zebrafish controls CRY-1a and WEE-1. Proc Natl Acad Sci USA 102:10194–10199

Hoffmann L, Kombrek J, Kastovski J (2004) System of Cyanoprokaryotes (Cyanobacteria) – state in 2004. Arch Hydrobiol Algolog Stud 117:95–115

Holtzendorff J, Partensky F, Jacquet S, Bruyant F, Marie D, Garczarek L, Mary I, Vaulot D, Hess WR (2001) Diel expression of cell cycle-related genes in synchronized cultures of *Prochlorococcus* sp. strain PCC 9511. J Bacteriol 183:915–920

Huang TC, Tu J, Chow TJ, Chen TH (1990) Circadian rhythm of the prokaryote *Synechococcus* sp. RF-1. Plant Physiol 92:531–533

Ishiura M, Kutsuna S, Aoki S, Iwasaki H, Andersson CR, Tanabe A, Golden SS, Johnson CH, Kondo T (1998) Expression of a gene cluster *kaiABC* as a circadian feedback process in cyanobacteria. Science 281:1519–1523

Ivleva NB, Bramlett MR, Lindahl PA, Golden SS (2005) LdpA: a component of the circadian clock senses redox state of the cell. EMBO J 24:1202–1210

Ivleva NB, Gao T, LiWang AC, Golden SS (2006) Quinone sensing by the circadian input kinase of the cyanobacterial circadian clock. Proc Natl Acad Sci USA 103:17468–17473

Iwasaki H, Williams SB, Kitayama Y, Ishiura M, Golden SS, Kondo T (2000) A *kaiC*-interacting sensory histidine kinase, SasA, necessary to sustain robust circadian oscillation in cyanobacteria. Cell 101:223–233

Jacquet S, Partensky F, Marie D, Casotti R, Vaulot D (2001) Cell cycle regulation by light in *Prochlorococcus* strains. Appl Environ Microbiol 67:782–790

Johnson CH (2005) Testing the adaptive value of circadian systems. Methods Enzymol 393:818–837

Jurgens U, Schneider S (1991) Cell wall and sheath constituents of the cyanobacterium *Gloeobacter violaceus*. Arch Microbiol 155:312–318

Katayama M, Tsinoremas NF, Kondo T, Golden SS (1999) *cpmA*, a gene involved in an output pathway of the cyanobacterial circadian system. J Bacteriol 181:3516–3524

Katayama M, Kondo T, Xiong J, Golden SS (2003) *ldpA* encodes an iron–sulfur protein involved in light-dependent modulation of the circadian period in the cyanobacterium *Synechococcus elongatus* PCC 7942. J Bacteriol 185:1415–1422

Kimura M, Ohta T (1974) On some principles governing molecular evolution. Proc Natl Acad Sci USA 71:2848–2852

Knudsen B, Miyamoto MM (2001) A likelihood ratio test for evolutionary rate shifts and functional divergence among proteins. Proc Natl Acad Sci USA 98:14512–14517

Knudsen B, Miyamoto MM, Laipis PJ, Silverman DN (2003) Using evolutionary rates to investigate protein functional divergence and conservation: a case study of the carbonic anhydrases. Genetics 164:1261–1269

Kondo T, Tsinoremas NF, Golden SS, Johnson CH, Kutsuna S, Ishiura M (1994) Circadian clock mutants of cyanobacteria. Science 266:1233–1236

Krasinsky GA (1999) Tidal effects in the earth–moon system and the earth's rotation. Celest Mech Dynam Astron 75:39–66

Krasinsky GA (2002) Dynamical history of the earth-moon system. Celest Mech Dynam Astron 84:27–55

Kucho K, Okamoto K, Tsuchiya Y, Nomura S, Nango M, Kanehisa M, Ishiura M (2005) Global analysis of circadian expression in the cyanobacterium *Synechocystis* sp. strain PCC 6803. J Bacteriol 187:2190–2199

Kutsuna S, Kondo T, Aoki S, Ishiura M (1998) A period-extender gene, *pex*, that extends the period of the circadian clock in the cyanobacterium *Synechococcus* sp. strain PCC 7942. J Bacteriol 180:2167–2174

Leipe DD, Aravind L, Grishin NV, Koonin EV (2000) The bacterial replicative helicase DnaB evolved from a RecA duplication. Genome Res 10:5–16

Lin Z, Kong H, Nei M, Ma H (2006) Origins and evolution of the *recA/RAD51* gene family: Evidence for ancient gene duplication and endosymbiotic gene transfer. Proc Natl Acad Sci USA 103:10328–10333

Liu Y, Tsinoremas NF, Johnson CH, Lebedeva NV, Golden SS, Ishiura M, Kondo T (1995) Circadian orchestration of gene expression in cyanobacteria. Genes Dev 9:1469–1478

Marchler-Bauer A, Anderson JB, Derbyshire MK, Weese-Scott C, Gonzales NR, Gwadz M, Hao L, He S, Hurwitz DI, Jackson JD, Ke Z, Krylov D, Lanczycki CJ, Liebert CA, Liu C, Lu F, Lu S, Marchler GH, Mullokandov M, Song JS, Thanki N, Yamashita RA, Yin JJ, Zhang D, Bryant SH (2007) CDD: a conserved domain database for interactive domain family analysis. Nucleic Acids Res 35:D237–D240

Meyer E, Leonard NJ, Bhat B, Stubbe J, Smith JM (1992) Purification and characterization of the *purE*, *purK*, and *purC* gene products: identification of a previously unrecognized energy requirement in the purine biosynthetic pathway. Biochemistry 31:5022–5032

Min H, Guo H, Xiong J (2005) Rhythmic gene expression in a purple photosynthetic bacterium, *Rhodobacter sphaeroides*. FEBS Lett 579:808–812

Miyagishima SY, Wolk CP, Osteryoung KW (2005) Identification of cyanobacterial cell division genes by comparative and mutational analyses. Mol Microbiol 56:126–143

Mori T, Johnson CH (2000) Circadian control of cell division in unicellular organisms. Prog Cell Cycle Res 4:185–192

Mori T, Johnson CH (2001) Independence of circadian timing from cell division in cyanobacteria. J Bacteriol 183:2439–2444

Mori T, Binder B, Johnson CH (1996) Circadian gating of cell division in cyanobacteria growing with average doubling times of less than 24 hours. Proc Natl Acad Sci USA 93:10183–10188

Mutsuda M, Michel KP, Zhang X, Montgomery BL, Golden SS (2003) Biochemical properties of CikA, an unusual phytochrome-like histidine protein kinase that resets the circadian clock in *Synechococcus elongatus* PCC 7942. J Biol Chem 278:19102–19110

Nair U, Ditty JL, Min H, Golden SS (2002) Roles for sigma factors in global circadian regulation of the cyanobacterial genome. J Bacteriol 184:3530–3538

Nakamura Y, Kaneko T, Sato S, Mimuro M, Miyashita H, Tsuchiya T, Sasamoto S, Watanabe A, Kawashima K, Kishida Y, Kiyokawa C, Kohara M, Matsumoto M, Matsuno A, Nakazaki N, Shimpo S, Takeuchi C, Yamada M, Tabata S (2003) Complete genome structure of *Gloeobacter violaceus* PCC 7421, a cyanobacterium that lacks thylakoids. DNA Res 10:137–145

Ouyang Y, Andersson CR, Kondo T, Golden SS, Johnson CH (1998) Resonating circadian clocks enhance fitness in cyanobacteria. Proc Natl Acad Sci USA 95:8660–8664

Pattanayek R, Williams DR, Pattanayek S, Xu Y, Mori T, Johnson CH, Stewart PL, Egli M (2006) Analysis of KaiA–KaiC protein interactions in the cyano-bacterial circadian clock using hybrid structural methods. EMBO J 25:2017–2028

Roca AI, Cox MM (1990) The RecA protein: structure and function. Crit Rev Biochem Mol Biol 25:415–456

Rossmann R, Sawers G, Bock A (1991) Mechanism of regulation of the formate-hydrogenlyase pathway by oxygen, nitrate, and pH – definition of the formate regulon. Mol Microbiol 5:2807–2814

Rye R, Holland HD (1998) Paleosols and the evolution of atmospheric oxygen: a critical review. Am J Sci 298:621–672

Sauter M, Bohm R, Bock A (1992) Mutational analysis of the operon (*hyc*) determining hydrogenase-3 formation in *Escherichia coli*. Mol Microbiol 6:1523–1532

Schmitz O, Katayama M, Williams SB, Kondo T, Golden SS (2000) CikA, a bacteriophytochrome that resets the cyanobacterial circadian clock. Science 289:765–768

Schopf JW (1993) Microfossils of the Early Archean Apex chert: new evidence of the antiquity of life. Science 260:640–646

Schopf JW (1996) Cyanobacteria: Pioneers of the early Earth. Nowa Hedwigia 112:13–32

Schopf JW (2000) The paleobiologic record of cyanobacterial evolution. In: Brun YV, Shimkets LJ (eds) Prokaryotic development. American Society for Microbiology, Washington, D.C., pp 105–129

Schopf JW, Packer BM (1987) Early Archean (3.3-billion to 3.5-billion-year-old) microfossils from Warrawoona Group, Australia. Science 237:70–73

Shalapyonok A, Olson RJ, Shalapyonok LS (1998) Ultradian growth in *Prochlorococcus* spp. Appl Environ Microbiol 64:1066–1069

Sonett CP, Kvale EP, Zakharian A, Chan MA, Demko TM (1996) Late Proterozoic and Paleozoic tides, retreat of the Moon, and rotation of the Earth. Science 273:100–104

Sweeney BM, Borgese BM (1989) A circadian rhythm in cell division in a prokaryote, the cyanobacterium *Synechococcus* WH7803. J Phycol 25:183–186

Takai N, Nakajima M, Oyama T, Kito R, Sugita C, Sugita M, Kondo T, Iwasaki H (2006) A KaiC-associating SasA-RpaA two-component regulatory system as a major circadian timing mediator in cyanobacteria. Proc Natl Acad Sci USA 103:12109–12114

Taniguchi Y, Katayama M, Ito R, Takai N, Kondo T, Oyama T (2007) *labA*: a novel gene required for negative feedback regulation of the cyanobacterial circadian clock protein KaiC. Genes Dev 21:60–70

Tomita J, Nakajima M, Kondo T, Iwasaki H (2005) No transcription–translation feedback in circadian rhythm of KaiC phosphorylation. Science 307:251–254

Tomitani A, Knoll AH, Cavanaugh CM, Ohno T (2006) The evolutionary diversification of cyanobacteria: molecular-phylogenetic and paleontological perspectives. Proc Natl Acad Sci USA 103:5442–5447

Tsinoremas NF, Ishiura M, Kondo T, Tanaka K, Takahashi H, Johnson CH, Golden SS (1996) A sigma factor that modifies the circadian expression of a subset of genes in cyanobacteria. EMBO J 15:2488–2495

Vakonakis I, LiWang AC (2004) Structure of the C-terminal domain of the clock protein KaiA in complex with a KaiC-derived peptide: implications for KaiC regulation. Proc Natl Acad Sci USA 101:10925–10930

Watanabe W, Sampei G, Aiba A, Mizobuchi K (1989) Identification and sequence analysis of *Escherichia coli* purE and purK genes encoding 5'-phosphoribosyl-5-amino-4-imidazole carboxylase for de novo purine biosynthesis. J Bacteriol 171:198–204

Wells JW (1963) Coral growth and geochronometry. Nature 197:948–950

Woelfle MA, Ouyang Y, Phanvijhitsiri K, Johnson CH (2004) The adaptive value of circadian clocks: an experimental assessment in cyanobacteria. Curr Biol 14:1481–1486

Zahnle K, Walker JC (1987) A constant daylength during the Precambrian era? Precambrian Res 37:95–105

Chapter 15
Circadian Clocks of *Synechocystis* sp. Strain PCC 6803, *Thermosynechococcus elongatus*, *Prochlorococcus* spp., *Trichodesmium* spp. and Other Species

Setsuyuki Aoki and Kiyoshi Onai

Abstract The cyanobacterium *Synechococcus elongatus* PCC 7942 has been established as the model system for studying the molecular mechanisms of the circadian clock in cyanobacteria. This chapter mainly focuses on other cyanobacteria, such as *Synechocystis* sp. strain PCC 6803, *Thermosynechococcus elongatus* and the genera *Trichodesmium* and *Prochlorococcus*. Here, we describe the research background, current status, possible problems and perspectives for studying circadian rhythms for each species/group and we summarize the related works of other cyanobacteria and plastids.

15.1 Introduction

Understanding the mechanisms of the cyanobacterial circadian clock has been greatly advanced by using *Synechococcus elongatus* sp. strain PCC 7942 (hereafter called *Synechococcus*). While the tractability of gene manipulation in this species prompted its use in many ways, the introduction of a luciferase reporter to monitor circadian gene transcription was a critical step in the advancement of discerning the circadian properties of *Synechococcus*. Importantly, this reporter enabled the isolation of many mutants with aberrant phenotypes in circadian gene expression (Kondo et al. 1994), which led to the identification of the cyanobacterial oscillator genes *kaiA*, *kaiB* and *kaiC* (Ishiura et al. 1998). These genes encode the *Synechococcus* oscillator; deletion of any or all of the *kai* genes results in a nonfunctional clock (Ishiura et al. 1998). A recent outcome of this line of study is the generation of a circadian oscillation in the phosphorylation state of KaiC in vitro by incubating the protein products of the three *kai* genes with ATP (Nakajima et al. 2005; see

S. Aoki(✉)
Graduate School of Information Science, Nagoya University, Furo-cho, Chikusa-ku, Nagoya 464-8601, Japan, e-mail: aoki@is.nagoya-u.ac.jp

K. Onai(✉)
Center for Gene Research, Nagoya University, Furo-cho, Chikusa-ku, Nagoya 464-8602, Japan, e-mail: onai@gene.nagoya-u.ac.jp

J.L. Ditty et al. (eds.), *Bacterial Circadian Programs*.

Chap. 5). Other clock-related genes, such as *pex*, *cpmA*, *sasA*, *cikA*, *ldpA*, *rpaA* and *labA*, have also been identified using real-time reporter technology; and their roles in the clock system have been analyzed (see Chaps. 8, 9; Kutsuna et al. 1998; Katayama et al. 1999; Iwasaki et al. 2000; Schmitz et al. 2000; Katayama et al. 2003; Takai et al. 2006; Taniguchi et al. 2007).

In parallel with the studies conducted on *Synechococcus*, the study of circadian clocks in other species of cyanobacteria has been attempted. Application of luciferase reporters to monitor gene expression rhythms has been published for three additional species: *Synechocystis* sp. strain PCC 6803 (Aoki et al. 1995, 1997), *Thermosynechococcus elongatus* (Onai et al. 2004b) and *Leptolyngbya boryana* (formerly *Plectonema boryanum*; Terauchi et al. 2005) which, based on their unique physiological characteristics, have experimental advantages that are not inherent to *Synechococcus*. Additional species, in which molecular genetics is not applicable, have been used to study clock-controlled growth and metabolism. The genera *Trichodesmium* and *Prochlorococcus* show physiologically unique diurnal rhythms of nitrogen fixation activity and cell cycle progress, respectively (Vaulot et al. 1995; Berman-Frank et al. 2001). These two groups of cyanobacteria are ecologically important because they are hugely abundant in the sea (Capone et al. 1997; Partensky et al. 1999). In this chapter, we describe circadian rhythm research on cyanobacteria other than *Synechococcus*, concentrating on *Synechocystis* sp. strain PCC 6803, *Thermosynechococcus elongatus*, *Trichodesmium* spp. and *Prochlorococcus* spp., as well as some other species of cyanobacteria and plastids. Finally, we discuss perspectives of studying multiple circadian systems from different standpoints.

15.2 *Synechocystis* sp. Strain PCC 6803

Synechocystis sp. strain PCC 6803 (hereafter called *Synechocystis*) is a unicellular cyanobacterium that inhabits fresh water. This species has many characteristics that allow for it to be widely used as the cyanobacterium of choice for many fields of research. Its genome was the first completed sequencing project in a photosynthetic organism (Kaneko et al. 1996a, b). A few years later, the *Synechocystis* genomic sequence was made available in a highly useful format on the website CyanoBase (http://www.kazusa.or.jp/cyano/cyano.html; Nakamura et al. 1998, 1999, 2000). Additionally, gene manipulation methods are well established in *Synechocystis*, and in particular, gene-targeting techniques based on homologous recombination are applicable to this species (Williams 1988).

An additional advantage that *Synechocystis* has over *Synechococcus* is that *Synechocystis* grows photoheterotrophically, without the need for a functional photosystem (PS) II, by using glucose as a source of reduced organic carbon (Anderson and McIntosh 1991). Anderson and McIntosh (1991) found that *Synechocystis* grows heterotrophically on glucose while maintained in the dark if cells are subjected to a 5-min light pulse every day. They named this mode of growth "light-activated heterotrophic growth (LAHG)", and demonstrated that *Synechocystis* can grow on glucose in these conditions for more than one week (Anderson and McIntosh 1991).

This growth property is advantageous to circadian researchers to study the photic input pathways of the clock, where the effect of light on features of a circadian rhythm such as phase, period, amplitude and sustainability is measured (Ninnemann 1979; Johnson 1994). To carry out such experiments in a straightforward way, it would be advantageous to observe a circadian rhythm in the dark and expose the cells to light pulses. *Synechococcus* is an obligate photoautotroph and cannot grow without light. In constant darkness (DD), the levels of transcription decrease rapidly to levels near those of background, and bioluminescence cannot be used as an effective indicator of circadian rhythms in this strain (Tomita et al. 2005). The ability of *Synechocystis* to grow heterotrophically in the dark, in addition to its ease of genetic manipulation, provides a potentially excellent system for the study of photic input pathways.

15.2.1 Bioluminescence Rhythms from **Synechocystis** *in Constant Light and Constant Darkness*

A luciferase reporter strain was generated using a glucose-tolerant strain of *Synechocystis* as the host (Aoki 1995). The *dnaK1* gene, which encodes the heat-shock protein (HSP) DnaK, was chosen as the gene to be tested because it is inducible by heat and mRNA accumulation of some eukaryotic HSPs was reported to be under the control of the clock (Cornelius and Rensing 1986; Otto et al. 1988; Rikin 1992). The resulting P*dnaK1::luxAB* reporter strain (CFC2) displayed a rhythm of bioluminescence that satisfied all three criteria of a circadian rhythm, i.e., persistence of oscillation with a period of about one day in constant conditions (in this case, constant light; LL; Fig. 15.1A), temperature compensation of the period length (at ambient temperatures between 25°C and 35°C) and entrainment to daily light/dark (LD) cycles (Aoki 1995).

Bioluminescence rhythms from CFC2 cells persist more stably when grown on agar plates than when grown in liquid culture (S. Aoki, T. Kondo, M. Ishiura, unpublished data); however, *Synechocystis*, whether wild-type cells or CFC2 cells, cannot undergo LAHG on agar plates, unless the organism is pre-adapted to the LAHG conditions, i.e., DD except for a brief light pulse every day, with glucose added in the medium (Anderson and McIntosh 1991). Therefore, wild-type *Synechocystis* cells were pre-adapted to LAHG by gradually increasing the period of the dark phase day by day; and these LAHG-adapted cells were then transformed with the same reporter construct used for generating CFC2 (Aoki et al. 1997). The resulting reporter strain CFC4 grew on agar plates under the LAHG conditions and exhibited circadian rhythms of bioluminescence in DD (Fig. 15.1B; Aoki et al. 1997). The amplitude of this circadian bioluminescence rhythm of CFC4 in DD was considerably lower than that of CFC2 (Fig. 15.1, compare part A to part B; Aoki et al. 1997).

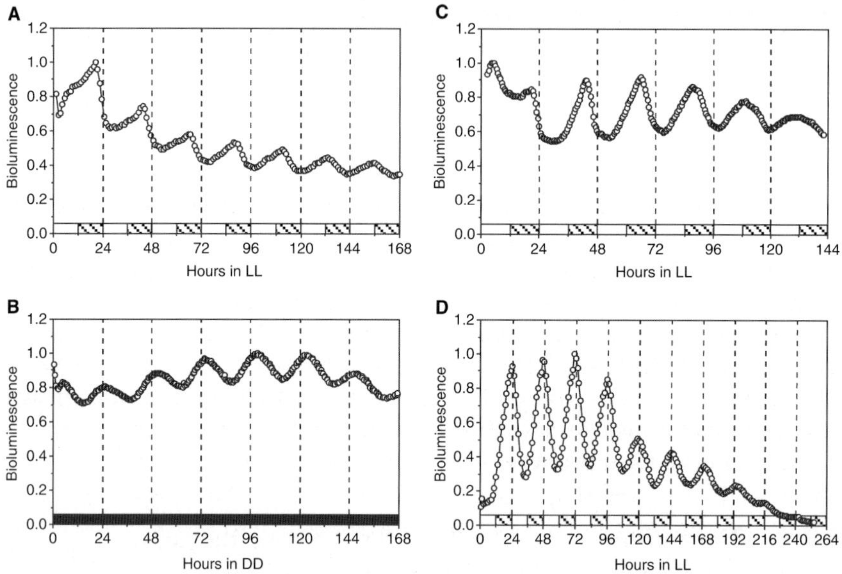

Fig. 15.1 Bioluminescence rhythms from luciferase reporter strains of *Synechocystis* and *Thermosynechococcus elongatus*. **A**, **B**, **C** Bioluminescence rhythms from *Synechocystis* reporter strains CFC2, CFC4 and P*dnaK::luxAB*(+), respectively. **D** Bioluminescence rhythm from *T. elongatus* reporter strain A205. In (**A**), (**C**) and (**D**), cells of CFC2, P*dnaK::luxAB*(+) and A205 were synchronized by a LD12:12 cycle before bioluminescence monitoring in LL. In these graphs, *open boxes* and *hatched boxes* on the horizontal axes represent subjective day and night phases, respectively. In (**B**), cells of CFC4 were entrained to three cycles of daily LAHG light pulses (15 min each), not to a LD12:12 cycle, before being monitored in DD. Therefore, we do not show subjective day and night phases on the *horizontal axis* in (**B**). Light intensities of LL were 46, 67 and 50 μmol m^{-2} s^{-1} for (**A**), (**C**) and (**D**), respectively. Ambient temperatures were 30°C and 41°C for *Synechocystis* and *T. elongatus* strains, respectively. *Vertical axes* show relative bioluminescence levels where the maximal bioluminescence levels are set to one. Panel (**B**) is used with permission from Aoki et al. (1997)

The interval between the last LAHG pulse and the first peak of bioluminescence in DD was always the same (~25.3 h; Aoki et al. 1997), which indicated that LAHG pulses synchronized the *Synechocystis* clock. Unexpectedly, an LAHG pulse suppressed the level of bioluminescence rapidly and irreversibly to background levels, when the daily timing of its application was shifted from those of preceding LAHG pulses to a certain extent (e.g., 6-h advance or delay; S. Aoki, T. Kondo, M. Ishiura, unpublished data). For this reason, it was impossible to estimate the phase response of the clock to a brief bright light pulse, such as that used during LAHG, after cells were entrained by daily cycles of LAHG pulses. By using longer (3 h) dim light pulses, instead of brief bright light pulses, a phase response curve (PRC) to light was obtained for *Synechocystis*; the PRC included relatively small phase shifts like that of a type 1 PRC (Winfree 1990; Aoki et al. 1997). Though long dim light pulses did not suppress bioluminescence levels as the brief bright pulses did, they did reduce the amplitude of the rhythms to varying degrees, which led to difficulties in determining the precise period and/or phase of the rhythms (S. Aoki, T. Kondo, M. Ishiura, unpublished data).

Bioluminescence intensity of CFC2 was much lower (~5%) than that of AMC149, the *psbA1* reporter strain of *Synechococcus* (Aoki et al. 1995). Although the bioluminescence intensity of CFC4 cells was comparable to that of AMC149, the amplitude of its rhythm was very low. For these reasons, neither of the *Synechocystis* reporter strains was used for further studies, such as screening of mutants that display altered rhythms in bioluminescence using a cooled-CCD camera system.

Kucho et al. (2005a) attempted to improve the *Synechocystis* bioluminescent reporter system by generating transgenic strains of *Synechocystis* with newly designed reporter constructs. These constructs carry a *dnaK* promoter fragment that is fused to the *luxA* coding sequence seamlessly, i.e., with no additional sequence between the end of the promoter fragment and the start codon of *luxA*. This new construct differs from that of CFC2 that contained an N-terminal part of the *dnaK* coding sequence immediately upstream of *luxA*, which resulted in the addition of a short peptide extension to the N-terminus of the LuxA protein. This extension may have contributed to the low bioluminescence levels from CFC2. Moreover, the selectable marker *aadA* gene cassette (spectinomycin resistance; Sp^R) was designed to be antiparallel to *luxAB* such that transcription of the Sp^R did not interfere with the access of RNA polymerase to the *dnaK* promoter. The resulting reporter strain P*dnaK::luxAB*(+) exhibited bioluminescence rhythms whose levels were comparable to those of AMC149, and the amplitude was 10% higher than that of CFC2 (Fig. 15.1C; Kucho et al. 2005a). The period length and phase of the rhythms produced by P*dnaK::luxAB*(+) were the same as those of CFC2.

15.2.2 Genome-Wide Search for Clock-Controlled Genes in Synechocystis

Using the *Synechococcus* model system, the circadian clock was shown to globally regulate gene transcription by introducing a promoterless *luxAB* gene throughout the *Synechococcus* genome (Liu et al. 1995). Each of the resulting transgenic colonies that produced bioluminescence did so with a circadian rhythm with a period near 24 h, although the waveform and phase of the expression varied. A modified method, in which the number of tested colonies was scaled down, was performed in *Synechocystis*. Instead of random insertion of *luxAB* into the genome by a single crossover event as was done in *Synechococcus*, *luxAB* fused downstream of various genomic DNA fragments was targeted to a specific neutral site in the *Synechocystis* chromosome by a double crossover event (Aoki et al. 2002). This experimental design was chosen for two reasons: (1) *cis*-regulatory elements responsible for the resulting bioluminescence pattern could be restricted to a certain DNA fragment that was targeted to the neutral site and (2) this DNA fragment could easily be cloned by PCR (Aoki et al. 2002). Each of the tested bioluminescent clones, except those showing rapid damping of bioluminescence, exhibited circadian rhythms (Aoki et al. 2002). Remarkably, the distribution of peak phase of the bioluminescence rhythms differed between the results of these "promoter trap" experiments in *Synechocystis* and those in *Synechococcus*. In *Synechocystis*, a majority of the transformant clones peaked at the late subjective night (47 out of 56 rhythm-expressing clones; Aoki et al. 2002), while in *Synechococcus*, transformant clones displayed bioluminescence rhythms that primarily peaked during the mid- to late subjective day (149 out of 318 clones; Liu et al. 1995). This difference may be due to the intracellular stability of the luciferase protein in each of the two strains; the phase of a given rhythmically transcribed gene is dependent on the half-life of its expression product (So and Rosbash 1997). Consistent with this idea, the average level of bioluminescence from *Synechocystis* clones was approximately 10% of that from *Synechococcus* clones (Aoki et al. 2002).

Kucho et al. (2005b) conducted a genome-wide analysis of clock-controlled gene transcription in *Synechocystis* using a DNA microarray. This microarray contained probes for 3,070 genes, which encompassed 94% of the *Synechocystis* genome (Kucho et al. 2005b). They identified 54 (2%) and 237 (9%) genes that exhibited clock-controlled expression under stringent and relaxed filtering conditions, respectively. These numbers are much lower than those from the results of the promoter trap experiments in which a majority (~80%) of the genes tested showed circadian bioluminescence profiles (Aoki et al. 2002). This apparent discrepancy could be largely due to lower sensitivity and larger experimental errors in the microarray method relative to the promoter trap method (Kucho et al. 2005b) because most genes that displayed circadian expression profiles in bioluminescence did so with very low amplitude (Aoki et al. 2002). Microarray analysis may not have detected these low amplitude rhythms, which may contribute to the underestimation of the number of clock-controlled genes. The difference in the results between the two methods could also be due to the fact that microarray analysis detects the amount of accumulated mRNA whereas the luciferase method detects transcriptional activity (Kucho et al. 2005b). Even when clock-controlled transcription is detected, the amplitude of the resulting rhythm in the amount of the accumulated mRNA could be lower than that of the transcription rhythm depending on the stability of the mRNA. In theory, an mRNA with a longer half-life shows an even lower amplitude rhythm (So and Rosbach 1997). Consistent with this idea, Gutiérrez et al. (2002) reported that clock-controlled genes were overrepresented in the population of unstable transcripts, compared to those expected to distribute in the entire genome of *Arabidopsis thaliana*.

Synechocystis transcripts that were shown to accumulate in a rhythmic fashion encode proteins predicted to be involved in various metabolic processes, membrane transport, transcription, translation and signal transduction (Kucho et al. 2005b), which suggests that the circadian clock influences a variety of cellular functions. Interestingly, the majority of cycling genes, including those involved in respiration, showed peak expression levels at the transition from subjective day to night (Kucho et al. 2005b). These respiration-related genes include those encoding enzymes in the pentose phosphate cycle, components of the respiratory electron transport chain and subunit C of ATP synthase. The *Synechocystis* clock also co-regulates genes involved in the synthesis of poly(3-hydroxyalkanoate) (PHA), which is used as a carbon and energy reserve in the cyanobacterial cell, with peak expression at the end of the subjective day. Collectively, these observations suggest that one of the main roles of the circadian clock of *Synechocystis* is to adjust the physiological state of the cell for the upcoming night environment. Circadian regulation of genes involved in respiration and PHA synthesis would help supply energy and a carbon source at night. In support of this idea, respiratory oxygen uptake peaks during subjective night in another species of cyanobacteria, *Synechococcus* sp. strain Miami (Mitsui et al. 1986). DNA replication and cell division should be candidate processes for which genes involved in respiration and PHA synthesis are co-regulated to supply enough energy and carbon source in *Synechocystis*, though there seems to be divergence in the timing for DNA replication and cell division among cyanobacteria (Mitsui et al. 1986; Vaulot et al. 1995; Mori et al. 1996). It has not yet been examined whether these processes are under the control of the clock in *Synechocystis*.

Genes involved in transcription and translation were expressed in circadian cycling patterns, e.g., genes encoding a sigma factor of RNA polymerase which confers promoter specificity, and a DNA-binding response regulator each produced cyclic profiles. A gene encoding prolyl-tRNA synthase as well as seven genes associated with various steps of translation, such as aminoacyl-tRNA synthesis, elongation and termination of the polypeptide chain, were detected to be cycling under relaxed filtering conditions (Kucho et al. 2005a). Expression of these genes was shown to be temporally co-regulated with peak expression occurring at early subjective day. These genes involved in gene expression may have regulatory functions in the output pathways of the *Synechocystis* circadian system.

15.2.3 kai Genes in Synechocystis

In *Synechococcus*, the *kaiA*, *kaiB* and *kaiC* genes are each present in single copy. *Synechocystis* also has one copy of the *kaiA* gene, but in contrast to that in *Synechococcus*, *kaiB* and *kaiC* are each present in three copies, i.e., *kaiB1*, *kaiB2*, *kaiB3* and *kaiC1*, *kaiC2*, *kaiC3*. These *kai* genes are located on four loci in the genome: *kaiAB1C1*, *kaiC2B2*, *kaiB3* and *kaiC3*. *In vivo* rhythm assay of the *kai* disrupted strains indicated that: (1) the *kaiAB1C1* cluster has critical functions in the clock oscillation, (2) *kaiB3* and *kaiC3* are important in fine-tuning of clock parameters such as period length and phase and (3) the *kaiC2B2* cluster does not have clock functions in the experimental conditions tested (K. Onai, K. Okamoto, M. Morishita, M. Ishiura, unpublished data). Therefore, it is likely that the *kaiAB1C1* gene cluster is an ortholog of the *Synechococcus kaiABC* cluster that encodes the oscillator components.

Phylogenetic analyses show that the gene product of *kaiC1* clusters with KaiC of *Synechococcus*, along with KaiC proteins from *Anabaena* sp. PCC 7120 and *T. elongatus* (Fig. 15.2; see Chap. 14). This group (Group A) containing *Synechocystis* KaiC1 consists of only cyanobacterial sequences and it is next clustered with a smaller group (Group B) that includes the *Synechocystis* KaiC3 sequence. Group B contains KaiC sequences not only from cyanobacteria but also from other groups of bacteria, such as *Chloroflexi* and Proteobacteria. *Synechocystis* KaiC2 belongs to another group (Group C) that is related to Groups A and B and consists of KaiC sequences from Archea, as well as those of some Proteobacteria. Group D with different clusters consists of unduplicated ancestral type sequences (Leipe et al. 2000; Dvornyk et al. 2003). Taken together, KaiC2 may represent an ancient type of KaiC, which probably does not have a clock-related function and might also hold true for similar *kaiC* sequences in Archaea. In this scenario, the clock (and clock-related) functions were probably acquired by the *kaiC1*- and *kaiC3*-type genes at or after the divergence from the *kaiC2*-type genes.

15.2.4 Current Problems and Perspectives

Initially, the ability to grow *Synechocystis* in the dark using the LAHG protocol hinted at the possibility of discovering the components necessary for light input to the circadian oscillator. However, the unexpected light sensitivity of the *Synechocystis* circadian clock system during LAHG hampered further exploitation of this species for the study of light input pathways. The LAHG pulses efficiently supported the growth of *Synechocystis* cells only when they were applied regularly at the same time every day; moreover, cells of *kaiA*-disrupted *Synechocystis* strains could no longer perform LAHG (Y. Obama, K. Fujii, M. Kis, H. Wada, unpublished data). These observations strongly suggest that some circadian regulation is involved in LAHG, and that an LAHG pulse might have a deleterious effect on the growth and/or viability of cells depending on the circadian phase of its application.

The advent of the P*dnaK::luxAB*(+) strain (Kucho et al. 2005a) now allows us to use *Synechocystis* for a large-scale screen for mutants with abnormal patterns of circadian gene expression. Therefore, *Synechocystis* may also serve as a cyanobacterial model system for molecular and genetic analyses of clock-related genes.

15.3 *Thermosynechococcus elongatus*

A unicellular thermophilic cyanobacterium *Thermosynechococcus elongatus* BP-1 was isolated at Beppu hot spring in Japan and grows optimally at around 57°C (Yamaoka et al. 1978). This species branched at a very early stage of evolution of cyanobacteria, based on the 16S rRNA sequence (Honda et al. 1999).

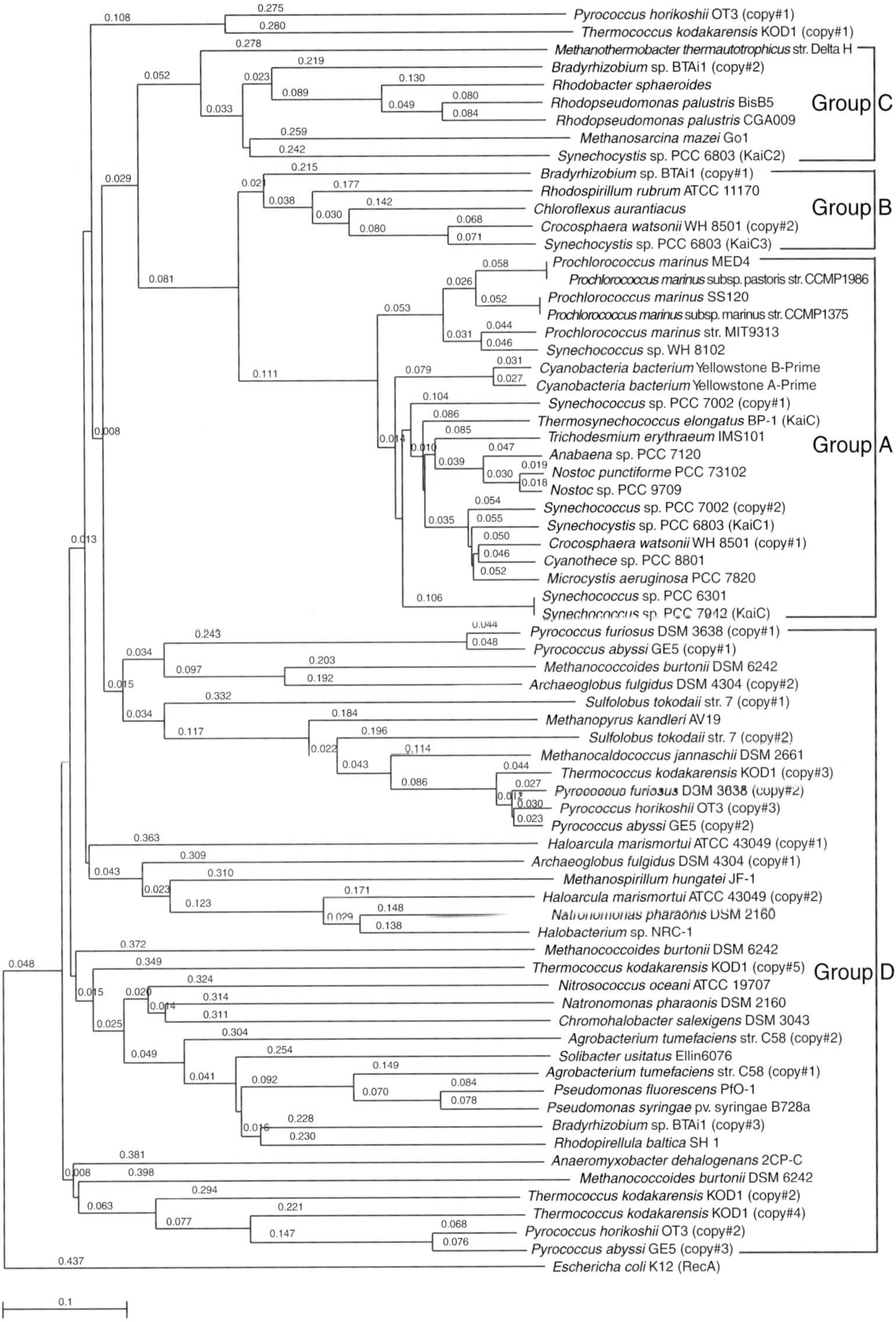
Pyrococcus horikoshii OT3 (copy#1)
Thermococcus kodakarensis KOD1 (copy#1)
Methanothermobacter thermautotrophicus str. Delta H
Bradyrhizobium sp. BTAi1 (copy#2)
Rhodobacter sphaeroides
Rhodopseudomonas palustris BisB5
Rhodopseudomonas palustris CGA009
Methanosarcina mazei Go1
Synechocystis sp. PCC 6803 (KaiC2)
Group C
Bradyrhizobium sp. BTAi1 (copy#1)
Rhodospirillum rubrum ATCC 11170
Chloroflexus aurantiacus
Crocosphaera watsonii WH 8501 (copy#2)
Synechocystis sp. PCC 6803 (KaiC3)
Group B
Prochlorococcus marinus MED4
Prochlorococcus marinus subsp. pastoris str. CCMP1986
Prochlorococcus marinus SS120
Prochlorococcus marinus subsp. marinus str. CCMP1375
Prochlorococcus marinus str. MIT9313
Synechococcus sp. WH 8102
Cyanobacteria bacterium Yellowstone B-Prime
Cyanobacteria bacterium Yellowstone A-Prime
Synechococcus sp. PCC 7002 (copy#1)
Thermosynechococcus elongatus BP-1 (KaiC)
Trichodesmium erythraeum IMS101
Anabaena sp. PCC 7120
Nostoc punctiforme PCC 73102
Nostoc sp. PCC 9709
Synechococcus sp. PCC 7002 (copy#2)
Synechocystis sp. PCC 6803 (KaiC1)
Crocosphaera watsonii WH 8501 (copy#1)
Cyanothece sp. PCC 8801
Microcystis aeruginosa PCC 7820
Synechococcus sp. PCC 6301
Synechococcus sp. PCC 7942 (KaiC)
Group A
Pyrococcus furiosus DSM 3638 (copy#1)
Pyrococcus abyssi GE5 (copy#1)
Methanococcoides burtonii DSM 6242
Archaeoglobus fulgidus DSM 4304 (copy#2)
Sulfolobus tokodaii str. 7 (copy#1)
Methanopyrus kandleri AV19
Sulfolobus tokodaii str. 7 (copy#2)
Methanocaldococcus jannaschii DSM 2661
Thermococcus kodakarensis KOD1 (copy#3)
Pyrococcus furiosus DSM 3638 (copy#2)
Pyrococcus horikoshii OT3 (copy#3)
Pyrococcus abyssi GE5 (copy#2)
Haloarcula marismortui ATCC 43049 (copy#1)
Archaeoglobus fulgidus DSM 4304 (copy#1)
Methanospirillum hungatei JF-1
Haloarcula marismortui ATCC 43049 (copy#2)
Natronomonas pharaonis DSM 2160
Halobacterium sp. NRC-1
Methanococcoides burtonii DSM 6242
Thermococcus kodakarensis KOD1 (copy#5)
Nitrosococcus oceani ATCC 19707
Natronomonas pharaonis DSM 2160
Chromohalobacter salexigens DSM 3043
Agrobacterium tumefaciens str. C58 (copy#2)
Solibacter usitatus Ellin6076
Agrobacterium tumefaciens str. C58 (copy#1)
Pseudomonas fluorescens PfO-1
Pseudomonas syringae pv. syringae B728a
Bradyrhizobium sp. BTAi1 (copy#3)
Rhodopirellula baltica SH 1
Anaeromyxobacter dehalogenans 2CP-C
Methanococcoides burtonii DSM 6242
Thermococcus kodakarensis KOD1 (copy#2)
Thermococcus kodakarensis KOD1 (copy#4)
Pyrococcus horikoshii OT3 (copy#2)
Pyrococcus abyssi GE5 (copy#3)
Group D
Eschericha coli K12 (RecA)
0.1

Benefits that a thermophilic organism provides to the researcher are that its proteins are highly stable, can be easily purified, characterized and crystallized for analysis of their biochemical and biophysical properties. *T. elongatus* has become a new model cyanobacterium for studying the clock. KaiA, KaiB and KaiC proteins derived from this strain have contributed to the studies on the clock molecular machinery at the atomic level (see Chaps. 6, 7). The pot-shaped structure of the KaiC-hexamer was determined by single particle analysis of electron microscopic images and X-ray crystal structure of the KaiA and KaiB proteins were determined (Hayashi et al. 2003; Uzumaki et al. 2004; Iwase et al. 2005). This organism also possesses its own internal biological clock; here, we focus on the genetic and physiological analyses of the circadian system in this thermophilic species.

15.3.1 Transformation of Thermophilic Cyanobacteria

Until very recently, there had been no reports of circadian rhythms from thermophilic cyanobacterial species. The luciferase reporters for circadian gene transcription were expected to be the method of choice for initiating these circadian-based studies. Gene transfer techniques had been accomplished in two thermophilic species, *T. elongatus* and *T. vulcanus*, by electroporation (Mühlenhoff and Chauvat 1996; Sugiura and Inoue 1999; Katoh et al. 2001), or by conjugation with *Escherichia coli* (Mühlenhoff and Chauvat 1996). These procedures had the following disadvantages: (1) transformants could be selected on solid medium only after amplification in liquid medium, (2) selection and maintenance of transformants with selective antibiotics was difficult at optimal bacterial growth temperatures due to a lack of a thermostable selectable marker, (3) electroporation may induce unexpected mutations, (4) the conjugation method required an extra step to eliminate *E. coli* from transformant cultures after conjugation, because the growth of recipient cells is inhibited by *E. coli* cells, (5) the efficiency and frequency of gene transfer was low, and could not been determined precisely (Mühlenhoff and Chauvat 1996; Sugiura and Inoue 1999; Katoh et al. 2001).

These problems were avoided when it was discovered that *T. elongatus* cells are able to take up exogenous DNA, i.e., they are naturally competent (Onai et al. 2004a). With the development of a new selectable marker gene through codon

Fig. 15.2 A phylogenetic tree constructed by sequences similar to KaiC from various prokaryotes. The tree was constructed by the neighbor-joining method using aligned amino acid residues of the sequences. The RecA sequence from *E. coli* strain K12 was used as the outgroup. The numbers on branches represent branch lengths (the number of substitutions every residue). Each sequence is represented by the species name and, if multiple copies of KaiC homolog are present in the species, the serial number of the gene copy in parentheses. As for *Synechococcus*, *Synechocystis* and *T. elongatus* sequences, protein names are also indicated in parentheses

optimization of a kanamycin nucleotidyltransferase gene, transformants could be selected for on agar plates at 52°C (Onai et al. 2004a).

15.3.2 Development of a Reporter System in T. elongatus

An automated bioluminescence real-time monitoring system was established in *T. elongatus* using a thermostable luciferase gene set (Xl *luxAB*) derived from the luminous terrestrial bacterium *Xenorhabdus luminescens* (Onai et al. 2004b; Okamoto et al. 2005). A promoter region of the *psbA1* gene of *T. elongatus* was fused to the Xl *luxAB* gene set and inserted into a specific targeting site in the genome of *T. elongatus*. The resulting reporter strain A205 exhibited circadian rhythms of bioluminescence for >10 days in LL (Fig. 15.1D). The rhythms were reset by an LD cycle (Onai et al. 2004b) or by a temperature cycle (K. Onai, M. Morishita, S. Itoh, M. Ishiura, unpublished data), and the period length of the rhythm remained nearly constant from 30°C to 60°C (Onai et al. 2004b). This temperature range is the widest for which temperature compensation of the period length has been investigated in any organism. These characteristics support the notion that a functional clock could exist in hot springs where the ambient temperature changes dramatically.

15.3.3 Genome-Wide Analysis of Clock-Controlled Genes in T. elongatus

The entire genome sequence of *T. elongatus* (2.6 Mbp) has been determined, and it is estimated to contain 2,524 genes (Nakamura et al. 2002a, b). Using DNA microarrays with unmodified oligonucleotide probes that encompass 95% of the *T. elongatus* genome, experiments were conducted using RNA samples from two different time points (2 h, LL2, early subjective day; 14 h, LL14, early subjective night) into LL, from cultures that had been previously synchronized (Kucho et al. 2004a, b). A total of 143 candidate clock-controlled genes were identified as having significantly different expression levels at LL2 as compared to LL14. The physiological functions of these genes were diverse, consistent with the microarray results in *Synechocystis* (Kucho et al. 2005b) and included predicted roles in metabolism, transcription, translation, membrane transport, DNA replication and repair, cell growth and cell death. Expression of 74 and 69 of these genes were enhanced at LL2 and LL14, respectively. Expression of several genes associated with photosynthesis was enhanced at LL2, i.e., the early subjective day, which suggests that they are timed such that photosynthesis is efficiently supported in the daytime. Expression of many genes involved in energy metabolism, e.g., several respiratory genes, was enhanced at LL14. This finding is consistent with the *Synechocystis* result, in which circadian regulation of respiratory genes also occurs with peak expression in the early subjective night (Kucho et al. 2005b). Of course, genes differentially regulated between the two time points, but not controlled by the clock, may also be included in the 143 genes; moreover, many clock-controlled genes likely were not identified using this protocol and require longer-term analyses with a higher time resolution. Nevertheless, these observations suggest that various genes involved in wide-range cellular physiology and metabolism may be under the control of the clock in *T. elongatus* as well as in *Synechocystis*.

15.3.4 Perspectives

Structural and biochemical studies on clock and clock-related proteins have been effectively conducted in *T. elongatus* (Hayashi et al. 2003, 2004a, b, 2006; Iwase et al. 2004, 2005; Uzumaki et al. 2004; Vakonakis et al. 2004; Pattanayek et al. 2006; Murakami et al. 2008). The establishment of the real-time reporter to examine bioluminescence rhythms adds great advantages to *T. elongatus* for studying the structure–function relationships of the clock and clock-related proteins. *T. elongatus* has a *kaiABC* cluster organized as in *Synechococcus*, and this cluster functions as the *T. elongatus* clock. Loss of the *kaiABC* cluster disrupted the circadian rhythms in *T. elongatus* (K. Onai, M. Morishita, S. Itoh, M. Ishiura, unpublished data) and both *T. elongatus kaiA* and *kaiB* genes functioned as the clock genes in *Synechococcus* cells (Uzumaki et al. 2004; Iwase et al. 2005). *T. elongatus* also has similar sequences to other clock-related genes of *Synechococcus* such as *sasA* and *pex*. Effective analyses that combine structural, biochemical, genetic and physiological methods are now applicable to *T. elongatus*, which makes this species one of the most promising model systems for unraveling the circadian clock mechanisms in cyanobacteria.

15.4 Nitrogen-Fixing, Nonheterocystous Cyanobacteria (*Trichodesmium* spp.)

Certain groups of cyanobacteria (diazotrophic cyanobacteria) fix atmospheric nitrogen when bioavailable forms of nitrogen are limited. An essentially anaerobic enzyme, nitrogenase, catalyzes nitrogen fixation, and this enzyme is irreversibly inhibited *in vitro* when exposed to O_2. Therefore, diazotrophic cyanobacteria must prevent nitrogenase from being damaged by oxygen produced as a by-product of oxygenic photosynthesis. In heterocystous cyanobacteria, such as *Anabaena* spp., a fraction of cells irreversibly differentiates into heterocysts, which lack PSII and do not evolve O_2. In these strains, nitrogen fixation occurs only in heterocysts, while photosynthesis is carried out in vegetative cells. This spatial separation of two seemingly incompatible processes, nitrogen fixation and photosynthesis, protects nitrogenase from O_2 evolved through photosynthesis (Wolk 1996).

Nonheterocystous, nitrogen-fixing cyanobacteria, such as the unicellular *Synechococcus* sp. strain RF-1 or the filamentous *Oscillatoria* sp. strain 23, undergo nitrogen fixation only during the night in daily LD cycles, thereby protecting nitrogenase from oxygen generated by photosynthesis that takes place only during the day. The circadian clock-controlled expression of nitrogenase is the molecular underpinning of this temporal separation of nitrogen fixation from photosynthesis (see Chap. 3).

Trichodesmium spp. are nitrogen-fixing, filamentous, nonheterocystous, marine cyanobacteria. This group is considered to be ecologically important because they are abundant in oligotrophic, tropical and subtropical oceans, and they contribute significantly to the annual input of new nitrogen to the surface waters of these areas (Carpenter 1983; Gallon et al. 1996; Capone et al. 1997; Zehr et al. 1998). Nearly 50 years ago, it was reported that *Trichodesmium* spp. fix nitrogen during the day, when photosynthesis is occurring, without the development of heterocysts (Dugdale et al. 1961); the mechanisms by which these two incompatible processes occurred has only recently been elucidated.

In populations of *Trichodesium* spp. maintained in their natural habitat as well as in laboratory conditions, temporal separation of photosynthesis and nitrogen fixation indeed occurs, though in a manner different from other nonheterocystous cyanobacteria (Berman-Frank et al. 2001). In *Trichodesmium* cells, these two metabolic processes are separated from each other during the photoperiod: High nitrogen fixation rates were measured for ~6 h in the midday, and the PSII activity and photosynthetic oxygen evolution varied inversely with nitrogen fixation. Not only did the activity patterns fluctuate in a diurnal manner, but also the accumulation of mRNA and protein levels for nitrogenase (*nifHDK*) and photosynthesis (*psaA*, *psbA*) were shown to oscillate with a period of about one day in constant conditions (Chen et al. 1999). Notably, there were phase differences between the net accumulation of these photosynthesis gene transcripts and the *nifHDK* gene transcripts: transcription of *nifHDK* reached the maximal level at 1–4 h into the light period, whereas that of *psbA* peaked near the end of the light period. The phase differences among transcription of the three genes were maintained in cultures grown in LL, indicating that this differential regulation is under the control of the clock (Chen et al. 1999). These data suggest that differential regulation of related genes by the circadian clock is, at least in part, involved in the separation of photosynthesis and nitrogenase activities in *Trichodesmium*.

More detailed analyses are needed to understand the mechanism by which the intricate temporal separation pattern between the two incompatible processes is precisely generated. It is also implied that *Trichodesmium* cells not only separate the two incompatible processes temporally but also spatially by performing a reversible and partial differentiation of cells. Photosynthetic activity is relatively low in some zones in a nonheterocystous cell filament (trichome) of *Trichodesmium*, and the occurrence of these zones increases during the hours of high nitrogen fixation (Berman-Frank et al. 2001; Küpper et al. 2004). These zones probably correspond to diazocytes, short stretches of cells in trichomes where higher levels of nitrogenase are observed (Bergman and Carpenter 1991; El-Shehawy et al. 2003). These observations imply that there is interplay between the temporal and spatial separation processes, the mechanism of which also remains unsolved.

Another point to address is the manner by which *Trichodesmium* obtained such a unique and complex strategy to perform nitrogen fixation. This strategy by *Trichodesmium* might reflect the evolutionary history of nitrogen fixation in cyanobacteria. Berman-Frank et al. (2001) developed phylogenetic trees of diazotrophic cyanobacteria based on *nifH* gene sequences to suggest that *Trichodesmium* branched out very early. Therefore, they inferred that the strategy used by *Trichodesmium* is likely a primitive one and that a complete temporal separation, in which nitrogen is fixed only at night, or a full spatial segregation based on heterocyst differentiation evolved later (Berman-Frank et al. 2001).

15.5 Nitrogen-Fixing Heterocystous Cyanobacteria

Nitrogen-fixing heterocystous cyanobacteria also show diurnal rhythms, including those of nitrogen fixation, though these daily fluctuations have not yet been conclusively demonstrated to be clock-controlled. For example, Kellar and Pael (1980) observed diurnal changes in N_2 and CO_2 fixation in *Anabaena spiroides* cells that were collected from a lake. Church et al. (2005) observed a diurnal rhythm in the abundance of *nifH* gene transcript, whose sequence clustered with *nifH* sequences of heterocystous cyanobacteria, from the natural population in the oligotrophic North Pacific Ocean. Therefore, there may also be interplay between the temporal and spatial separation strategies of heterocystous cyanobacteria. Sinha et al. (2001) reported that an *Anabaena* sp. showed a diurnal rhythm in the induction of ultraviolet (UV)-absorbing mycosporine-like amino acids (MAAs), which are thought to protect cyanobacteria from harmful UV radiation.

Transgenic reporter strains were generated using *Anabaena* sp. strain PCC 7120 by introducing a construct that contains a bacterial luciferase gene (*luxAB*) fused downstream of promoters for genes related to heterocyst patterning (H. Iwasaki, personal communication) or photosynthesis (T. Kondo, M. Ishiura, personal communication); the resulting strains exhibited bioluminescence rhythms in LL. *Anabaena* PCC 7120 is a cyanobacterial species in which various genetic tools have been well established. The complete genomic sequence is also available for this species (Kaneko et al. 2001). Therefore, this heterocystous cyanobacterium will also provide an excellent model system for studying the molecular mechanisms of the circadian clock in a multi-cellular species.

15.6 *Prochlorococcus* spp.

The tiny marine cyanobacteria of the genus *Prochlorococcus* (0.5–0.8 mm across) populate the oceans from the surface waters down to ~175 m and are predicted to be the dominant photosynthetic organisms in the intertropical areas of oceans (Chisholm et al. 1988; Partensky et al. 1999). *Prochlorococcus* spp. are also unique in photosynthetic pigment composition, i.e., they contain both chlorophylls *a* and *b* as antenna pigments and lack phycobilisomes (Chisholm et al. 1988; Partensky et al. 1999).

Although genuine circadian rhythms that meet the three criteria have not yet been demonstrated for *Prochlorococcus*, some studies reported diurnal rhythms that could well be explained by the involvement of circadian regulation. Vaulot et al. (1995) showed that the cell cycle of *Prochlorococcus* spp. in the equatorial eastern Pacific progressed in phase with the daily LD cycle: DNA replication occurred in the afternoon and cell division at night. They estimated the growth rate of *Prochlorococcus* cells to be about one division per day at the maximum. Shalapyonok et al. (1998) showed that *Prochlorococcus* exhibited, under optimal conditions, ultradian growth (faster than one division per day) both under natural conditions and in culture, while the timing of DNA synthesis and cell division is still strictly phased to the LD cycle. The first round of DNA synthesis and cell division are phased, similar to data by Vaulot et al. (1995), to late afternoon and early night, respectively. A fraction of cells then immediately undergo a second round of DNA synthesis followed by cell division that finishes before the onset of the next day. These observations suggest a possibility that a circadian clock gates the timing of the cell cycle of *Prochlorococcus*, independent of the average growth rates of cells, as was reported for *Synechococcus* (Mori et al. 1996). Jacquet et al. (2001) recorded cell size and chlorophyll fluorescence as well as cell cycle progress with higher time resolution in laboratory cultures. They demonstrated that these parameters showed not only diurnal rhythms in LD cycles, but also persistent rhythms in LL, though with significantly reduced amplitudes and phase disturbances. In addition, when the timing of LD cycles shifted with 4-h advance or delay, the rhythms also shifted their phases accordingly. These data strongly suggest that the parameters measured are under the control of a circadian clock.

Holtzendorff et al. (2001) examined two genes involved in either DNA replication (*dnaA*) or cell division (*ftsZ*) to understand the underlying mechanisms of the *Prochlorococcus* cell cycle rhythms. *Prochlorococcus* cells were axenically cultured in a turbidostat with a LD12:12 cycle and mRNA levels of *dnaA* and *ftsZ* were measured at 4-h intervals, while cell cycle synchronization was monitored. Both genes exhibited clear diurnal rhythms of expression, with mRNA maxima during the replication (S) phase. Western blot experiments indicated that the peak concentration of FtsZ protein occurred at night, i.e., at the time of cell division. These results indicated that the expression rhythms of key cell cycle-associated genes and proteins, which are well synchronized to ambient LD cycles, might be crucial for determining the timing of DNA replication and cell division rhythms. Holtzendorff et al. (2002) further examined the expression patterns of the *ftsZ* gene in natural *Prochlorococcus* populations in the northern Red Sea, using quantitative reverse transcriptase-coupled real-time PCR. They demonstrated that *Prochlorococcus* cells under natural day and night cycles also showed a diurnal expression rhythm of *ftsZ* expression, which was highly synchronized to the replication (S) phase. Garczarek et al. (2001) examined the expression patterns of four photosynthesis genes of *Prochlorococcus*, *psbA*, *psbC*, *psbD* and *pcbA*, under LD cycles. The first three *psb* genes showed diurnal rhythms similar to those of photosynthetic genes in higher plants, with anticipatory rises before the onset of light followed by peak expressions in light.

Sequences similar to *kaiB* and *kaiC* genes were found in the genomes of *P. marinus* sp. strains MED4, MIT 9313 and SS120 and *Prochlorococcus* sp. NATL2A, while no ortholog of the *kaiA* gene was found in their genomes (Ditty et al. 2003; Williams 2007). In *Synechococcus*, KaiA protein has a critical function in the molecular mechanisms of the circadian oscillator (Ditty et al. 2003; Williams 2007). *Prochlorococcus* might provide us with critical insight about the molecular mechanism for the generation of the circadian oscillation and its evolution in cyanobacteria.

15.7 Plastids

Plastids and extant cyanobacteria are phylogenetically closely related because plastids descended from a cyanobacterium that was endosymbiotically incorporated into the ancestral plant cell. Plastids contain their own genome, though the majority of original symbiont genes were lost or transferred to the nuclear genome; plastid genomes (~100 kbp) are much smaller in size than those of extant cyanobacteria (Sugiura 1992). In plants, no homologs of *kai* genes have been found in the plastid or nuclear genomes, which suggests that the *kai*-based clock was eliminated through the evolution of plastids.

Despite the lack of *kai* genes, there is a possibility that the output systems of the cyanobacterial clock might still be functional in plastids. Sigma factors are subunits of bacteria-type multi-subunit RNA polymerase, and they confer promoter specificity to the RNA polymerase holoenzyme. Cyanobacteria contain group 2 sigma factors, which are not essential for growth, in addition to the indispensable housekeeping sigma, RpoD1 (Wösten 1996). In *Synechococcus*, a group 2 sigma factor mutant (*rpoD2*) showed a low-amplitude phenotype in clock-controlled expression from only a certain group of luciferase reporters, including that driven by the *psbAI* promoter (Tsinoremas et al. 1996). Moreover, Nair et al. (2002) examined the effects of inactivation of four known group 2 sigma factor genes and demonstrated that there is functional division between the four sigma factor genes in conveying output signals. For example, inactivation of *rpoD2*, *rpoD3* or *rpoD4* resulted in low-amplitude phenotypes, whereas inactivation of *sigC* resulted in a high-amplitude phenotype. These observations indicate that sigma factors are regulatory components of the output pathways of the cyanobacterial circadian system.

In plants, multiple sigma factors are encoded in the nuclear genome which reflect past transfer from the plastid genome. Their protein products are transported into plastids, where they perform transcriptional regulation (Shiina et al. 2005). Some sigma factor genes have been reported to be under the control of the plant circadian clock, suggesting that they might also be involved in output regulation in the plant circadian system (Morikawa et al. 1999). Clock-controlled plastid genes have been reported, although the number is much less than that of reported clock-controlled nuclear genes. Hwang et al. (1996) reported that the circadian clock controls the *tufA* gene encoding the elongation factor Tu and some other plastid genes in the green alga *Chlamydomonas reinhardtii*. Matsuo et al. (2006) observed circadian transcription of *psbD* and *tufA* genes as bioluminescence rhythms, using a firefly luciferase reporter gene whose codons were optimized to the *C. reinhardtii* chloroplast. Although it was suggested that a sigma factor might be involved in the rhythmic regulation of *psbD* in wheat (Nakahira et al.1998; Morikawa et al. 1999), direct evidence of this idea has not yet been obtained. Recently, Matsuo et al. (2008) screened 16,000 insertional mutants for defects in chloroplast bioluminescence rhythms of *C. reinhardtii*, of which 105 mutants were isolated, and 30 genes were identified as being involved in causing rhythm defects. The genes they cloned might include output regulators and other factors that transmit timing information from the clock to chloroplast.

The moss *Physcomitrella patens* has four sigma factor genes, and only one of them (*PpSig5*) is under the control of the clock (Ichikawa et al. 2004). When *PpSig5* is disrupted, the amplitude of rhythmic expression of *psbD* decreased in LD cycles, indicating that sigma factor SIG5 (encoded by *PpSig5*) controls the diurnal pattern of the *psbD* gene (Ichikawa et al. 2008). A plausible explanation is that SIG5 transmits circadian regulation from the nucleus to plastids; however, because the amplitude of rhythms of moss *psbD* was so low in LL or DD, it was difficult to examine this idea clearly (Ichikawa et al. 2008). As mentioned above, robust circadian rhythms of *psbD* expression in constant conditions was observed in wheat (Nakahira et al. 1998) and *C. reinhardtii* (Matsuo et al. 2006). It should be investigated whether *psbD* is also rhythmically expressed in *A. thaliana* because, if this were the case, it would be possible to examine conclusively whether SIG5 is an output regulator of the circadian expression of *psbD*, by using the *AtSIG5*-disrupted tag insertion lines (Nagashima et al. 2004; Tsunoyama et al. 2004).

15.8 Concluding Remarks

The circadian system is composed of three regulatory parts that interact with each other, i.e., input pathways, the core oscillator and output pathways. To understand the mechanisms of this complex system, it is beneficial to study multiple model systems from different standpoints.

The development of the bioluminescent reporter strain in *T. elongatus* now allows us to conduct both structural and physiological analyses on the same species. These experiments will promote studies of structure–function relationships of the clock proteins in fine detail, which are critical in understanding the Kai-based oscillator machinery.

Although reporter strains that exhibited bioluminescence rhythms in DD were established in *Synechocystis*, they showed unexpected light sensitivities when undergoing LAHG. Therefore, it still remains a problem to establish a reporter strain that can be used for studying the effect of light signals on the cyanobacterial clock. Generating a *Synechocystis* mutant that is insensitive to the negative effect of light may be a possible solution to this problem, although there could be some overlap between signal transduction systems involved in these light sensitivities and the light input pathways of the clock. An attempt is underway to use another species, *Leptolyngbya boryana*, which grows heterotrophically in complete darkness, as a model system for observation of cyanobacterial circadian rhythms in the dark (Terauchi et al. 2005). *Trichodesmium* and *Prochlorococcus* can be regarded as unique models for studying the molecular mechanisms of the output pathways. Moreover, the ecological impact of their rhythms would also be of interest, given their abundance in the natural environment.

Cyanobacteria form a group with great variety in their phylogeny, morphology, physiology and strategy to adapt to different environments (see Chap. 2). Studies on various cyanobacterial species will also help to understand the diversity and evolution of cyanobacterial circadian systems (see Chap. 14). For example, studies on *Trichodesmium* species suggest that there are at least two strategies in the clock-driven temporal separation of nitrogen fixation and photosynthesis. Future studies should answer questions from an evolutionary point of view, such as: (1) whether these two strategies have the same origin and (2) what selective pressures or physiological requirements allowed *Trichodesmium* to evolve their seemingly complex strategy. Comparative genomics approaches will help to understand the diversities and evolution of the cyanobacterial circadian systems at the gene level. For example, the *pex* gene, which extends the endogenous period of the clock in *Synechococcus* (Kutsuna et al. 1998; see Chap. 8), is not found in *Synechocystis* (Kutsuna et al. 1998). This indicates that different genes may be exploited in the circadian systems among different species of cyanobacteria. *Prochlorococcus* would be a more interesting case: it has no *kaiA* homologs, suggesting that there might be divergence even in the core mechanisms of the circadian oscillators among different cyanobacteria. *Synechocystis* has a more complex organization of the *kai* genes relative to those in *Synechococcus* and *T. elongatus*, which provides an experimental disadvantage for it makes analyses more complicated in *Synechocystis*. However, additional copies of the *kaiB* or *kaiC* gene of *Synechocystis* are not recently duplicated paralogs with similar functions, but seem to be functionally divergent, reflecting the evolution of the *kai* genes. Therefore, studying the precise function of each copy should provide us with important clues about the evolution of the *kai* genes and the oscillator machineries. Unraveling the functions of the second *kaiC* gene cluster, *kaiC2B2*, might provide us a hint on the enzymatic activities of KaiB and KaiC and the origin of the prokaryotic circadian clock machinery.

By studying various species, it would be possible to understand the cyanobacterial circadian system and its evolution more comprehensively.

Acknowledgements We thank Drs. Masahiro Ishiura, Hideo Iwasaki, Takao Kondo, Ken-ichi Kucho, Shinsuke Kutsuna, Takuya Matsuo, Kazuhisa Okamoto, Tokitaka Oyama, Kazuki Terauchi and Hajime Wada for unpublished data, valuable advice and discussion. We also thank Dr. Masahiro Ishiura for giving us the opportunity to write this manuscript.

References

Anderson SL, McIntosh L (1991) Light-activated heterotrophic growth of the cyanobacterium *Synechocystis* sp. strain PCC 6803: a blue-light-requiring process. J Bacteriol 173:2761–2767

Aoki S, Kondo T, Ishiura M (1995) Circadian expression of the *dnaK* gene in the cyanobacterium *Synechocystis* sp. strain PCC 6803. J Bacteriol 177:5606–5611

Aoki S, Kondo T, Wada H, Ishiura M (1997) Circadian rhythm of the cyanobacterium *Synechocystis* sp. strain PCC 6803 in the dark. J Bacteriol 179:5751–5755

Aoki S, Kondo T, Ishiura M (2002) A promoter-trap vector for clock-controlled genes in the cyanobacterium *Synechocystis* sp. PCC 6803. J Microbiol Methods 49:265–274

Bergman B, Carpenter EJ (1991) Nitrogenase confined to randomly distributed trichomes in the marine cyanobacterium *Trichodesmium thiebautii*. J Phycol 27:158–165

Berman-Frank I, Lundgren P, Chen YB, Küpper H, Kolber Z, Bergman B, Falkowski P (2001) Segregation of nitrogen fixation and oxygenic photosynthesis in the marine cyanobacterium *Trichodesmium*. Science 294:1534–1537

Buikema WJ, Haselkorn R (1991) Characterization of a gene controlling heterocyst differentiation in the cyanobacterium *Anabaena* 7120. Genes Dev 5:321–330

Cai Y, Wolk CP (1997) *Anabaena* sp. strain PCC 7120 responds to nitrogen deprivation with a cascade-like sequence of transcriptional activations. J Bacteriol 179:267–271

Capone DG, Zehr JP, Paerl HW, Bergman B, Carpenter E (1997) *Trichodesmium*, a globally significant marine cyanobacterium. Science 276:1221–1229

Carpenter EJ (1983) Physiology and ecology of marine *Oscillatoria* (*Trichodesmium*). Mar Biol Lett 4:69–85

Chen YB, Zehr, JP, Mellon MT (1996) Growth and nitrogen fixation of the diazotrophic filamentous nonheterocystous cyanobacterium *Trichodesmium* sp. IMS 101 in defined media: Evidence for a circadian rhythm. J Phycol 32:916–923

Chen YB, Dominic B, Mellon MT, Zehr JP (1998) Circadian rhythm of nitrogenase gene expression in the diazotrophic filamentous nonheterocystous cyanobacterium *Trichodesmium* sp. strain IMS 101. J Bacteriol 180:3598–3605

Chen YB, Dominic B, Zani S, Mellon MT, Zehr JP (1999) Expression of photosynthesis genes in relation to nitrogen fixation in the diazotrophic filamentous nonheterocystous cyanobacterium *Trichodesmium* sp. IMS 101. Plant Mol Biol 41:89–104

Chisholm SW, Olson RJ, Zettler ER, Goericke R, Waterbury JB, Welschmeyer NA (1988) A novel free-living prochlorophyte abundant in the oceanic euphotic zone. Nature 334:340–343

Church MJ, Short CM, Jenkins BD, Karl DM, Zehr JP (2005) Temporal patterns of nitrogenase gene (*nifH*) expression in the oligotrophic North Pacific Ocean. Appl Environ Microbiol 71:5362–5370

Cornelius G, Rensing L (1986) Circadian rhythm of heat shock protein synthesis of *Neurospora crassa*. Eur J Cell Biol 40:130–132

Debus RJ, Barry BA, Babcock GT, McIntosh L (1988a) Site-directed mutagenesis identifies a tyrosine radical involved in the photosynthetic oxygen-evolving system. Proc Natl Acad Sci USA 85:427–430

Debus RJ, Barry BA, Sithole I, Babcock GT, McIntosh L (1988b) Directed mutagenesis indicates that the donor to P+680 in photosystem II is tyrosine-161 of the D1 polypeptide. Biochemistry 27:9071–9074

Ditty JL, Williams SB, Golden SS (2003) A cyanobacterial circadian timing mechanism. Annu Rev Genet 37:513–543

Dugdale RC, Menzel DW, Ryther JH (1961) Nitrogen fixation in the Sargasso Sea. Deep Sea Res 7:297–300

Dvornyk V, Vinogradova O, Nevo E (2003) Origin and evolution of circadian clock genes in prokaryotes. Proc Natl Acad Sci USA 100:2495–2500

El-Shehawy R, Lugomela C, Ernst A, Bergman B (2003) Diurnal expression of *hetR* and diazocyte development in the filamentous non-heterocystous cyanobacterium *Trichodesmium erythraeum*. Microbiology 149:1139–1146

Gallon JR, Jones DA, Page TS (1996) *Trichodesmium*, the paradoxical diazotroph. Algol Stud 83:215–243

Garczarek L, Partensky F, Irlbacher H, Holtzendorff J, Babin M, Mary I, Thomas JC, Hess WR (2001) Differential expression of antenna and core genes in *Prochlorococcus* PCC 9511 (Oxyphotobacteria) grown under a modulated light-dark cycle. Environ Microbiol 3:168–175

Golden SS, Ishiura M, Johnson CH, Kondo T (1997) Cyanobacterial circadian rhythms. Annu Rev Plant Physiol Plant Mol Biol 48:327–354

Gutierrez RA, Ewing RM, Cherry JM, Green PJ (2002) Identification of unstable transcripts in *Arabidopsis* by cDNA microarray analysis: rapid decay is associated with a group of touch- and specific clock-controlled genes. Proc Natl Acad Sci USA 99:11513–11518

Hayashi F, Suzuki H, Iwase R, Uzumaki T, Miyake A, Shen JR, Imada K, Furukawa Y, Yonekura K, Namba K, Ishiura M (2003) ATP-induced hexameric ring structure of the cyanobacterial circadian clock protein KaiC. Genes Cells 8:287–296

Hayashi F, Ito H, Fujita M, Iwase R, Uzumaki T, Ishiura M (2004a) Stoichiometric interactions between cyanobacterial clock proteins KaiA and KaiC. Biochem Biophys Res Commun 316:195–202

Hayashi F, Itoh N, Uzumaki T, Iwase R, Tsuchiya Y, Yamakawa H, Morishita M, Onai K, Itoh S, Ishiura M (2004b) Roles of two ATPase-motif-containing domains in cyanobacterial circadian clock protein KaiC. J Biol Chem 279:52331–52337

Hayashi F, Iwase R, Uzumaki T, Ishiura M (2006) Hexamerization by the N-terminal domain and intersubunit phosphorylation by the C-terminal domain of cyanobacterial circadian clock protein KaiC. Biochem Biophys Res Commun 348:864–872

Holtzendorff J, Partensky F, Jacquet S, Bruyant F, Marie D, Garczarek L, Mary I, Vaulot D, Hess WR (2001) Diel expression of cell cycle-related genes in synchronized cultures of *Prochlorococcus* sp. strain PCC 9511. J Bacteriol 183:915–920

Holtzendorff J, Marie D, Post AF, Partensky F, Rivlin A, Hess WR (2002) Synchronized expression of *ftsZ* in natural *Prochlorococcus* populations of the Red Sea. Environ Microbiol 4:644–653

Honda D, Yokota A, Sugiyama J (1999) Detection of seven major evolutionary lineages in cyanobacteria based on the 16S rRNA gene sequence analysis with new sequences of five marine *Synechococcus* strains. J Mol Evol 48:723–739

Hwang S, Kawazoe R, Herrin DL (1996) Transcription of *tufA* and other chloroplast-encoded genes is controlled by a circadian clock in *Chlamydomonas*. Proc Natl Acad Sci USA 93:996–1000

Ichikawa K, Sugita M, Imaizumi T, Wada M, Aoki S (2004) Differential expression on a daily basis of plastid sigma factor genes from the moss *Physcomitrella patens*. Regulatory interactions among *PpSig5*, the circadian clock, and blue light signaling mediated by cryptochromes. Plant Physiol 136:4285–4298

Ichikawa K, Shimizu A, Okada R, Satbhai SB, Aoki S (2008) The plastid sigma factor SIG5 is involved in the diurnal regulation of the chloroplast gene *psbD* in the moss *Physcomitrella patens*. FEBS Lett 582:405–409

Ishiura M, Kutsuna S, Aoki S, Iwasaki H, Andersson CR, Tanabe A, Golden SS, Johnson CH, Kondo T (1998) Expression of a gene cluster *kaiABC* as a circadian feedback process in cyanobacteria. Science 281:1519–1523

Iwasaki H, Williams SB, Kitayama Y, Ishiura M, Golden SS, Kondo T (2000) A KaiC-interacting sensory histidine kinase, SasA, necessary to sustain robust circadian oscillation in cyanobacteria. Cell 101:223–233

Iwase R, Imada K, Hayashi F, Uzumaki T, Namba K, Ishiura M (2004) Crystallization and preliminary crystallographic analysis of the circadian clock protein KaiB from the thermophilic cyanobacterium *Thermosynechococcus elongatus* BP-1. Acta Crystallogr D Biol Crystallogr 60:727–729

Iwase R, Imada K, Hayashi F, Uzumaki T, Morishita M, Onai K, Furukawa Y, Namba K, Ishiura M (2005) Functionally important substructures of circadian clock protein KaiB in a unique tetramer complex. J Biol Chem 280:43141–43149

Jacquet S, Partensky F, Marie D, Casotti R, Vaulot D (2001) Cell cycle regulation by light in *Prochlorococcus* strains. Appl Environ Microbiol 67:782–790

Jansson C, Debus RJ, Osiewacz HD, Gurevitz M, McIntosh L (1987) Construction of an obligate photoheterotrophic mutant of the cyanobacterium *Synechocystis* 6803: inactivation of the *psbA* gene family. Plant Physiol 85:1021–1025

Johnson CH (1994) Illuminating the clock: circadian photobiology. Semin Cell Biol 5:355–362

Jordan P, Fromme P, Witt HT, Klukas O, Saenger W, Krauss N (2001) Three-dimensional structure of cyanobacterial photosystem I at 2.5 A° resolution. Nature 411:909–917

Kamiya N, Shen JR (2003) Crystal structure of oxygen-evolving photosystem II from *Thermosynechococcus vulcanus* at 3.7-A° resolution. Proc Natl Acad Sci USA 100:98–103

Kaneko T, Sato S, Kotani H, Tanaka A, Asamizu E, Nakamura Y, Miyajima N, Hirosawa M, Sugiura M, Sasamoto S, Kimura T, Hosouchi T, Matsuno A, Muraki A, Nakazaki N, Naruo K, Okumura S, Shimpo S, Takeuchi C, Wada T, Watanabe A, Yamada M, Yasuda M, Tabata S (1996a) Sequence analysis of the genome of the unicellular cyanobacterium *Synechocystis* sp. strain PCC6803. II. Sequence determination of the entire genome and assignment of potential protein-coding regions. DNA Res 3:109–136

Kaneko T, Sato S, Kotani H, Tanaka A, Asamizu E, Nakamura Y, Miyajima N, Hirosawa M, Sugiura M, Sasamoto S, Kimura T, Hosouchi T, Matsuno A, Muraki A, Nakazaki N, Naruo K, Okumura S, Shimpo S, Takeuchi C, Wada T, Watanabe A, Yamada M, Yasuda M, Tabata S (1996b) Sequence analysis of the genome of the unicellular cyanobacterium *Synechocystis* sp. strain PCC6803. II. Sequence determination of the entire genome and assignment of potential protein-coding regions (supplement). DNA Res 3:185–209

Kaneko T, Nakamura Y, Wolk CP, Kuritz T, Sasamoto S, Watanabe A, Iriguchi M, Ishikawa A, Kawashima K, Kimura T, Kishida Y, Kohara M, Matsumoto M, Matsuno A, Muraki A, Nakazaki N, Shimpo S, Sugimoto M, Takazawa M, Yamada M, Yasuda M, Tabata S (2001) Complete genomic sequence of the filamentous nitrogen-fixing cyanobacterium *Anabaena* sp. strain PCC 7120. DNA Res 8:205–213, 227–253

Katayama M, Tsinoremas NF, Kondo T, Golden SS (1999) *cpmA*, a gene involved in an output pathway of the cyanobacterial circadian system. J Bacteriol 181:3516–3524

Katayama M, Kondo T, Xiong J, Golden SS (2003) *ldpA* encodes an iron-sulfur protein involved in light-dependent modulation of the circadian period in the cyanobacterium *Synechococcus elongatus* PCC 7942. J Bacteriol 185:1415–1422

Katoh H, Itoh S, Shen JR, Ikeuchi M (2001) Functional analysis of *psbV* and a novel c-type cytochrome gene *psbV2* of the thermophilic cyanobacterium *Thermosynechococcus elongatus* strain BP-1. Plant Cell Physiol 42:599–607

Kellar PE, Paerl HW (1980) Physiological adaptations in response to environmental stress during an $N_{(2)}$-fixing *Anabaena* bloom. Appl Environ Microbiol 40:587–595

Kondo T, Tsinoremas NF, Golden SS, Johnson CH, Kutsuna S, Ishiura M (1994) Circadian clock mutants of cyanobacteria. Science 266:1233–1236

Kucho K, Yoneda H, Harada M, Ishiura M (2004a) Determinants of sensitivity and specificity in spotted DNA microarrays with unmodified oligonucleotides. Genes Genet Syst 79:189–197

Kucho K, Tsuchiya Y, Okumoto Y, Harada M, Yamada M, Ishiura M (2004b) Construction of unmodified oligonucleotide-based microarrays in the thermophilic cyanobacterium *Thermosynechococcus elongatus* BP-1: screening of the candidates for circadianly expressed genes. Genes Genet Syst 79:319–329

Kucho K, Aoki K, Itoh S, Ishiura M (2005a) Improvement of the bioluminescence reporter system for real-time monitoring of circadian rhythms in the cyanobacterium *Synechocystis* sp. strain PCC 6803. Genes Genet Syst 80:19–23

Kucho K, Okamoto K, Tsuchiya Y, Nomura S, Nango M, Kanehisa M, Ishiura M (2005b) Global analysis of circadian expression in the cyanobacterium *Synechocystis* sp. strain PCC 6803. J Bacteriol 187:2190–2199

Küpper H, Ferimazova N, Setlík I, Berman-Frank I (2004) Traffic lights in *Trichodesmium*. Regulation of photosynthesis for nitrogen fixation studied by chlorophyll fluorescence kinetic microscopy. Plant Physiol 135:2120–2133

Kutsuna S, Kondo T, Aoki S, Ishiura M (1998) A period-extender gene, *pex*, that extends the period of the circadian clock in the cyanobacterium *Synechococcus* sp. strain PCC 7942. J Bacteriol 180:2167–2174

Leipe DD, Aravind L, Grishin NV, Koonin EV (2000) The bacterial replicative helicase DnaB evolved from a RecA duplication. Genome Res 10:5–16

Liu Y, Tsinoremas NF, Johnson CH, Lebedeva NV, Golden SS, Ishiura M, Kondo T (1995) Circadian orchestration of gene expression in cyanobacteria. Genes Dev 9:1469–1478

Matsuo T, Onai K, Okamoto K, Minagawa J, Ishiura M (2006) Real-time monitoring of chloroplast gene expression by a luciferase reporter: evidence for nuclear regulation of chloroplast circadian period. Mol Cell Biol 26:863–870

Matsuo T, Okamoto K, Onai K, Niwa Y, Shimogawara K, Ishiura M (2008) A systematic forward genetic analysis identified components of the *Chlamydomonas* circadian system. Genes Dev 22:918–930

Mitsui A, Kumazawa S, Takahashi A, Ikemoto H, Cao S, Arai T (1986) Strategy by which nitrogen-fixing unicellular cyanobacteria grow photoautotrophically. Nature 323:720–722

Mori T, Binder B, Johnson CH (1996) Circadian gating of cell division in cyanobacteria growing with average doubling times of less than 24 hours. Proc Natl Acad Sci USA 93:10183–10188

Morikawa K, Ito S, Tsunoyama Y, Nakahira Y, Shiina T, Toyoshima Y (1999) Circadian-regulated expression of a nuclear-encoded plastid sigma factor gene (*sigA*) in wheat seedlings. FEBS Lett 451:275–278

Mühlenhoff U, Chauvat F (1996) Gene transfer and manipulation in the thermophilic cyanobacterium *Synechococcus elongatus*. Mol Gen Genet 252:93–100

Murakami R, Miyake A, Iwase R, Hayashi F, Uzumaki T, Ishiura M (2008) ATPase activity and its temperature compensation of the cyanobacterial clock protein KaiC. Genes Cells 13:387–395

Nagashima A, Hanaoka M, Shikanai T, Fujiwara M, Kanamaru K, Takahashi H, Tanaka K (2004) The multiple-stress responsive plastid sigma factor, SIG5, directs activation of the *psbD* blue light-responsive promoter (BLRP) in *Arabidopsis thaliana*. Plant Cell Physiol 45:357–368

Nair U, Ditty JL, Min H, Golden SS (2002) Roles for sigma factors in global circadian regulation of the cyanobacterial genome. J Bacteriol 184:3530–3538

Nakahira Y, Baba K, Yoneda A, Shiina T, Toyoshima Y (1998) Circadian-regulated transcription of the *psbD* light-responsive promoter in wheat chloroplasts. Plant Physiol 118:1079–1088

Nakajima M, Imai K, Ito H, Nishiwaki T, Murayama Y, Iwasaki H, Oyama T, Kondo T (2005) Reconstitution of circadian oscillation of cyanobacterial KaiC phosphorylation in vitro. Science 308:414–415

Nakamura Y, Kaneko T, Hirosawa M, Miyajima N, Tabata S (1998) CyanoBase, a www database containing the complete nucleotide sequence of the genome of *Synechocystis* sp. strain PCC 6803. Nucleic Acids Res 26:63–67

Nakamura Y, Kaneko T, Miyajima N, Tabata S (1999) Extension of CyanoBase. CyanoMutants: repository of mutant information on *Synechocystis* sp. strain PCC 6803. Nucleic Acids Res 27:66–68

Nakamura Y, Kaneko T, Tabata S (2000) CyanoBase, the genome database for *Synechocystis* sp. strain PCC 6803: status for the year 2000. Nucleic Acids Res 28:72

Nakamura Y, Kaneko T, Sato S, Ikeuchi M, Katoh H, Sasamoto S, Watanabe A, Iriguchi M, Kawashima K, Kimura T, Kishida Y, Kiyokawa C, Kohara M, Matsumoto M, Matsuno A, Nakazaki N, Shimpo S, Sugimoto M, Takeuchi C, Yamada M, Tabata S (2002a) Complete genome structure of the thermophilic cyanobacterium *Thermosynechococcus elongatus* BP-1. DNA Res 9:123–130

Nakamura Y, Kaneko T, Sato S, Ikeuchi M, Katoh H, Sasamoto S, Watanabe A, Iriguchi M, Kawashima K, Kimura T, Kishida Y, Kiyokawa C, Kohara M, Matsumoto M, Matsuno A, Nakazaki N, Shimpo S, Sugimoto M, Takeuchi C, Yamada M, Tabata S (2002b) Complete genome structure of the thermophilic cyanobacterium *Thermosynechococcus elongatus* BP-1 (supplement). DNA Res 9:135–148

Ninnemann H (1979) Photoreceptors for circadian rhythms. Photochem Phobiol Rev 4:207–266
Okamoto K, Onai K, Furusawa T, Ishiura M (2005) A portable integrated automatic apparatus for the real-time monitoring of bioluminescence in plants. Plant Cell Environ 28:1305–1315
Onai K, Morishita M, Kaneko T, Tabata S, Ishiura M (2004a) Natural transformation of the thermophilic cyanobacterium *Thermosynechococcus elongatus* BP-1: a simple and efficient method for gene transfer. Mol Genet Genomics 271:50–59
Onai K, Morishita M, Itoh S, Okamoto K, Ishiura M (2004b) Circadian rhythms in the thermophilic cyanobacterium *Thermosynechococcus elongatus*: compensation of period length over a wide temperature range. J Bacteriol 186:4972–4977
Otto B, Grimm B, Ottersbach P, Kloppstech K (1988) Circadian control of the accumulation of mRNAs for light- and heat-inducible chloroplast proteins in pea (*Pisum sativum* L.). Plant Physiol 88:21–25
Partensky F, Hess WR, Vaulot D (1999) *Prochlorococcus*, a marine photosynthetic prokaryote of global significance. Microbiol Mol Biol Rev 63:106–127
Pattanayek R, Williams DR, Pattanayek S, Xu Y, Mori T, Johnson CH, Stewart PL, Egli M (2006) Analysis of KaiA–KaiC protein interactions in the cyanobacterial circadian clock using hybrid structural methods. EMBO J 25:2017–2028
Rikin A (1992) Circadian rhythm of heat resistance in cotton seedlings: synthesis of heat-shock proteins. Eur J Cell Biol 59:160–165
Schmitz O, Katayama M, Williams SB, Kondo T, Golden SS (2000) CikA, a bacteriophytochrome that resets the cyanobacterial circadian clock. Science 289:765–768
Shalapyonok A, Olson RJ, Shalapyonok LS (1998) Ultradian growth in *Prochlorococcus* spp. Appl Environ Microbiol 64:1066–1069
Shiina T, Tsunoyama Y, Nakahira Y, Khan MS (2005) Plastid RNA polymerases, promoters, and transcription regulators in higher plants. Int Rev Cytol 244:1–68
Sinha RP, Klisch M, Helbling EW, Häder D (2001) Induction of mycosporine-like amino acids (MAAs) in cyanobacteria by solar ultraviolet-B radiation. J Photochem Photobiol B 60:129–135
So WV, Rosbash M (1997) Post-transcriptional regulation contributes to *Drosophila* clock gene mRNA cycling. EMBO J 16:7146–7155
Sugiura M (1992) The chloroplast genome. Plant Mol Biol 19:149–168
Sugiura M, Inoue Y (1999) Highly purified thermo-stable oxygenevolving photosystem II core complex from the thermophilic cyanobacterium *Synechococcus elongatus* having His-tagged CP43. Plant Cell Physiol 40:1219–1231
Takai N, Nakajima M, Oyama T, Kito R, Sugita C, Sugita M, Kondo T, Iwasaki H (2006) A KaiC-associating SasA-RpaA two-component regulatory system as a major circadian timing mediator in cyanobacteria. Proc Natl Acad Sci USA 103:12109–12114
Taniguchi Y, Katayama M, Ito R, Takai N, Kondo T, Oyama T (2007) *labA*: a novel gene required for negative feedback regulation of the cyanobacterial circadian clock protein KaiC. Genes Dev 21:60–70
Terauchi K, Katayama M, Fujita Y, Kondo T (2005) Circadian rhythm of the facultative cyanobacterium *Plectonema boryanum*. In: van der Est A, Bruce D (eds) Photosynthesis: fundamental aspects to global perspectives, vol 2. Allen, New York, pp 729–731
Tomita J, Nakajima M, Kondo T, Iwasaki H (2005) No transcription-translation feedback in circadian rhythm of KaiC phosphorylation. Science 307:251–254
Tsinoremas NF, Ishiura M, Kondo T, Andersson CR, Tanaka K, Takahashi H, Johnson CH, Golden SS (1996) A sigma factor that modifies the circadian expression of a subset of genes in cyanobacteria. EMBO J 15:2488–2495
Tsunoyama Y, Ishizaki Y, Morikawa K, Kobori M, Nakahira Y, Takeba G, Toyoshima Y, Shiina T (2004) Blue light-induced transcription of plastid-encoded *psbD* gene is mediated by a nuclear-encoded transcription initiation factor, AtSig5. Proc Natl Acad Sci USA 101:3304–3309
Uzumaki T, Fujita M, Nakatsu T, Hayashi F, Shibata H, Itoh N, Kato H, Ishiura M (2004) Crystal structure of the C-terminal clock-oscillator domain of the cyanobacterial KaiA protein. Nat Struct Mol Biol 11:623–631

Vakonakis I, LiWang AC (2004a) Structure of the C-terminal domain of the clock protein KaiA in complex with a KaiC-derived peptide: implications for KaiC regulation. Proc Natl Acad Sci USA 101:10925–10930

Vakonakis I, Sun J, Wu T, Holzenburg A, Golden SS, LiWang AC (2004b) NMR structure of the KaiC-interacting C-terminal domain of KaiA, a circadian clock protein: implications for KaiA–KaiC interaction. Proc Natl Acad Sci USA 101:1479–1484

Vaulot D, Marie D, Olson RJ, Chisholm SW (1995) Growth of *Prochlorococcus*, a photosynthetic prokaryote, in the equatorial Pacific Ocean. Science 268:1480–1482

Vermaas WF, Williams JG, Rutherford AW, Mathis P, Arntzen CJ (1986) Genetically engineered mutant of the cyanobacterium *Synechocystis* 6803 lacks the photosystem II chlorophyll-binding protein CP-47. Proc Natl Acad Sci USA 83:9474–9477

Vermaas WFJ, Rutherford AW, Hansson O (1988) Site-directed mutagenesis in photosystem II of the cyanobacterium *Synechocystis* sp. PCC 6803: donor D is a tyrosine residue in the D2 protein. Proc Natl Acad Sci USA 85:8477–8481

Williams JGK (1988) Construction of specific mutations in photosystem II photosynthetic reaction center by genetic engineering methods in *Synechocystis* 6803. In: Packer L, Glazer A (eds) Methods in Enzymology, vol 167. Academic, San Diego, pp 766–778

Williams SB (2007) A circadian timing mechanism in the cyanobacteria. Adv Microb Physiol 52:229–296

Winfree AT (1990) The geometry of biological time. Springer, Heidelberg

Wolk CP (1996) Heterocyst formation. Annu Rev Genet 30:59–78

Wösten MM (1998) Eubacterial sigma-factors. FEMS Microbiol Rev 22:127–150

Yamaoka T, Satoh K, Katoh S (1978) Photosynthetic activities of a thermophilic blue-green alga. Plant Cell Physiol 19:943–954

Zehr JP, Dominic B, Chen YB, Mellon MT, Meeks JC (1998) Nitrogen fixation in the marine cyanobacteria *Trichodesmium*: a challenging model for ecology and molecular biology. In: Peschek GA, Loffelhardt W, Schmetterer G (eds) Phototrophic prokaryotes, Plenum, New York, pp 485–500

Zouni A, Witt HT, Kern J, Fromme P, Krauss N, Saenger W, Orth P (2001) Crystal structure of photosystem II from *Synechococcus elongatus* at 3.8 A° resolution. Nature 409:739–743

Chapter 16
Mathematical Modeling of the In Vitro Cyanobacterial Circadian Oscillator

Mark Byrne

Abstract This chapter describes recent attempts to formulate and validate mathematical models of the in vitro KaiABC oscillator from cyanobacteria. A variety of proposed mathematical models are discussed and compared, with a brief overview of recent experimental data relevant to the construction of these models and the constraints they must satisfy. A generic model is introduced which accounts for the hexameric structure of KaiC, intermediate states of hexamer phosphorylation, site- dependent reactions, stoichiometry, and monomer exchange.

16.1 Introduction

Previous studies have described the motivation for studying the circadian oscillator in the cyanobacterium *Synechococcus elongatus* PCC 7942: the discovery of oscillations in KaiC phosphorylation levels in the absence of transcription and translation (Tomita et al. 2005), the reconstruction of the circadian oscillator in vitro using only KaiA, KaiB, KaiC, and ATP (Nakajima et al. 2005), and the associated structural characteristics and conformational dynamics that are implicated in the in vitro circadian cycle (see Chap. 5; Kageyama et al. 2006; Mori et al. 2006; Ito et al. 2007; Rust et al. 2007). This chapter discusses current efforts to integrate the experimental findings on the KaiABC system within the framework of mathematical models describing the time-dependent dynamics of the participating molecular species. The primary purpose of creating such mathematical models is at least twofold: (a) to determine whether a specific interpretation of the experimental data (a proposed biological "mechanism") is consistent with oscillatory behavior, and (b) within a specific model mechanism, to unambiguously predict what happens to the oscillator under various experimental perturbations. The KaiABC oscillator represents an excellent opportunity to rigorously characterize the kinetics of the individual

M. Byrne
Department of Physics, 4000 Dauphin Street, Spring Hill College, Mobile, AL 36608, USA,
e-mail: mbyrne@shc.edu

J.L. Ditty et al. (eds.) *Bacterial Circadian Programs.*

molecular interactions, to represent these reactions with mass action kinetics, and to deduce from multiple (combinatorial) reactions how a stable oscillation of the appropriate timescale (~24 h) might be achieved. As such, these molecular processes of the KaiABC oscillator approximate an ideal model within systems biology – a molecular "machine" which can be deconstructed experimentally, whose dynamics may (presumably) be reconstructed entirely with chemical kinetics and represented by coupled differential equations, and which can then be studied in vivo to determine how the oscillator functions with other cellular components.

Given the simplicity of the oscillator (relative to typical mammalian transcription/translation feedback loop models), it might be assumed that creating stable oscillations in phosphorylation from only three proteins would be a trivial task. However, there are several possible mechanisms for generating the population oscillations in the phosphorylation levels of KaiC. Experimentally observed reactions in the system include KaiC monomer exchange (Kageyama 2006; Ito et al. 2007; Mori et al. 2007), site-dependent phosphorylation (Nishiwaki 2007; Rust et al. 2007) and the formation of different stable (or semi-stable) molecules (Iwasaki et al. 1999) including KaiA–KaiB, KaiA–KaiC, KaiB–KaiC, KaiA–KaiB–KaiC. Furthermore, each of these different complexes may have different effects on the rate of phosphorylation of KaiC hexamers, and the probabilities of association/disassociation may depend on the phosphorylation status of KaiC. It is therefore not surprising that there are a variety of proposed mathematical models for the KaiC oscillator (Emberly and Wingreen 2006; Kurosawa et al. 2006; Mehra et al. 2006; Mori et al. 2006; Clodong et al. 2007; Imamura et al. 2007; Li and Fang 2007; Miyoshi et al. 2007; Rust et al. 2007; Van Zon et al. 2007; Yoda et al. 2007). These models typically include a mechanism for generating oscillations in the phosphorylation of individual hexamers and usually also contain explicitly or implicitly a synchronization mechanism for the population of hexamers. These models may include some form of effective negative feedback and employ conformational changes and monomer exchange, autocatalytic mechanisms for phosphorylation, site-dependent cyclic phosphorylation, phenomenological inactivation of KaiA for single-hexamer cycling, and sequestration of KaiA in the complex. Independent of these mathematical models, I first reiterate the salient experimental findings that have been previously described and outline a generic ("bare bones") model of the KaiC system which includes most published models as particular examples. I then describe specific realizations with different model mechanisms, emphasizing their strengths and potential weaknesses of each. In the last section I comment on future work for the role of mathematical modeling in understanding this system.

16.2 Experimental Basis for Constructing a Mathematical Model

Any given mathematical model may or may not include all the known experimental information available about a system, depending on the desired simplicity of the model and the type of predictions that specific model is intended to produce. In the particular case discussed here, for example, the fact that KaiC forms a hexamer in the presence of ATP (Mori et al. 2002) may or may not need to be explicitly included in a mathematical model to describe the relevant oscillation in population phosphorylation levels or for generating predictions related to the oscillatory dynamics. On smaller timescales most population models (consisting of a sufficiently large number of molecules) do not use the detailed molecular structure of the molecules to predict, ab initio, the dynamics and kinetics of interaction. For example, simulations with structural dynamics might be useful over a timescale on the order of 10^{-9} s for suggesting potential semi-stable complexes and characteristic lifetimes of those states. The case of the KaiC system is particularly interesting because some of the characteristic effective timescales for reactions are minutes to hours; whereas the molecular dynamics and diffusive interactions take place at least six orders of magnitude faster. In this chapter, the discussion is limited to mathematical models of the oscillation process on the appropriate circadian timescale so that the models are approximations of the actual molecular dynamics that take place in the test tube.

A mathematical model of a system consists of stating the components of the system and defining their interactions. Fortunately, and in contrast to most other processes under investigation in cells, the in vitro oscillator forms a closed, isolated system. The molecular constituents of the oscillator are monomers of KaiC (58.4 kDa), which form a barrel-like hexamer in the presence of ATP (Pattanayek 2004), KaiA monomers (33.0 kDa), which form homodimers (Vakonakis et al. 2004; Ye et al. 2004), and KaiB (11.8 kDa), which forms either a dimer or a tetramer (Kageyama et al. 2003, 2006; Hitomi et al. 2005). The KaiC hexamer consists of two barrel-shaped domains, CI and CII, and is not entirely symmetric with respect to rotations by multiples of 60°. The CI and CII domains are posited to have different functional properties in terms of binding KaiA and KaiB (Pattanayek et al. 2004, 2006). The molecular abundance per cell is approximately (in order of magnitude): 10^4 KaiC monomers, 10^4 KaiB monomers, and 10^2 KaiA monomers (Kitayama et al. 2003). Given these typical molecular abundances and the fact that the abundance of KaiC can fluctuate by as much as ~50% during constant light (LL) conditions in vivo, it may be useful to perform stochastic simulations of the oscillator to determine the effects of noise on the circadian phosphorylation rhythm in vivo (Gillespie 1976).

In terms of reactions, each hexamer can autophosphorylate and autodephosphorylate, with a preference for autodephosphorylation in the absence of KaiA at room temperature. Starting with roughly 100% phosphorylated KaiC, the steady-state phosphorylation level falls to about 10% (Kageyama et al. 2006), which indicates that the rate of autodephosphorylation is roughly ten times greater than autophosphorylation in the absence of KaiA. Each KaiC monomer has at least two phosphorylation sites, S431 and T432, so that the entire hexamer likely has at least 12 phosphorylation sites available (Nishiwaki et al. 2004; Xu et al. 2004). There is the possibility of dynamic shuffling of phosphates if the free energy barriers are sufficiently low at room temperature. Interestingly, there are also 12 ATP binding sites in the KaiC hexamer, which serve to bridge adjacent monomers. In the case of the CII domain, these sites very likely provide the phosphates for the phosphorylation of S431 and T432.

In terms of interactions, KaiA, KaiB, and KaiC may each form separate complexes (KaiA–KaiB, KaiA–KaiC, KaiB–KaiC, KaiA–KaiB–KaiC) although the relative quantities of KaiA–KaiB appear to be much smaller than the other three (Kageyama et al. 2006; Mori et al. 2007). The phosphorylation rates of KaiC, presumably the rates at which each of the two monomer sites are phosphorylated, are apparently affected by association with other proteins, such that KaiA enhances the rate of autophosphorylation, suppresses autodephosphorylation, or both (Nishiwaki et al. 2002; Xu et al. 2003). Recent data for KaiA–KaiC mixing suggests that the added gamma phosphate preferentially binds to the T432 site, and during autodephosphorylation (in the absence of KaiA or KaiB), the phosphate is preferentially transferred to a longer-lived S431, which slowly dephosphorylates (Nishiwaki et al. 2007; Rust et al. 2007). This differential stability of T432 and S431 suggests that a mathematical model attempting to describe the dynamics of site-dependent phosphorylation should include different kinetics for the different phosphorylation sites on each monomer. Presumably these two different rates of autophosphorylation and autodephosphorylation (for S431 and T432) are roughly the same for the six monomers on the hexamer. It may be the case that the distribution of phosphates between the two sites depends on the association of KaiC with KaiA–KaiB and/or conformational dynamics that the hexamer may undergo as a result of association.

This rather intricate story of interactions is usually phenomenologically described by saying that KaiA increases the rate at which the KaiC population phosphorylation rises and that, for typical cellular stoichiometry, essentially saturates the KaiC molecules with phosphates (close to 100% phosphorylation). Thus we can say that KaiA is the driving force behind the "phosphorylation phase" of the oscillation. Also phenomenologically, KaiB counteracts the effect of KaiA by some unknown mechanism that is most likely a result of the formation of bound states with the KaiC hexamer (Kitayama et al. 2003). Kageyama et al. (2006) noted that KaiB preferentially binds to hyper-phosphorylated KaiC, where hyper-phosphorylation implies a predominance of phosphates on the hexamer. In some unknown way, the binding of KaiB is required to produce a net dephosphorylation of the hexamer and is correlated with a change in the preferred site of phosphorylation on the monomer. KaiB preferentially binds KaiC with a predominance of S431 phosphorylated monomers (Rust et al. 2008). There are several interesting possibilities related to the mechanism whereby KaiB mitigates the hyper-phosphorylated state to dephosphorylate the hexamer, even in the presence of KaiA. A variety of alternate mechanisms for the effect of KaiB on the oscillator may be checked for consistency with mathematical models: (a) phosphorylation-dependent competitive binding of KaiA versus KaiB, (b) "inactivation" of KaiA correlated with KaiB binding, either by association (removing KaiA from reactions with KaiC) or conformational inactivation of KaiA, and (c) conformational changes in the KaiC hexamer as a result of KaiB binding that render the usual action of KaiA ineffective. There is reason to suspect that KaiC is undergoing conformational changes because the probability of KaiB–KaiC complex formation increases as the number of phosphates on KaiC increases. However, a large number of phosphates on the hexamer alone is obviously insufficient to generate the dephosphorylation phase and the binding of KaiB apparently serves as a hyper-phosphorylation "sensor" to signal the dephosphorylation of monomers (or rather to turn off the enhanced autophosphorylation due to the presence of KaiA). This appears to occur in tandem with a transition in the preferential sites of bound phosphates and is likely a reflection of another conformational transition in the KaiC molecule. Specifically, it appears that phosphorylation on S431 is correlated with the dephosphorylation phase of the oscillations (Nishiwaki et al. 2007; Rust et al 2007).

In the scenarios outlined above, option (a) is the simplest and most economical explanation but is not favored by kinetics despite the larger number of KaiB molecules. Let us consider the case in which the phosphorylation-enhancing activities of KaiA are considerably diminished when KaiB–KaiC is formed. KaiB binding probabilities decrease as the number of phosphates decrease, suggesting KaiA would be more likely to win the competition for the binding site and the dephosphorylation phase would rapidly return to preferential phosphorylation; this mechanism would occur before dephosphorylation of a hexamer was complete. Given this scenario it is hard to see how oscillations of 20–80% (population) phosphorylation would be possible.

Option (b) is a phenomenological statement where KaiA is inactivated and this inactivation is correlated with hyper-phosphorylation of KaiC, a "minimum" threshold of S431 phosphates, and KaiB–KaiC binding (which is perfectly acceptable in a minimal mathematical model and is readily imposed). There is currently no experimental evidence that two forms of KaiA ("active" and "inactive" conformations) are present in solution. It is important to note that Rust et al. (2007) have added exogenous KaiA at various phases of the oscillation and the addition of KaiA perturbed the oscillator as expected at all phases – the KaiC molecules during the dephosphorylation phase are sensitive to any free KaiA that happens to be available. This finding implies that KaiA association with KaiB–KaiC may be sufficient to "trap" enough KaiA (a stoichiometry-dependent statement) and render it "inactive." In order for this to occur effectively, the unbinding rate of KaiA would have to be reduced so that the long-lived KaiA–KaiC (and KaiA–KaiB–KaiC) states last a sufficient time such that the dephosphorylation phase of the oscillation is not disrupted. Gel filtration experiments indicate that most of the dimerized KaiA is in complexes during the dephosphorylation phase of the cycle, which provides the most likely mechanism for the "inactivation" of KaiA. It is worthwhile to examine this idea in detail, as it sheds light on the clock's sensitive dependence on the relative molecular abundance of KaiA and KaiC. The monomer stoichiometry used in the in vitro studies of Kageyama et al. (2006), KaiA = 1.2 μM, KaiB = 3.5 μM, KaiC = 3.5 μM, implies an approximate 1:1 stoichiometry of dimerized KaiA:(hexamer) KaiC. Interestingly, the fraction of free KaiC hexamers at any time during the oscillation is 60–70% while most of the dimerized KaiA (>90%) appears to be in bound KaiC complexes (Kageyama et al. 2006). These KaiA–KaiC and KaiA–KaiB–KaiC bound states account for only 25% of the total KaiC, which suggests that, on average, there may be three or four KaiA proteins bound to a single hexamer. This important mechanism implies that the in vitro oscillator is surprisingly sensitive to the relative ratio of KaiA:KaiC, such that if there is a large enough ratio of KaiA:KaiC present (certainly 6:1), the inactivation of KaiA by association is not possible; the oscillation ceases and steady-state high levels of phosphorylated KaiC result. A potential reason for multiple KaiA binding to a single hexamer is to increase the autophosphorylation reactions maximally and to simultaneously sequester as much KaiA as possible in complex during the dephosphorylation phase. However, it should be noted that the in vitro data of Mori et al. (2007) does not provide evidence that multiple KaiA dimers bind to single KaiC hexamers. Option (c) appears to be ruled out by the experimental verification that KaiC can be phosphorylated by adding exogenous KaiA at any time during the cycle, so that whatever conformational changes KaiC undergoes, these changes do not render the hexamers unphosphorylatable by KaiA (Rust et al. 2007).

The preceding remarks are intended to describe the phosphorylation kinetics of single hexamers for one cycle: rapid association and dissociation of KaiA leading to hyper-phosphorylation, conformational change and KaiB binding, inactivation of KaiA (perhaps by sequestration), and dephosphorylation until the hexamer is hypo-phosphorylated (Fig. 16.1 is a pictorial description of this process, adapted from Mori et al. 2007). This phosphorylation cycle may be taking place for individual KaiC hexamers in a population, but if the hexamers are not synchronized across the population there would be no population circadian rhythm. Rather, there would be a set of clocks independently ticking with random phases and no net cycling, even

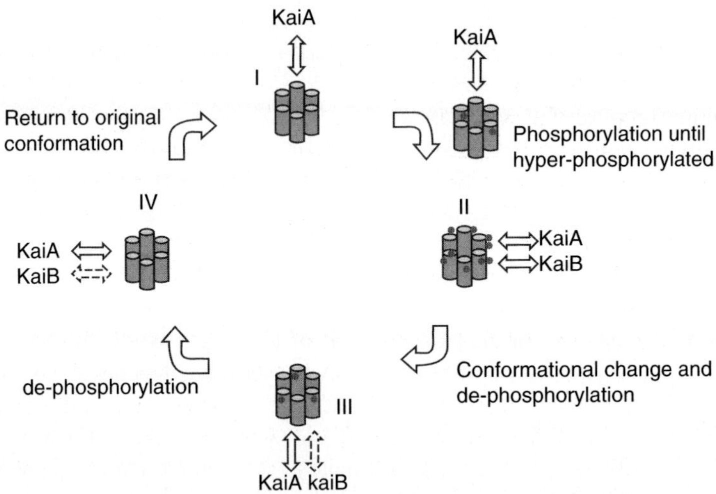

Fig. 16.1 The KaiC phosphorylation cycle: a proposed pictorial representation of the phosphorylation cycle of an individual KaiC hexamer. The *dots* represent phosphates at S431 or T432 on individual monomers. Starting from a non-phosphorylated state (*I*) rapid binding and unbinding of KaiA facilitates phosphorylation until the hexamer is hyper-phosphorylated (state *II*). KaiB is assumed to preferentially associate and disassociate from hyper-phosphorylated KaiC; there is a simultaneous conformational change to a new state (KaiC*). The KaiC* hexamer (state *III*) may sequester KaiA and dephosphorylate at the auto-dephosphorylation rate until it is no longer phosphorylated (state *IV*). Adapted from Mori et al. (2007)

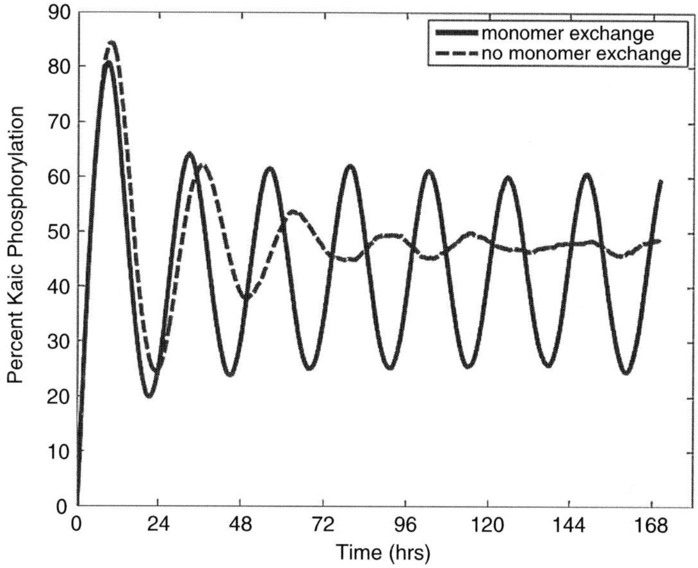

Fig. 16.2 The effect of monomer exchange in synchronizing the hexamer population. Turning off monomer exchange results in a decaying oscillation toward a stable steady-state phosphorylation distribution. *hrs* Hours. Adapted from Mori et al. (2007)

if the individual clocks have the same period. The last potentially relevant mechanism is related to synchronization of hexamers and involves the experimental indication that monomers may be exchanged between different hexamers during the reaction (Kageyama et al. 2006; Ito et al. 2007; Mori et al. 2007). The probability of this exchange appears to depend on the state of the KaiC hexamer as there are differences in the exchange rates during the phosphorylation versus dephosphorylation phases. The rate of exchange may depend on whether KaiA or KaiB is present. Kageyama and co-workers (2006) found that KaiA inhibited monomer exchange among hexamers, while Mori and co-workers (2007) found that neither KaiA nor KaiB significantly reduced monomer exchange. In either case monomer exchange implies that, even with arbitrary initial conditions (hexamer preparations), the hexamers shuffle their monomers at the relevant rate until the population is synchronized. A particular dynamic example of turning off the exchange of monomers results in damped oscillations in the population phosphorylation levels (Fig. 16.2, adapted from Mori et al. 2007).

16.3 Generic Mathematical Model of the KaiC In Vitro System

In this section, a very general formulation for describing the in vitro system is described, either for implementation with coupled differential equations or for stochastic algorithms. The following sections then specialize to specific cases considered in the literature, including assumptions, methods, and predictions. Let $C_k(m_1, m_2, \ldots m_6)$ represent the *k*th hexamer in a population and $m_1, m_2, \ldots m_6$ represent the six monomers in the hexamer. Including site dependence, we can label each monomer by two (or more) labels: for example, $m_1 = m_1(S,T)$ where *S* and *T* refer to the S431 and T432 labels for this monomer, so that there are really four potentially functionally significant states, *U* (unphosphorylated) $= m_1$ (0,0), $S = m_1$ (1,0), $T = m_1$ (0,1), and $S,T = m_1$ (1,1). If we assume, to first order, that all monomers in a hexamer are equivalent, then we can simplify the description of a hexamer to $C_k(S',T')$ where S' and T′ now each take on values from zero to six. A pictorial representation of a symmetric single hexamer with 12 total sites is given in Fig 16.3A, with potential reactions indicated in Fig. 16.3B, and a hexamer in a potential intermediate state of phosphorylation in Fig. 16.3C. Based on the indication in the experimental data that multiple KaiA molecules may bind to KaiC, but a single KaiB appears to bind to the KaiC hexamer, we assume the only legitimate hexamer reaction is KaiB association and disassociation from the KaiC hexamer (with or without multiple KaiA proteins bound to the monomers):

$$C_{(S,T)} + B \leftrightarrow BC_{(S,T)} \qquad (16.1)$$

In Eq. 16.1, the forward and backward rates might be some complicated functions of *S* and *T*, but should be chosen so that hyper-phosphorylated complexes (large *S*, large *T*) are more likely to bind KaiB. Another possibility is that the number and location of monomers with KaiA bound could affect the above rates; each monomer and its association (or lack thereof) with KaiA would need to be tracked with a label. That is, in the worst possible case, using the above nomenclature, we would write $m_1 = m_1(S,T,A)$, using *A* as a label to indicate whether or not that monomer had KaiA bound. For simplicity, we might label KaiA bound with a value of 1, and KaiA unbound with zero. In principle, if the hexamer reactions are *not symmetric* with respect to the different monomers, and the rates in Eq. 16.1 are also KaiA-dependent, then hexamer reaction rates take on values labeled by the set of numbers (*S*1, *T*1, *A*1; *S*2, *T*2, *A*2; … *S*6, *T*6, *A*6), all either 0 or 1, and there would be 2^{18} (262,144) rate parameters which would need to be chosen! It is clearly not useful to consider such complications further since the mathematical models with the most power have the fewest rate parameters (for a desired level of predictive power), are clearly interpretable and are falsifiable. The latter is difficult to assess in models with numerous freely adjustable rate parameters. A simpler case to consider is where the monomers are assumed equivalent and the rate constants are labeled by the set (S', T', A') which states that only the net number of S431 phosphates, the net number of T432 phosphates, and the net number of KaiA bound affect the rates. Even in this simplified case there are in principle 7^3 (343) possible different rates depending on the values of S', T', and A'. Clearly the simplest and cleanest approach to interpreting the preference for KaiB binding to hyper-phosphorylated KaiC is to simplify this further so that the rates are monomer-independent, KaiA-independent, and only depend on the sum of phosphates $S'+T'$. The simplest implementation for KaiB binding is a single rate when $S'+T'$ is above some threshold value.

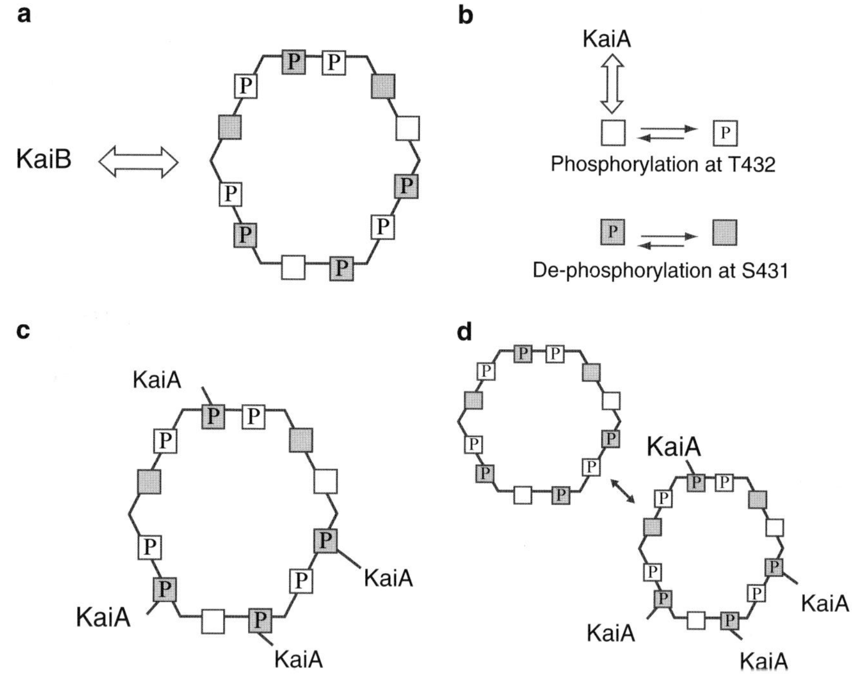

Fig. 16.3 Site-dependent model with 12 phosphorylation sites per hexamer, S431 and T432 for each monomer. **A** KaiB binding above some threshold phosphorylation of the hexamer that likely depends on the total number of S431 phosphates bound. **B** Monomer phosphorylation reactions (auto-phosphorylation, auto-dephosphorylation) modulated by KaiA. **C** A particular hexamer with four KaiA bound in an intermediate stage of phosphorylation. **D** The potential exchange of monomers across two different hexamers

Monomer reactions include phosphorylation, dephosphorylation, and KaiA association and disassociation. Again, we assume monomers may be treated equivalently to avoid a profusion of rates. The phosphorylation reactions for each monomer may be indicated by:

$$m_{(k)} \rightarrow m_{(k)} + 1 \; (k = 0, m_{\max} - 1), \qquad (16.2)$$

where $m_{(k)}$ refers to the phosphorylation level of the kth monomer and $m_{\max} = 2$; and the dephosphorylation reactions may be indicated by:

$$m_{(k)} \rightarrow m_{(k)} - 1 \; (k = 1, m_{\max}). \qquad (16.3)$$

For the rate-dependence of phosphorylation, an interpretation of the experimental data is that the rates depend on whether KaiA is bound to the monomer (or perhaps they may depend on whether KaiA is bound to any monomer in the hexamer), whether KaiB is bound to the hexamer, and whether the S431 (S) or T432 (T) site is being phosphorylated for that monomer. We can indicate such rate dependencies schematically with $r = r(A, B, S, T)$ and use + and – to indicate phosphorylation and dephosphorylation, respectively. For example the monomer auto-phosphorylation rate is then $r_+ = r_+ (0,0,S,T)$; and $r_- = r_- (0,0,S,T)$ is the rate of monomer auto-dephosphorylation. It does not appear that KaiB has much effect on the effective dephosphorylation rate, so if this assumption is made we write $r_- (0,0,S,T) = r_- (0,B,S,T)$.

KaiA association and disassociation is treated similarly:

$$A + m_{(k)} \leftrightarrow \mathrm{Am}_{(k)} \tag{16.4}$$

where the forward and backward rates may depend on the extent of phosphorylation, or more precisely, the four states the monomer may be in at any given time (S and T site-dependence as described above). Monomer exchange (with a single exchange between two hexamers indicated in Fig. 16.3D) may be implemented in a number of ways. A simple method is to treat the exchange of monomers across any two hexamers (C_l, C_r) as a "reaction:"

$$C_l(m_{(k)}) \leftrightarrow C_r(m_{(j)}) \tag{16.5}$$

Again we can imagine that the rates of such exchange may depend on a number of variables associated with each hexamer as previously discussed (net degree of S or T phosphorylation, differences in state of association of the two different hexamers, etc.). In summary, Eqs. 16.1–16.5 represent the reactions for the oscillator (hexamer and monomer) with all of the interesting dynamics encoded in the rate dependencies of each reaction. An alternative version of Eqs. 16.1–16.5 is to assume KaiA binds to the hexamer (rather than monomers) so that Eq. 16.4 is replaced by the equivalent of Eq. 16.1. The formation of hexamers and the phosphorylation reactions require ATP. For this chapter, I am not focusing on the dynamics of hexamer formation, essentially assuming that such hexamers readily form in non-limiting ATP (1 mM in typical experimental runs), and that the phosphorylation reactions also proceed at a fixed rate (for given values of S, T, A) in non-limiting ATP. This does not imply the ATPase activity would be constant but would rather directly reflect the oscillatory dynamics resulting from the effective (overall) rate at which the phosphorylation kinetics proceeds as a result of all the reactions occurring simultaneously in the entire population of hexamers.

16.4 Summary of Proposed Mathematical Models for the In Vitro Oscillator

16.4.1 Model 1 (Emberly and Wingreen)

After the surprising announcement of in vitro circadian oscillations (Nakajima et al. 2005), the model of Emberly and Wingreen (2006) was the first of several proposed mathematical models for the KaiABC system. The authors' proposed primary mechanism of oscillation was the exchange of monomers during the KaiC phosphorylation phase (for population synchronization), followed by the formation of extended clusters during the phase of dephosphorylation. Both KaiA and KaiB were assumed to be non-limiting, and the authors assumed constant single effective rates of KaiC monomer phosphorylation and dephosphorylation. As such, the mathematical model was minimal, assuming it was sufficient to track the population phosphorylation levels (0, 1 or 2) for monomers (assuming the hexamers were appropriately randomized), the concentration of completely phosphorylated clusters, and the concentration of clusters of some net phosphorylation level above a minimum threshold (a free parameter of the model). Monomer exchange was implicitly assumed in the model because all monomers were, by construction, randomly mixed at all times, whereas the authors noted that modeling the hexamer phosphorylation concentrations separately (in the effective forms C_0, C_1, ... C_{12}) without assuming a randomized monomer population yielded no oscillatory solutions. This model suggested the importance of monomer exchange among hexamers for maintenance of the oscillations prior to experimental evidence for such exchange (Kageyama et al. 2006; Mori et al. 2007). However, experimental evidence for the dephosphorylation mechanism – extended clustering by hexamers – has not been reported. This particular model was not constructed for explicit stoichiometry studies for the effects of KaiA and KaiB, but rather for suggesting potential viable oscillatory mechanisms.

16.4.2 Model 2 (Mehra, Hong, Shi et al.)

Another potential oscillatory mechanism, which explicitly included KaiA and KaiB, was suggested by Mehra et al. (2006) to describe an autocatalytic KaiA KaiC interaction in which the formation of phosphorylated KaiA–KaiC complexes forms an auto-feedback loop. To simplify the model, the authors assumed two effective KaiC states, a "low" phosphorylated form of KaiC (C) and a "high" phosphorylated form (C*). In this case, the formation of (highly) phosphorylated KaiA–KaiC complex (labeled AC*) increases the rate of complexation and phosphorylation reactions of further AC* complexes from free A and free C:

$$\mathrm{A+C+AC^*} \rightarrow \mathrm{2AC^*} \qquad (16.6)$$

Therefore the time-dependence of AC* complexes takes the form:

$$\mathrm{d}/\mathrm{d}t[\mathrm{AC^*}]=k_{p1}[\mathrm{AC}]+k_{p2}[\mathrm{AC}]^*[\mathrm{A}][\mathrm{C}]+... \qquad (16.7)$$

where brackets have been used to denote concentrations, k_{p1} is the "normal" rate of KaiC phosphorylation from KaiA–KaiC complexes and k_{p2} is the rate associated with the autocatalysis assumption. The ellipses represent additional reactions of KaiA–KaiC* not specifically relevant for the proposed mechanism. The autocatalysis step involves both association with KaiA and phosphorylation of KaiC from the

KaiA–KaiC* complexes, perhaps by molecular scaffolding or some other structural mechanism. An additional assumption is that KaiA–KaiB–KaiC* created from the reaction KaiA–KaiC* + KaiB selectively generates KaiB–KaiC* (and free KaiA) so that the reaction cycle for the hexamer is approximately represented by the cycle: C → AC → AC* → ABC* → BC* → C* → C. Essentially, by speeding up the phosphorylation reaction with auto-feedback and selecting appropriate rates for the other reactions, the hexamers roughly cycle through the individual states almost in synchrony so that a stable steady-state of C and C* is not possible; instead the system shows limit cycle behavior with the appropriate oscillation timescale for various rate parameter choices (Mehra et al. 2006). KaiA is also effectively rendered inoperable during the dephosphorylation phase since ABC* and the selective disassociation into BC* complexes (excluding ABC* → AC* + B as a possibility) results in the dephosphorylation of C* with limited simultaneous phosphorylation reactions. This particular model is useful for predicting the complexes that form as a function of time and making testable hypotheses about the effect of KaiA and KaiB abundance perturbations on the oscillator. In addition, the authors studied the effects of temperature on the system and indicated how such a mechanism could be temperature compensated. It is not clear to what extent the model misses essential aspects of the dynamics of the system in the simplifying assumption of only two possible phosphorylation states. By not including multiple states of phosphorylation for hexamers (C_0, C_1, … C_{12}), an explicit synchronization of hexamers was not needed for obtaining oscillatory dynamics (similar to the initial Emberly–Wingreen model above) since hexamers, in this case, instantaneously switch from "low" phosphorylation to "high" phosphorylation without proceeding through intermediate states. Of course, this does not imply the autocatalytic mechanism could not, in principle, sufficiently synchronize the hexamers for specific ranges of rate parameters with the inclusion of such intermediate states.

16.4.3 Model 3 (Kurosawa, Aihara and Iwasa)

The mathematical model of Kurosawa et al. (2006) also uses an effective two-state approximation in which phosphorylated KaiC (C* using the previous nomenclature) and non-phosphorylated KaiC ($C = C_{total} - C^*$) is described as a function of time. The authors include an attempt to model the clock both in vivo under LL conditions and in vitro, corresponding to constant darkness (DD) in vivo. Focusing on the in vitro model the authors assume a completely effective description of the oscillator and assume active and inactive states of KaiA (A, A*) and KaiB (B, B*) with the dynamics of KaiC phosphorylation and dephosphorylation described by:

$$\frac{dC^*}{dt} = k_+ A^* C - k_{1-} C^* B^* - k_{2-} C^* \tag{16.8}$$

where k_+ can be interpreted as the rate of phosphorylation of C associated with A*C complexes, k_{1-} describes dephosphorylation due to KaiB* and k_{2-} is the rate of auto- dephosphorylation of C*. Similarly the concentration of active KaiB is described by the following equation:

$$\frac{dB^*}{dt} = k_B B - \frac{kB^*(C^*)^n}{[(C^*)^n + K^n]} \tag{16.9}$$

where active KaiB is generated from KaiB and inactivated proportional to the abundance of B* and the concentration of phosphorylated KaiC assuming a Hill-type functional form. Active KaiA regulation is similarly described by a phenomenological equation, which includes both a rate for increasing active KaiA and decreasing active KaiA as the degree of phosphorylation increases:

$$dA^*/dt = k_3(A^*_{max} - A^*) - k_4 A^*(1 + aC^*) \tag{16.10}$$

In the above A^*_{max}, k_3, k_4, and α are constants. This particular model shares the drawbacks of other mathematical models of the KaiABC system – the model does not include intermediate states of phosphorylation, neglects the hexamer nature of KaiC and synchronization of hexamers, and assumes states of KaiA and KaiB, which have not been observed experimentally. However it does share similar aspects with other mathematical models of the oscillator, including the effective inactivation of KaiA as the degree of phosphorylation increases (this could occur by sequestration of KaiA, as the authors note) and incorporates many of the experimental findings, but in a generalized form not specifically motivated by mass action kinetics of the molecular species. By using such phenomenological models, the authors were able to test alternative simple mechanisms of regulation (assuming these active and inactive forms) that indicated the regulation of active KaiB by C* (Eq. 16.9) was the most robust mechanism for obtaining sustained high amplitude oscillations. Alternative mechanisms tended to produce low amplitude or decaying oscillations as solutions.

16.4.4 Model 4 (Clodong, Düring, Kronk et al.)

A more extensive analysis in terms of robustness of the circadian oscillator was provided by Clodong et al. (2007) who examined various feedback mechanisms built on a core cyclic mechanism of effective hexamer states ($C_0 \rightarrow C_1 \rightarrow \ldots \rightarrow C_6 \rightarrow BC^*_6 \rightarrow \ldots BC^*_0 \rightarrow C_0$) and systematically determined what types of feedback (negative and positive) generated the most robust rhythms for the oscillator. These authors found several candidate mechanisms consistent with oscillations, including autocatalysis in the phosphorylation phase as previously discussed in Mehra et al. (2006). Another such feedback "topology" mechanism was the interpretation of monomer exchange as a particular type of negative feedback mechanism among complexes in the phosphorylation phase ($C_0 \rightarrow C_1, \ldots \rightarrow C_{12}$) used implicitly in the Emberly–Wingreen model described above. However, using this negative feedback topology alone in the phosphorylation phase, without a mechanism to maintain synchrony in the dephosphorylation phase, does not generate stable oscillations.

However, a form of negative feedback from the KaiB–KaiC complexes generated, computationally, the most robust oscillations (amplitude, frequency, phase) with respect to stoichiometric perturbations of the oscillator as measured by Kageyama et al. (2006). This negative feedback was specifically interpreted as the sequestration of KaiA by KaiB–KaiC complexes (BC_n) and was used by the authors to construct an explicit biochemical model for the time-dependent formation of the complexes which roughly matched that time-dependence of complexes and the stoichiometry data of Kageyama and Mori et al. (2007), although with some differences in the apparent oscillatory behavior of KaiA–KaiC complexes.

16.4.5 Model 5 (Mori, Saveliev, Xu et al.)

An alternative mechanism which also attempted to match the time-dependence of the complexes and the perturbations under abundance variation was proposed in Mori et al. (2007), who also reported electron microscopy (EM) imagery for the stable complexes and estimated time-dependencies for these complexes which were similar to those in Kageyama et al. (2006) – for example, see Fig. 16.4. The mathematical model mechanism of Mori et al. (2007) included a core KaiC cyclic phosphorylation mechanism similar to Clodong et al. (2007) which can be roughly represented by $C_0 \rightarrow C_1 \rightarrow \ldots C_6 \rightarrow C^*_6 \ldots \rightarrow C_0^* \rightarrow C_0$, where the step $C_6 \rightarrow C^*_6$ is a result of KaiB association and the star indicates conformationally altered KaiC, presumably correlated with the hyper-phosphorylation of the hexamer. The mathematical model explicitly simulated the kinetics of complex formation and disassociation of hexamers (KaiA–KaiC, KaiB–KaiC, KaiA–KaiB–KaiC) and monomer phosphorylation/dephosphorylation reactions. The conformational dynamics of KaiC hexamers and their degree of phosphorylation was assumed to be responsible for the asymmetric binding of KaiB and subsequent de-phosphorylation phase, while explicit sequestration of KaiA was not assumed a necessary ingredient for sustained oscillations. The authors used monomer exchange as a synchronization mechanism among hexamers in both the phosphorylation and dephosphorylation phases. If these exchanges were selectively de-coupled so that dephosphorylation phase monomers (in C^* hexamers) did not exchange with phosphorylation phase monomers (in C hexamers), the population phosphorylation levels oscillated in a sustained manner and the complexes roughly matched those of the EM data and the pull-down data of Kageyama et al. (2006). The dephosphorylation phase was defined by assuming that the C^* hexamers were essentially dephosphorylating independent of whether KaiA was bound or not. A very similar mathematical model incorporating allosteric transitions and selective monomer exchange of the two phases was proposed shortly thereafter by Yoda et al. (2007), who studied time-dependent complex formation, stoichiometry, and temperature compensation of the oscillator. Whereas the model of Mori et al. (2007) simulated hexamer kinetics with a stochastic (probabilistic) matrix model, the Yoda et al. (2007) model used deterministic equations to implement the reactions, including monomer exchange.

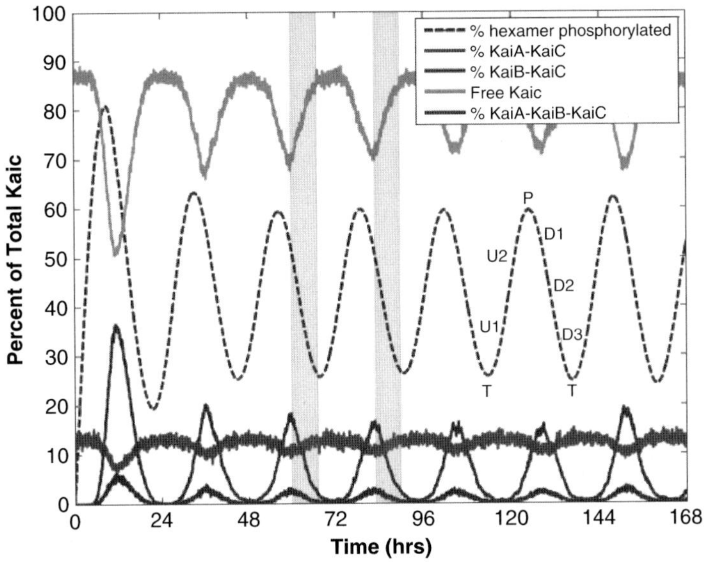

Fig. 16.4 Simulated oscillations in relative levels of complexes in the KaiABC oscillator with the population phosphorylation oscillation indicated by the dashed line. *T, U1, U2, P, D1, D2, D3* Various phases of the oscillator (*T* trough, *U* up, *D* down, *P* peak) labeled for ease of comparison with experimental measurements. Adapted from Mori et al. (2007)

16.4.6 Model 6 (Van Zon, Lubensky, Altena and ten Wolde)

Along similar lines, the work of Van Zon et al. (2007) uses a similar core mechanism with allosteric transitions in KaiC (although they allow for small flipping probabilities between each of the inactive states, C^*_j, and the active states, C_j). The essential ingredients for synchronous hexamer dynamics and sustained oscillations in the mathematical model of Van Zon et al. (2007) includes: (a) an inactive form of KaiC (indicated by C^* as in previous models), (b) differential affinity of KaiA for KaiC at different degrees of phosphorylation – KaiA is assumed to prefer binding to hexamers with lower numbers of phosphates than to those with more phosphates allowing those hexamers lagging behind to "catch up" with the hexamers in the lead, and (c) sequestration of KaiA by KaiB–KaiC* complexes to prevent the dephosphorylation phase hexamers from desynchronizing. Without the sequestration of KaiA, the first hexamers which transition to the active form ($C^*_0 \rightarrow C_0$) may quickly phosphorylate and destroy the population rhythm for the hexamers which lag behind and are still dephosphorylating, resulting in overall population damping to a stable steady state. The authors also present, for a specific choice of rates, the effects of abundance changes on the period and amplitude, and a rough agreement for temperature compensation of the period with some additional assumptions. It is important to note that the mechanism of KaiA implies that, if too much KaiA is present, the population oscillation de-synchronizes.

16.4.7 Model 7 (Rust, Markson, Lane et al.)

The mathematical model of Rust et al. (2007) attempts to reproduce their experimental findings related to the circadian cycle in the population dynamics of the net amount of phosphoryl groups at different sites in the population of KaiC monomers (net T432 phosphates; T-KaiC, net S431 phosphates; S-KaiC, phosphorylated at both sites; ST-KaiC). Rust and co-workers noticed that the amount of T432 tracked the phosphorylation phase while the amount of phosphoryl groups on S431 tracked the dephosphorylation phase of the oscillations. Their mathematical model consists of describing the population dynamics of these phosphoforms using phenomenological inactivation of KaiA by S-KaiC, and assuming specific KaiA-dependent transition rates between the four states in the model. Impressively, by fitting the rates to partial reactions the authors were able to generate a circadian oscillation in the population phosphoforms roughly consistent with their experimental data. The primary drawback is that the model does not make contact with the individual KaiC hexamer kinetics and the explicit biochemical interactions of the hexamer (and monomer constituents) with KaiA and KaiB. Thus, inactivation of KaiA is treated phenomenologically in the model and cannot be used to test alternate mechanisms of inactivation. Since these KaiA–KaiC and KaiB–KaiC interactions are presumably partially responsible for an individual hexamer cycling through the distinct phosphoform states, a detailed stoichiometric analysis and examination of hexamer–hexamer interactions is lacking in the simplified version of the model.

16.4.8 Other Mathematical Models

There are also other recent additional mathematical models for the KaiC system which reiterate many of these themes incorporating and gene expression and hypothetical "states" (Miyoshi et al. 2007), include delays for the generation of oscillations (Li and Fang 2007), and attempt to explain the oscillations by theoretically studying various types of feedback (Imamura et al. 2007).

16.5 Conclusions

The discussion of what may be drawn from such models is that there are two distinct processes for which different mechanisms are required, and that both mechanisms are required to adequately explain the circadian oscillations in a population of KaiC hexamers. The first process is the cyclic phosphorylation/dephosphorylation reactions that occur on a single hexamer. The mathematical model must contain a mechanism for the oscillation in the phosphorylation levels of the single hexamer. The second process is the synchronization of this process across the population of hexamers. Within each process there is roughly a phosphorylation and dephosphorylation phase, as experimentally observed.

While the essential core structure for the single hexamer phosphorylation cycle is almost identical in many of the mathematical models discussed above, most differ in terms of how the hexamers are synchronized. There are currently only a few proposed mechanisms and the synchronization may occur in either the phosphorylation phase, the dephosphorylation phase, or both, and may include one or more mechanisms in each phase (or none in one and several in the other). That the two population phases must be in "weak" or non-existent "communication" is obvious, for otherwise the oscillations would not occur (e.g., Fig. 16.2). Those mathematical models that neglect (or implicitly include) randomized synchronized "states" may appear to lack such population synchronizing mechanisms. In general, potential mechanisms of hexamer synchronization appear to be at least in three forms: (a) differential reaction rates – phosphorylation or dephosphorylation reaction rates depend on the number of phosphoryl groups on the monomer or hexamer (e.g., autocatalytic phosphorylation or autocatalytic dephosphorylation, either when unbound bound to KaiA or bound to KaiB or both), (b) differential affinities – association/disassociation rates of KaiA and/or KaiB depend on the number of phosphates on the hexamer, and (c) direct exchange of monomers between hexamers. Each of these mechanisms allows for the possibility of the population moving as a group (with some standard deviation in number of phosphates per hexamer that is mechanism-dependent) rather than as individuals.

It is clear that there is both interesting experimental and theoretical work involved in understanding the functioning of this particular in vitro clock and considerably more work involved in determining how the clock functions in vivo. The integration of experimental work with mathematical models on this and other similar biophysical systems should provide us with a clear understanding of which fundamental mechanisms are possible for a given phenomena under investigation and which mechanisms are more likely (have verified predictions and stimulate further experimental work). It should also provide a clear basis for reasoning quantitatively about these complex systems within a well defined framework.

References

Clodong S, Düring U, Kronk L, Axmann I, Wilde A, Herzel H, Kollmann M (2007) Functioning and robustness of a bacterial circadian clock. Mol Sys Bio 3:90

Emberly E, Wingreen NS (2006) Hourglass model for a protein-based circadian oscillator. Phys Rev Lett 96:038303

Gillespie DT (1976) A general method for numerically simulating the stochastic time evolution of coupled chemical reactions. J Comput Phys 22:403–444

Hitomi K, Oyama T, Han S, Arvai AS, Getzoff ED (2005) Tetrameric architecture of the circadian clock protein KaiB: a novel interface for intermolecular interactions and its impact on the circadian rhythm. J Biol Chem 280:19127–19135

Ito H, Kageyama H, Mutsuda M, Nakajima M, Oyama T, Kondo T (2007) Autonomous synchronization of the circadian KaiC phosphorylation rhythm. Nat Stat Mol Biol 14:11

Iwasaki H, Taniguchi Y, Ishiura M, Kondo T (1999) Physical interactions among circadian clock proteins KaiA, KaiB and KaiC in cyanobacteria. EMBO J 18:1137–1145

Kageyama H, Nishiwaki T, Nakajima M, Iwasaki H, Oyama T, Kondo T (2006) Cyanobacterial circadian pacemaker: Kai protein complex dynamics in the KaiC phosphorylation cycle in vitro. Mol Cell 23:161–171

Kitayama Y, Iwasaki H, Nishiwaki T, Kondo T (2003) KaiB functions as an attenuator of KaiC phosphorylation in the cyanobacterial circadian clock system. EMBO J 22:2127–2134

Kurosawa G, Aihara K, Iwasa Y (2006) A model for the circadian rhythm of cyanobacteria that maintains oscillation without gene expression. Biophys J 91(6):2015–2023

Li S, Fang YH (2007) Modelling circadian rhythms of protein KaiA, KaiB and KaiC interactions in cyanobacteria. Biol Rhythm Res 38:43–53

Mehra A, Hong C, Shi M, Loros J, Dunlap J, Ruoff P (2006) Circadian rhythmicity by autocatalysis. PLoS Comput Biol 2:e96

Miyoshi F, Nakayama Y, Kaizu K, Iwasaki H, Tomita M (2007) A mathematical model for the Kai-protein–based chemical oscillator and clock gene expression rhythms in cyanobacteria. J Biol Rhythms 22:69–80

Mori T, Saveliev SV, Xu Y, Stafford WF, Cox MM, Inman RB, Johnson CH (2002) Circadian clock protein KaiC forms ATP-dependent hexameric rings and binds DNA. Proc Natl Acad Sci USA 99:17203–17208

Mori T, Williams DR, Byrne MO, Qin X, Egli M, Mchaourab HS, Stewart PL, Johnson CH (2007) Elucidating the ticking of an in vitro circadian clockwork. PLoS Biol 4:e93

Nakajima M, Imai K, Ito H, Nishiwaki T, Murayama Y, Iwasaki H, Oyama T, Kondo T (2005) Reconstitution of circadian oscillation of cyanobacterial KaiC phosphorylation in vitro. Science 308:414–415

Nishiwaki T, Satomi Y, Kitayama Y, Terauchi K, Kiyohara R, Takao T, Kondo T (2007) A sequential program of dual phosphorylation of KaiC as a basis for circadian rhythm in cyanobacteria. EMBO J 26:4029–4037

Pattanayek, R, Wang J, Mori T, Xu Y, Johnson CH, Egli M (2004) Visualizing a circadian clock protein: crystal structure of KaiC and functional insights. Mol Cell 15:375–388

Pattanayek R, Williams DR, Pattanayek S, Xu Y, Mori T, Johnson CH, Stewart PL, Egli M (2006) Analysis of KaiA-KaiC protein interactions in the cyano-bacterial circadian clock using hybrid structural methods. EMBO J 25:2017–2028

Rust MJ, Markson JS, Lane WS, Fisher DS, O'Shea EK (2007) Ordered phosphorylation governs oscillation of a three-protein circadian clock. Science 318:809–812

Takigawa-Imamura H, Mochizuki A (2006) Predicting regulation of the phosphorylation cycle of KaiC clock protein using mathematical analysis. J Biol Rhythms 21(5):405–416

Terauchi K, Kitayama Y, Nishiwaki T, Miwa K, Murayama Y, Oyama T, Kondo T (2007) ATPase activity of KaiC determines the basic timing for circadian clock of cyanobacteria. Proc Natl Acad Sci USA 104:16377–16381

Tomita J, Nakajima M, Kondo T, Iwasaki H (2005) No transcription-translation feedback in circadian rhythm of KaiC phosphorylation. Science 307:251–254

Vakonakis I, LiWang AC (2004) Structure of the C-terminal domain of the clock protein KaiA in complex with a KaiC-derived peptide: implications for KaiC regulation. Proc Natl Acad Sci USA 101:10925–10930

Van Zon JS, Lubensky DK, Altena PR, ten Wolde PR (2007) An allosteric model of circadian KaiC phosphorylation. Proc Natl Acad Sci USA 104:7420

Xu Y, Mori T, Pattanayek R, Pattanayek S, Egli M, Johnson CH (2004) Identification of key phosphorylation sites in the circadian clock protein KaiC by crystallographic and mutagenetic analyses. Proc Natl Acad Sci USA 101:13933–13938

Ye S, Vakonakis I, Ioerger TR, LiWang AC, Sacchettini JC (2004) Crystal structure of circadian clock protein KaiA from *Synechococcus elongatus*. J Biol Chem 279:20511–20518

Yoda M, Eguchi K, Terada TP, Sasai M (2007) Monomer-shuffling and allosteric transition in KaiC circadian oscillation. PLoS ONE 2:e408

Chapter 17
A Synthetic Biology Approach to Understanding Biological Oscillations: Developing a Genetic Oscillator for *Escherichia coli*

Alexander J. Ninfa, Mariette R. Atkinson, Daniel Forger, Stephen Atkins, David Arps, Stephen Selinsky, Donald Court, Nicolas Perry, and Avraham E. Mayo

Abstract Our goals are to construct a simple genetic clock that will stably oscillate in *Escherichia coli* and to identify the design principles and parameters responsible for oscillations. We previously described a simple genetic circuit of linked activator and repressor operons that produced damped oscillations. Here, we altered the repression of the activator operon and identified an oscillator that produces improved oscillations over our initial system. We also explored mathematical models of the oscillator. Toy models were used to investigate the behaviors that may be obtained from our clock circuitry. Depending on parameters, the circuitry produced a wide array of oscillatory systems, including sinusoidal and relaxation oscillators. We also attempted to explicitly model all known interactions that affect the oscillator, producing a 32-dimensional ODE model. This model can produce results similar to those obtained in experiments, and we have begun attempts to fit experimental data to the model.

A.J. Ninfa(✉), M.R. Atkinson, S. Atkins, D. Arps, S. Selinsky, and A.E. Mayo

Department of Biological Chemistry, University of Michigan Medical School, Ann Arbor, MI 48109-0606, USA, e-mails: aninfa@umich.edu, sjatkins@umich.edu

D. Forger

Department of Mathematics, University of Michigan, Ann Arbor, MI 48109-1043, USA, e-mail: forger@umich.edu

D. Court

National Cancer Institute–Frederick, Frederick, MD 21702-1201, USA, e-mail: court@ncifcrf.gov

N. Perry

Department of Biophysics, University of Michigan, Ann Arbor, MI 48109, USA, e-mail: nperryp@umich.edu

A.E. Mayo

Current address: Weizmann Institute of Science, Rehovot, Israel, e-mail: avimayo@weizmann.ac.il

J.L. Ditty et al. (eds.), *Bacterial Circadian Programs*.

17.1 Introduction

The goal of our synthetic biology approach is not to imitate the circuit topology or regulatory mechanisms of any natural genetic clock. Rather, we attempt to construct a small model oscillatory system in which all parameters are identified and can be manipulated. This construction allows direct comparisons between models and experiments. The synthetic genetic oscillator that we study here reproducibly displays damped oscillations of gene expression in large *Escherichia coli* cell populations (Atkinson et al. 2003; Ninfa et al. 2007). In the experiments shown here and before (Atkinson et al. 2003; Ninfa et al. 2007), we observe rhythms of *lacZ* expression in populations of $\sim 10^{11}$ cells over a time frame of ~50 cell generations, with the period of the damped oscillations being about 10 generations. Thus, the individual cells in the population inherit information regarding the time since the release from induction as they grow and divide. The availability of this oscillator naturally leads to further efforts to understand and improve its function. These efforts do not come easily. As part of our oscillatory mechanism, we require the dilution of transcription factors by cell growth as the means for reduction of transcription factor concentration. That is, after a bolus of transcription factor (activator and repressor) synthesis, our clock is designed to use feedback to block further synthesis of the transcription factors, after which cell growth is required to reduce the level of the repressor to the concentration at which the next round of activator and repressor synthesis occurs (see below).

Our initial goal was to develop a system that produced oscillations that would occur over many generations, as occurs naturally in the prokaryotic *Synechococcus elongatus* PCC 7942 model system (Kondo et al. 1997). The result of this engineering decision was a double-edged sword, as very dramatic oscillations were obtained, but one must maintain a bacterial culture in a turbidostat and monitor gene expression for many cell generations, which requires several days of continuous monitoring. A consequence of using a transcription factor that regulates other cellular genes (NRI; see Sect. 17.2) is that our oscillator inhibited cellular growth in its "activated" phase relative to its "repressed" phase. Thus, during fermentation, significant changes in the speed of the nutrient pumps were required to maintain constant culture turbidity. One solution to this predicament was to automate the experiment such that constant human maintenance of the experiment was not required. This advance occurred recently (Ninfa et al. 2007), and our current and future experiments will be automated. Here, we present data that were generated before the era of automated experiments, where continuous human maintenance of the turbidostats limited the number of experiments that could be performed. For most of the experiments shown below, we used a standard laboratory fermentor and operated the device as a turbidostat by manually measuring the culture turbidity and adjusting the nutrient pump by hand to maintain constant culture turbidity.

17.2 Background: the Initial Synthetic Oscillator

Our synthetic genetic oscillator for *E. coli* consists of interacting activator and repressor operons (Fig. 17.1), which are located within "landing pads" on the *E. coli* chromosome referred to as the activator and repressor modules. "Landing pads" refer to sections of the *E. coli* chromosome that have been modified to allow for simple placement of genes onto the chromosome. Typically they contain a selectable marker, such as drug resistance, near a multiple cloning site that is flanked by transcriptional terminator sequences; regions homologous to the chromosome flank these elements. As described previously (Ninfa et al. 2007), devices such as synthetic operons that are cloned into landing pad plasmids may be recombined onto the *E. coli* chromosome, after which the chromosomal genes may be transferred between strains

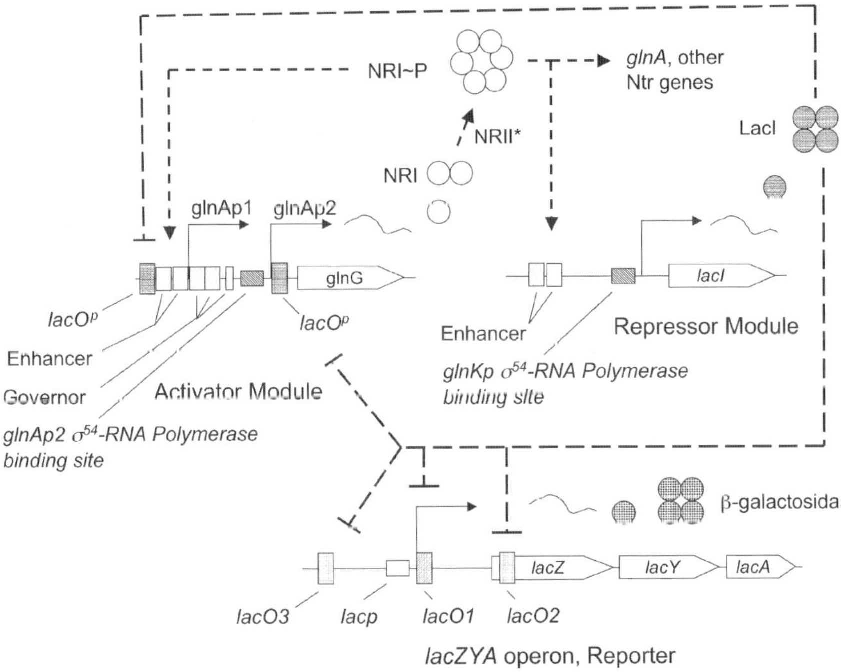

Fig. 17.1 Structure of the synthetic genetic clock. The activator module is shown *left*, the repressor module is shown *right*, and the *lacZYA* operon that serves as the reporter in many experiments is shown *bottom*. *Bent arrow* Site of transcription initiation for gene promoters, *wavy line* mRNA species, *circle* protein subunit, *heavy dashed line* regulatory interaction, *arrowhead* gene activation, *flat ending* gene repression, *small boxes* DNA-binding sites for activator and repressor. The enhancer site consists of two repeated binding sites for NRI; these may be strong or weak binding sites, designated by *empty* or *stippled boxes*, respectively. The enhancer of the repressor module consists of one strong and one weak NRI-binding site. The governor sites of the activator module are sites that weakly bind activator. For further description, see text

by standard phage-mediated generalized transduction. For the synthetic genetic clock discussed here, the activator operon was positioned within a chromosomal landing pad within the *rbs* operon; this landing pad contains a gentamycin-resistance marker and results in the loss of the ability of the cells to use ribose as the sole carbon source. The repressor operon was positioned within a chromosomal landing pad in the *glnK* operon (Atkinson et al. 2003); this landing pad contains a chloramphenicol-resistance marker but does not result in a simple nutritional phenotype. The two landing pads are separated by about 25% of the *E. coli* chromosome.

The activator module promoter is a modified version of the *glnA* control region (Fig. 17.1). This control region contains two promoters, *glnAp1* and *glnAp2*, resulting in the transcriptional start sites designated by bent arrows in Fig. 17.1. The *glnAp1* promoter overlaps the binding sites for the activator (designated "Enhancer" in Fig. 17.1); this promoter is repressed upon binding of the activator. The activator is the phosphorylated form of the NRI protein (NRI~P); when phosphorylated the dimeric NRI~P forms an oligomer (most likely a hexamer) that stimulates transcription of its own structural gene, *glnG*, by binding to the enhancer and causing RNA polymerase containing sigma 54 (σ^{54}) to utilize the *glnAp2* promoter (Fig. 17.1). The NRII protein is responsible for phosphorylating NRI, but has been modified such that it is no longer capable of dephosphorylating NRI (depicted as NRII*). The activator works by binding to high-affinity enhancer sequences upstream from the activator gene; this activator also stimulates the transcription of the repressor gene by binding to a lower-affinity enhancer upstream of the repressor gene promoter (Ninfa et al. 1987; Atkinson et al. 2002a, b). This design was intended to provide a delay between the activation of the two module promoters.

The repressor, LacI, whose own expression is driven by the *glnKp* promoter, blocks transcription of the activator gene by binding to the operator site just downstream from the site of *glnAp2* transcription initiation. The activator module of our oscillator also contains a second operator for LacI-binding located upstream from the enhancer, and we anticipated that this distal operator site would enhance repression by allowing the formation of a repression loop, as occurs in the repression of the *lacZYA* operon (Oehler et al. 1990). For our initial clock, we used "perfect" *lac* operators (*lacOp*) that bind repressor significantly better than the strongest natural *lac* operator, *lacO1* (Sadler et al. 1983). As noted, the activator module also contained a second, minor promoter (*glnAp1*) that overlaps with the enhancer sequences. This promoter provides a low level of expression in the absence of NRI~P, which allows for priming of the system in the absence of activator and is repressed by NRI~P (Reitzer and Magasanik 1985). The initial activator module also contained three weak activator-binding sites, known as the governor sites, which lie between the enhancer and promoter (Fig. 17.1). Occupancy of these sites by NRI~P has been shown to repress expression from the promoter at very high concentrations of NRI~P (Atkinson et al. 2002a, b).

To build the bacterial strain that had all necessary elements of the synthetic genetic clock, the activator and repressor modules were introduced into the chromosome of a bacterial strain that was deleted for *glnL* and *glnG*, which encode the native NRII and NRI proteins, respectively, and contained a null mutation of *lacI* (Atkinson et al. 2003). The function of NRII was partially restored by introducing a plasmid that encodes the modified NRII* (unable to dephosphorylate NRI). This mutant form of NRII was used to bypass the normal cellular regulation of NRI phosphorylation by nitrogen status, so that high levels of activation could be obtained in cultures growing in nitrogen-rich conditions (Ninfa and Magasanik 1986). To monitor oscillator function, repression of the *lacZYA* operon (Fig. 17.1) was measured to determine the level of repressor. The activity of β-galactosidase, product of *lacZ*, was measured using the standard Miller assay or using a miniaturized version of the assay (Ninfa et al. 2007); this activity is reported in Miller units. In some cases, the expression of the chromosomal glutamine synthetase (GS) *glnA* gene was also measured to determine activator function.

To conduct the oscillator experiments, cells were synchronized by induction with IPTG, washed to remove IPTG, and introduced into a turbidostat where optical density was held constant by manual adjustment of a nutrient pump. Samples were periodically removed to assay the reporters. This initial oscillator that we earlier described (Atkinson et al. 2003) produced damped oscillations in *E. coli*, in which large populations of *E. coli* demonstrated synchronous waves of *lacZ* expression for three or four cycles with periods of about 10 cell generations or more (where the doubling time was about 1 h), in experiments lasting around a total of 60 h (Fig. 17.2). As expected, activator and repressor module expression were out of phase with one another (by about two generations), as indicated by monitoring GS and β-galactosidase expression in a single experiment. Although the magnitude of the phasing difference between the two modules was not anticipated, a delay in the transcription of the repressor module promoter was the engineering goal of using the *glnK* promoter for this part of the system (Atkinson et al. 2003). A detailed description of the oscillator experimental protocols and assay procedures has been previously published, along with a description of a homemade turbidostat device that can be used to automate the procedure (Ninfa et al. 2007).

17.3 Synthetic Oscillator with Altered Repression of the Activator Module Promoters

17.3.1 Altered Governor Sites

Based on our experience with the natural *glnA* promoter region, we anticipated that the governor sequences in our initial activator module could become problematic if our system was able to operate at high activator concentrations. This is because, at high activator concentrations, activator would result in a reduction in expression of the activator module. Such a biphasic activity of the activator was undesirable both from an engineering perspective and for simplification of the modeling of clock function. Furthermore, in the future we plan to develop clocks that function at high activator concentrations. Thus, the governor NRI~P-binding sequences were

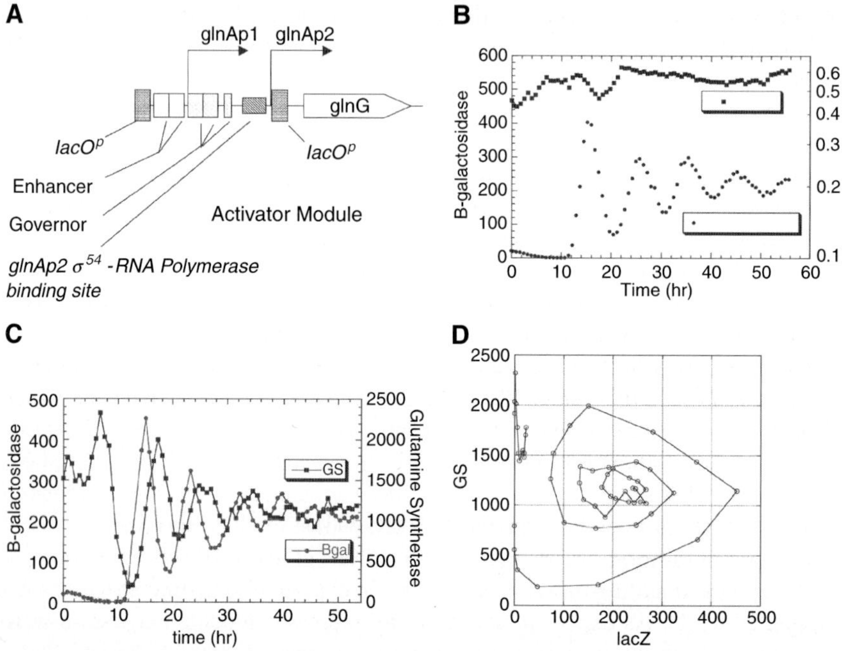

Fig. 17.2 Function of the initial synthetic genetic clock. **A** Structure of the activator module of the clock. Symbols are as in Fig. 17.1. The activator module contained two copies of the *lacOp* operator and the natural complement of governor sequences from the *glnA* control region. **B** Example of a clock experiment using the initial synthetic genetic clock. ▪ Optical density of the culture, • β-galactosidase activity, hr hours. As shown, there were four obvious waves of *lacZ* expression over a time-course of 50 h. **C** Another clock experiment, in which both β-galactosidase (*Bgal*) and glutamine synthetase (*GS*) were measured. As shown, both activities oscillated and were not in phase with one another. **D** Phase diagram depicting the *lacZ* (β-galactosidase) and GS data from the same experiment shown in panel (**C**)

altered to maintain the length and base composition of the DNA between enhancer and promoter, but the sequences of the two most upstream of the three governor elements were scrambled (Atkinson et al. 2002a, b). Direct measurement of activator abundance during oscillator experiments by immunoblotting indicated that activator was quite low and thus below the level where the governor sites should be functional (Atkinson et al. 2003). As expected, we observed that removal of two upstream governor sites did not greatly influence function of the genetic clock (compare Figs. 17.2A, B, 17.3B). Details of the experimental methods of these clock experiments have been published (Ninfa et al. 2007).

17.3.2 Altered Activator Module Operator Sites

Our initial activator module contained two *lacOp* sites, separated by about 16 turns of the DNA helix (Fig. 17.1). We expected that these operators should permit very tight repression of the activator module promoter and that this repression would be highly cooperative due to the tetrameric repressor simultaneously binding to both operators and forming a repression loop, as in the natural *lacZYA* context (Oehler et al. 1990). We examined the effects of altering the identity and position of the operators in the activator module that contains only the most proximal of the three governor sequences. The operator sequence *lacOp* was replaced with that of either *lacO1* or *lacO3*. Previous studies showed that *lacOp* binds the repressor about 10-fold better than *lacO1*, which in turn binds to Lac repressor at least 10-fold better than *lacO3* (Sadler et al. 1983). The effect of removing the distal operator was also examined. For each experiment, each altered activator module was placed into the chromosome in the same landing pad of their respective strain. All conditions for this series of experiments were similar to those described earlier (Atkinson et al. 2003) except that, in this series of experiments (Figs. 17.3, 17.4, 17.5), the *E. coli* cells contained a mutation of the *nac* gene enhancer on the *E. coli* chromosome.

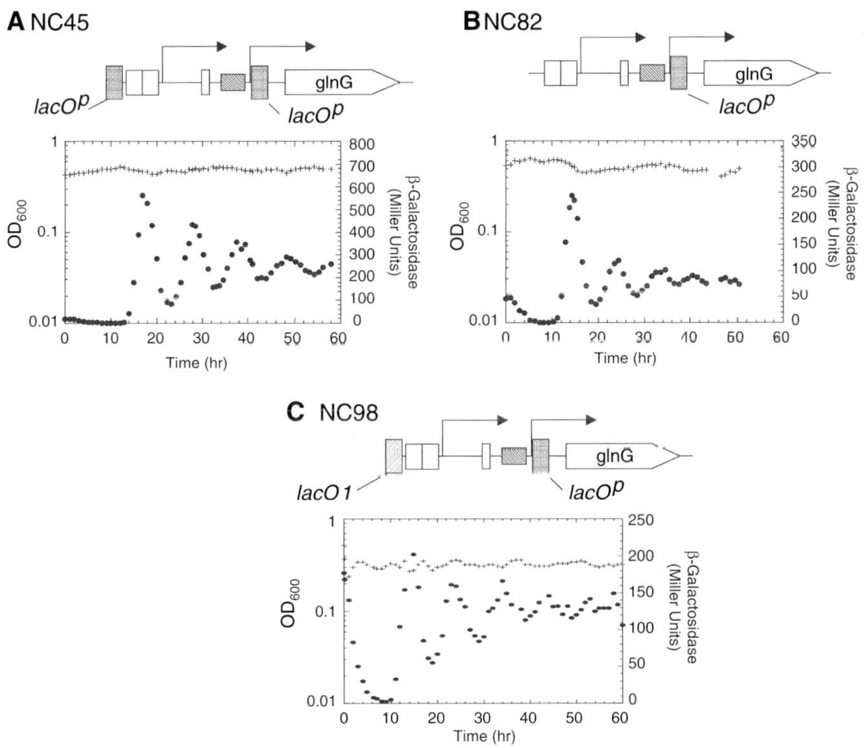

Fig. 17.3 Synthetic genetic clocks containing a proximal *lacOp* and various distal operators. For each panel, the structure of the activator module is shown along with the results of a clock experiment. + Optical density of the culture, • β-galactosidase activity. **A** NC45 clock strain with activator module containing a *lacOp* distal operator. **B** NC82 clock strain with activator module lacking a distal operator. **C** NC98 clock strain with activator module containing a *lacO1* distal operator

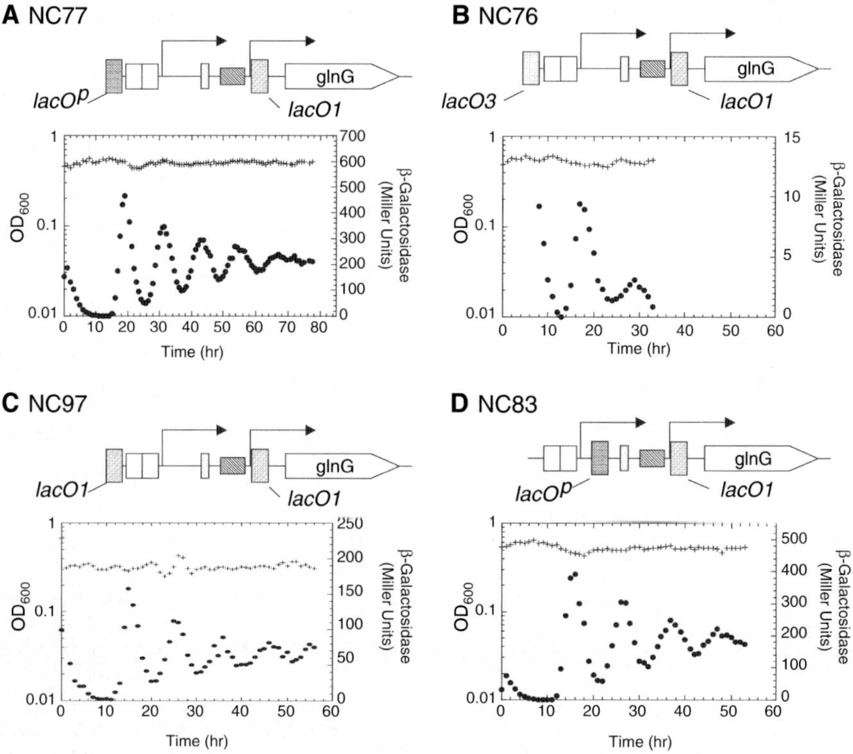

Fig. 17.4 Synthetic genetic clocks containing a proximal *lacO1* and various distal operators. For each panel, the structure of the activator module is shown along with the results of a clock experiment. Symbols are as in Fig. 17.3. **A** NC77 clock strain with activator module containing a *lacOp* distal operator. **B** NC76 clock strain with activator module containing a *lacO3* distal operator. **C** NC97 clock strain with activator module containing a *lacO1* distal operator. **D** NC83 clock strain with activator module containing a *lacOp* operator positioned at −78 relative to the transcription start site of the *glnAp2* promoter, where the governor sites are normally located (compare to Fig. 17.1)

This mutation, introduced by recombination (Yu et al. 2000), was used to help disconnect the clock from cellular physiology. The *nac* gene encodes a transcription factor that helps to slow cell growth during periods of nitrogen limitation (Blauwkamp and Ninfa 2002). Mutation of the *nac* enhancer, while not having a major effect on oscillator function (data not shown), did minimize the effects of the modules on cell growth rate, making it easier for the experimenters to maintain constant culture optical density during the experiments.

Figure 17.3 shows results for the NC45, NC82, and NC98 clock strains. These strains contain activator modules that have *lacOp* as the proximal operator and either *lacOp* (NC45), *lacO1* (NC98), or no upstream operator (NC82). The NC45 results show that removal of the two distal governor sequences did not discernibly alter performance of the clock relative to the initial clock (compare Fig. 17.2B with Fig. 17.3A), as four well defined peaks of *lacZ* expression were obtained in a clock

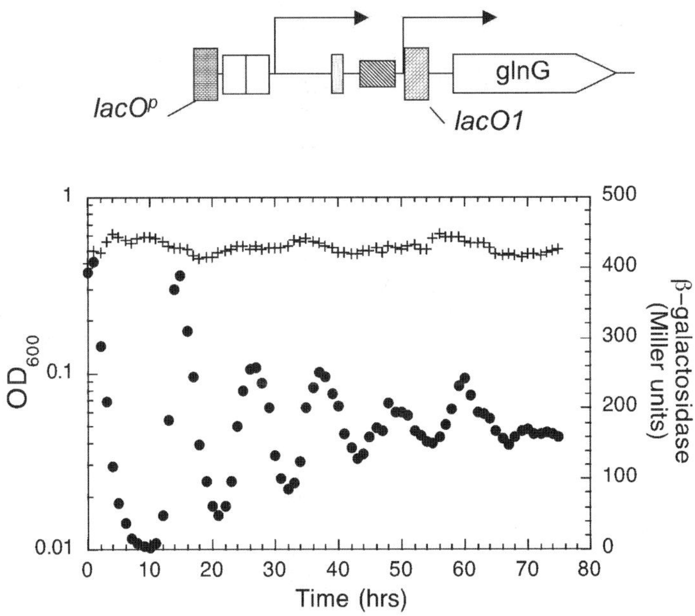

Fig. 17.5 A separate clock experiment using the NC77 clock strain. Symbols are as in Fig. 17.3

experiment lasting about 60 h. The NC82 results show that elimination of the distal operator (in addition to the two governor sites) resulted in increased damping of the oscillations. Thus, the distal operator plays some beneficial role in maintaining the amplitude of the oscillation. The NC98 results show that oscillations were observed when the upstream operator was *lacO1*, but the damping of oscillations seemed to be greater than when the upstream operator was *lacOp* (compare Fig. 17.3A with Fig. 17.3C).

Figure 17.4 shows results for the NC77, NC76, NC97, and NC83 clock strains. These strains all contain clock activator modules that have the *lacO1* sequence as the proximal activator. The distal operators were either *lacOp* (NC77), *lacO3* (NC76), *lacO1* (NC97), or *lacOp* positioned at –78 (Fig. 17.4). The NC76 (Fig. 17.4B) results show that when the distal operator was *lacO3* and the proximal operator was *lacO1,* as in the *lacZ* promoter, the clock barely oscillated at all with two weak peaks of *lacZ* expression (note the scale of the axis for this experiment is expanded so that the very small peaks in *lacZ* expression can be observed). Because of these results with the NC76 strain, we did not bother to examine the results of completely deleting the distal operator when the proximal operator was *lacO1*, as it would be expected to oscillate even less than that of the NC78 strain. The results obtained for the NC77 strain, where the distal operator was *lacOp* and the proximal operator was *lacO1*, oscillator function was improved (Fig. 17.4A). Indeed, this NC77 strain is our best oscillator (Fig. 17.4A). By comparison of the NC77 and NC76 results, it is again clear that the upstream operator is playing a significant role (Fig. 17.4A, B). Reasonably good oscillations were also obtained when both operators were *lacO1*

(NC97; Fig. 17.4C). Finally, when the proximal operator was *lacO1* and the distal *lacOp* operator was moved to a position centered at −78 where the governor sequence formerly was located in the original clock, reasonable oscillations were also obtained (NC83; Fig. 17.4D). Thus, the distal operator was playing a beneficial role, *lacOp* was best at this role, *lacO1* but not *lacO3* could substitute for the distal role, and the distal operator could be located either at its original position or at position −78. It should be noted that the last module mentioned, with *lacOp* at −78, might function by a different mechanism than the other arrangements.

The NC77 module arrangement was used in several additional repeated experiments, where its function again appeared to be improved over the initial system. Specifically, damping appeared to be reduced and in some experiments a clear fifth cycle could be detected. One such experiment is shown in Fig. 17.5, where five cycles of *lacZ* expression were obtained in 70 h. It should be pointed out that the initial levels of *lacZ*, which reflect the levels of expression in the IPTG-induced cells used at the beginning of each experiment, were observed to be highly variable. This variability is currently under investigation; conditions for the induction of the cultures with IPTG apparently determine these properties.

17.3.3 Improved Function of the Activator Module Promoter Did Not Correlate with Improved Repression

In order to get a sense of why the various arrangements of operators functioned as they did, we made a series of *lacZYA* fusions to the set of activator module promoters, and by measuring the levels of β-galactosidase activity, we examined repression in cells that had a constant level of NRI~P and a low, constant level of LacI (Fig. 17.6). It should be noted that the fixed level of activator in these experiments was lower than that required to trigger expression of *nac* and thus the cells grew rapidly at a constant rate during the experiments. In each of these experiments, the activator module promoters were linked to *lacY*$^+$ and thus were fully induced with 25 mM IPTG (Fig. 17.6, 25 mM IPTG column). Conversely, in the *lac* operon control, a *lacY*$^-$ version was used and induction was only partial at 25 mM IPTG (see Fig. 17.6 legend). Of the various activator module promoters tested, best repression was obtained with the promoter containing a single proximal *lacOp* (Fig. 17.6B). This promoter, analogous to that used in the clock activator module of strain NC82, had the lowest basal expression level in the absence of inducer (12 Miller units, Fig. 17.6B) and the highest repression ratio of 219 (maximum expression divided by the fully repressed level; Fig. 17.6B). The promoter with a pair of *lacOp*, at both proximal and distal locations (analogous to the activator module of the NC45 clock strain), displayed less repression, with a basal level of 18 Miller units and a repression ratio of 149 (Fig. 17.5A). However, the NC45 clock strain seemed to produce stronger (less damped) oscillations in the clock experiments (Fig. 17.3A, B). Thus, there was a lack of correspondence between the function of the activator module in clock experiments and the strength of the repression of the activator mod-

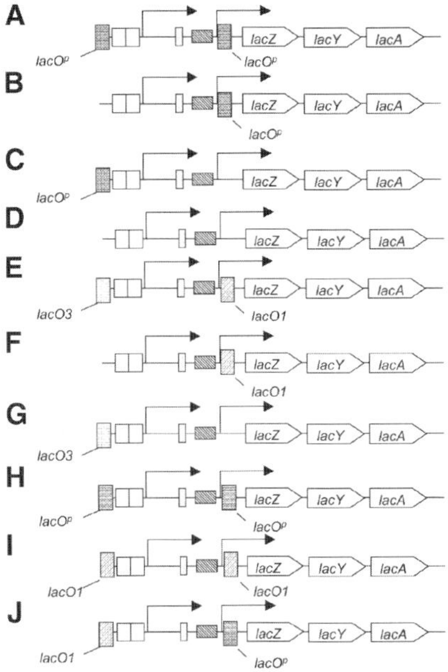

related to clock	0 IPTG	25 μM IPTG	Repression
NC45	18	2680	149.0
NC82	12	2630	219.0
----	2550	3600	1.4
----	3670	3850	1.1
NC76	57	4490	78.0
----	47	4215	90.0
----	3370	3700	1.1
NC77	90	4170	46.0
NC97	104	4010	39.0
NC98	14	2819	201.0

Fig. 17.6 Repression of the activator module promoters as measured when fused to a *lacZYA* reporter. Each of the promoter-*lacZYA* fusions was present in single copy within the *trp* landing pad on the bacterial chromosome (Atkinson et al. 2003 and references therein). The host cells were deleted for the natural *lacZYA* operon and contained a wild-type *lacI* gene. Cells were grown to mid-log phase under nitrogen-limiting conditions and β-galactosidase activity is expressed in Miller units. The basal expression level was determined at 0 mM IPTG, while the full expression level was obtained at 25 mM IPTG (see text). The repression (*right column*) is the ratio of expression in the presence and absence of IPTG. Activator module promoters that are analogous to those used in clock activator modules are indicated by showing the name of the clock strain

ule promoter when measured directly, as in Fig. 17.6. Another operator arrangement that demonstrated strong repression consisted of *lacO1* as the distal operator and *lacOp* as the proximal operator, analogous to the activator module promoter of the NC98 clock strain (Fig. 17.6J). This promoter demonstrated a basal level of 14 Miller units and an induction ratio of 201, which means that it is intermediate in these properties between the promoters in Fig. 17.6A, B. This result is logical, and in the clock experiment the strain NC98 appeared to produce oscillations with damping intermediate between the NC45 and NC82 clocks (Fig. 17.3). Together, the set of results in Fig. 17.6A, B, J and the corresponding clock experiments in Fig. 17.3 suggest that the distal operator causes the level of basal expression to increase and the ratio of repression to decrease; that is, it reduces repression. Furthermore, the stronger the distal operator, the more it reduces repression. Thus, *lacOp* was more effective than was *lacO1* in reducing repression. Finally and unexpectedly, one of the best oscillations in the clock experiments seemed to be obtained with the NC45 strain, that is, with the strain with the weakest repression.

In the set of promoters that we examined (Fig. 17.6), all of the promoters where the proximal operator was *lacO1* (Fig. 17.6E, F, H, I) displayed higher basal levels of expression in the fully repressed state, when compared to the repression observed with *lacOp* as the proximal operator (Fig. 17.6). This is reasonable, as tighter binding translates into tighter repression in vivo. Among the set of promoters with a proximal *lacO1*, best repression was obtained when there was no distal operator and the proximal *lacO1* was the only operator. In this case, a basal level of 47 Miller units was observed along with a repression ratio of 90 (Fig. 17.6F). Nearly the same repression properties were displayed by a promoter that contained a distal *lacO3* and a proximal *lacO1*, analogous to the NC76 clock activator module promoter (Fig. 17.6E). In the latter case, the promoter displayed a basal level of 57 Miller units and a repression ratio of 78 (Fig. 17.6E). Thus, the presence of the very weak *lacO3* as the distal operator did not have a great effect on repression. By contrast, the *lacOp* operator had a dramatic effect on repression when it was the distal operator, resulting in a basal level of 90 Miller units and a repression ratio of 46 (Fig. 17.6H). This arrangement is analogous to the activator module promoter of the NC77 clock, which produced the best oscillations of the clocks studies so far. By contrast, the NC76 clock barely produced detectable oscillations at all (Fig. 17.4). Thus, again, the promoter arrangements that produced best oscillations in the clock activator modules did not correspond to the promoter arrangements that demonstrated best repression, when measured directly, as in Fig. 17.6. The promoter with *lacO1* in both proximal and distal positions, analogous to the activator module of the NC97 clock strain, displayed repression properties that were nearly the same as the NC77 arrangement when measured directly; and the NC97 clock strain produced strong oscillations in a clock experiment (Fig. 17.4), although these were judged to be somewhat more damped than were oscillations from the NC77 strain (Fig. 17.4). Thus, NC97 has less repression, yet weaker oscillations than NC77, in discordance with the emerging pattern that weaker repression results in improved oscillations. Nevertheless, both the NC97 and NC77 clocks were far superior to the NC76 clock, and this corresponded to weaker repression of the activator module promoter of NC77 and NC97 relative to NC76.

We also examined three promoter arrangements that did not contain a proximal operator (Fig. 17.6C, D, G) and contained either no distal operator (Fig. 17.6D), a *lacOp* distal operator (Fig. 17.6C), or a *lacO3* distal operator (Fig. 17.6G). All three of these promoters displayed essentially no repression (repression ratios of 1.1–1.4; Fig. 17.6). These results suggest that the binding of Lac repressor to the distal operator, in the absence of a proximal operator, had no influence on either of the promoters of the activator module.

To summarize the data of Fig. 17.6, the expected improvement in repression that should have been obtained by including a distal operator in the system, based on the study of DNA looping in the natural *lacZYA* operon repression (Oehler et al. 1990), was not observed. If DNA looping were occurring as in the *lacZYA* system, allowing a repressor tetramer to occupy the proximal and distal operators at the same time, then the distal operators should have lowered the basal level of expression and increased the repression ratio of the system (Oehler et al. 1990). Instead, the opposite behavior was observed: the distal operators functioned to increase the basal level of expression and decrease the repression ratio relative to promoters that lacked these sites. Thus, it seems that repressor is not simultaneously contacting both operators in our systems with looping out of the intervening DNA. Yet, the distal operators clearly had a beneficial effect on the functioning of the synthetic genetic clock. One possibility is that the distal operator serves as a "sink" for repressor and thus weakens repression at low concentrations of repressor. Perhaps by serving as sinks, these operators may bring about an increase in the apparent kinetic order of repression. This, in turn, could be aiding clock function (Atkinson et al. 2003). It was fortuitous that we included a distal operator in our design; the distal operator aids the oscillator function, but not for the reason we imagined when we designed the clock.

17.4 The Synthetic Genetic Clock Appears to be Very Sensitive to Small Changes in the Host Cell Genotype

In the course of our studies, we observed that the initial clock shown in Fig. 17.1 performed significantly worse when placed into cells deleted for the *lac* region and containing a fusion of the *glnK* promoter linked to *lacZYA* in the *trp* landing pad (Atkinson et al. 2002a, b). The relevant differences between the two strain backgrounds are as follows: deletion of the chromosomal *lacZYA* operon removes the three natural *lac* operators, and the presence of the *glnK* promoter-*lacZYA* fusion in the *trp* landing pad results in the presence of an additional target for activator that is identical to the repressor module promoter. The other differences in the strains, such as the defect in the *trp* genes in the strain containing the *glnKp-lacZYA* fusion are not anticipated to be relevant, as we have already used this landing pad for clock module placement in functional clocks that were early in the development of the current system (Atkinson et al. 2003). Thus, it seems that the relevant differences between the two strain backgrounds is the presence of an additional activator target and the loss of a repressor target. Our purpose in building the strain was to be able to monitor activation of the repressor module by measuring the parallel activation of the *glnKp-lacZYA* fusion. Measuring both GS and β-galactosidase in this clock experiment provides two indirect measurements of the activator level, and as observed, these are not expected to be out of phase (Fig. 17.7). Note that this experiment differs from that shown in Fig. 17.2, as here both GS and β-galactosidase are directly reporting on the level of activator. The clock performed significantly worse in this strain background than in the original strain background (compare Fig. 17.2B with Fig. 17.7). Specifically, the clock that produced at least four waves of oscillations in Fig. 17.2B only produced two or three waves of oscillations in Fig 17.7. Further studies are needed to define the reasons.

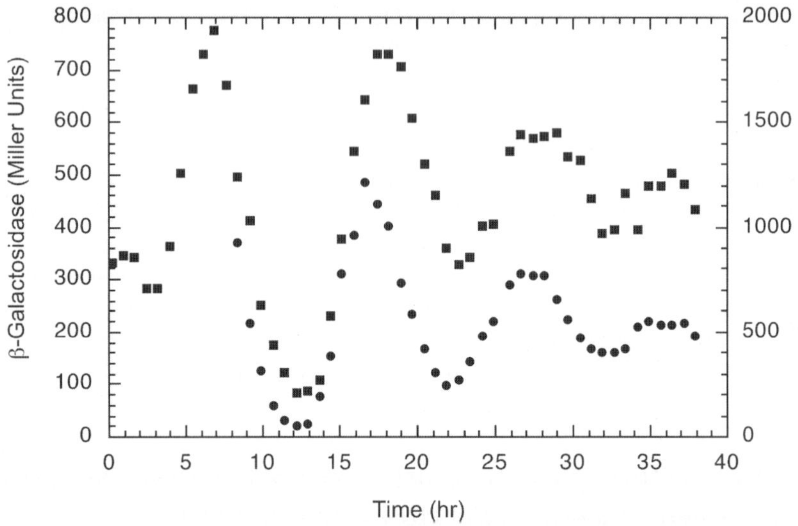

Fig. 17.7 Effect of host strain on oscillator function. Results of clock experiment using the same clock strain as in Figs. 17.1, 17.2, except that the clock was placed into a different host strain. For further details, see text

17.5 Modeling the Synthetic *E. coli* Oscillator

17.5.1 Exploring the Potential of the Regulatory Network with Mathematical Models

Figure 17.8 shows an example of a mathematical model that captures the essential features of the regulatory network, while not attempting to explicitly model all interactions that form the network. The underlying assumptions of this model are that: (i) the activator (x) is a hexamer, (ii) the repressor (y) is a tetramer, (iii) the activator and repressor are never bound to the activator module simultaneously, and (iv) the activator module has a basal level of expression while the repressor module has no basal rate of expression. In addition, it was also assumed that all protein species are perfectly stable and their concentrations are only reduced by dilution as cells grow and divide. To minimize the number of variables, parameters are combined. For example, production rates of the activator and repressor proteins consist of all factors involved in production of mRNA and protein, divided by the turnover. The ratio of affinity of activator for its two targets in the network is defined as s (the activator module enhancer and the repressor module enhancer), while b is defined as the ratio of the activated and basal rates from the activator module promoter. G is defined as the ratio of half-lives. Furthermore, it is possible to express proteins in terms of their association constant equivalents and time in terms of generations. Using these criteria and parameters, a simple model is obtained that should capture the essence of the network design (Fig. 17.8).

P-P

$$K_A A^6 = CA,$$
$$K_R R^4 = CR,$$

P-DNA

$$K_{S_aCA} S_a CA = S_{aCA},$$
$$K_{S_rCA} S_r CA = S_{rCA},$$
$$K_{S_aCR} S_a CR = S_{aCR}.$$

Mass Conservation

$$1 = S_a + S_{aCA} + S_{aCR},$$
$$1 = S_r + S_{rCA},$$

Transcription

$$\frac{da}{dt} = P_a S_{aCA} + p_a S_a - \gamma_a a,$$
$$\frac{dr}{dt} = P_r S_{rCA} - \gamma_r r.$$

Translation

$$\frac{dA}{dt} = P_A a - \gamma A,$$
$$\frac{dR}{dt} = P_R r - \gamma R.$$

$$\pi_a = \frac{P_A P_a}{\gamma \gamma_a}\left(K_{S_rCA} K_A\right)^{1/6}, \quad \pi_r = \frac{P_R P_r}{\gamma \gamma_r}\left(K_{S_aCR} K_R\right)^{1/4},$$
$$\sigma = \frac{K_{S_aCA}}{K_{S_rCA}}, \quad \beta = \frac{p_a}{P_a}, \quad \Gamma = \frac{\gamma_a}{\gamma_r}$$

$$x = \left(K_{S_rCA} K_A\right)^{1/6} A, \quad y = \left(K_{S_aCR} K_R\right)^{1/4} R, \quad \tau = \gamma_r t.$$

$$\boxed{\begin{aligned} \frac{1}{\Gamma}\frac{dx}{d\tau} &= \pi_a \frac{\beta + \sigma x^6}{1 + \sigma x^6 + y^4} - x, \\ \frac{dy}{d\tau} &= \pi_r \frac{x^6}{1 + x^6} - y. \end{aligned}}$$

Fig. 17.8 A simple model for the synthetic genetic clock. *P-P* Protein–protein interactions, P-DNA protein–DNA interactions. K_A Association constant for formation of hexameric activator, K_R association constant for formation of tetrameric repressor. *CA, CR* Hexamer activator and tetramer repressor species, respectively. S_a Site in the activator module that can be bound by either activator or repressor (for simplicity, see text), S_r enhancer site in the repressor module that is only bound by repressor. K_{SaCA} Association constant for activator binding the activator module, K_{SrCA} association constant for activator binding to enhancer of the repressor module, K_{SaCR} association constant for repressor binding to the activator module. In the model, the activator module may be free, bound by activator, or bound by repressor. The repressor module may be either free or bound by activator. Transcription and translation are described by differential equations as shown; lower case *a* and *r* refer to the mRNA species, while upper case *A* and *R* refer to the protein species. P_a Activated production rate, p_a basal expression rate from the activator module promoter. There is no term for basal expression of the repressor module, as this module has no basal expression in the absence of activator. *g* Decay rate, a summation of turnover of species and their dilution by cell growth. Simplifications used in the model are shown at the *top right*, and these are discussed in the text. The *boxed equations* are the model that was simulated using *Mathematica*

Exploring the model shown in Fig. 17.8 revealed that a large number of different oscillating systems were obtained, which included what we typically think of as sinusoidal or relaxation oscillators, along with a variety of irregular-shaped phase patterns (Fig. 17.9). That is, the parameters determined the type of oscillating system, as opposed to the system topology. Further work on this issue has failed to

Fig. 17.9 Phase diagrams for 100 different oscillating systems produced using the model in Fig. 17.8. In each phase diagram, repressor value forms the y-axis while activator value forms the x-axis. To identify the system, numerical simulations were performed for 500 simulated cell generations after which systems displaying a strong peak in Fourier space were identified. Thus, systems that produced damped oscillations (that would have reached the steady state within 500 cell generations) are not included here

alter this conclusion (Conrad et al. 2008). One may think of the oscillating systems as forming a "cloud" within the five-dimensional parameter space considered in the model. We investigated whether it was possible to "morph" one oscillating system into another by traveling from one set of parameter values to another within this "cloud", and found that 10 of the systems shown in Fig. 17.9 were connected in this manner. For this experiment, we chose 10 systems with very different shapes of the phase plots. This connection does not prove that there is a single "cloud" and that it is contiguous, but we suspect that this is the case.

Interestingly, the boundary between oscillating and non-oscillating systems in parameter space displays both a Hopf bifurcation and SNIC bifurcation. The practical result of these two types of bifurcations is that, as one "flies into the cloud" from one direction, one observes damped oscillations near the cloud that become stable oscillations with the same period as one crosses the boundary of the cloud through a Hopf bifurcation. But, as one "flies into the cloud" from another direction, one observes no hint of oscillations until one enters the cloud through a SNIC bifurcation. In one region of parameter space outside the cloud (near the Hopf bifurcation), we observed systems that produce damped oscillations similar to our experimental system.

These simulations modeled one idealized cell within the culture. Thus any effects of possible coupling were not included because our system was not designed with a coupling mechanism. However, indirect coupling could be present in our experiments by unintended mechanisms. The actual oscillator within an individual cell is probably noisy due to the small number of molecules of activator and repressor. Without coupling, this noise would cause cells to drift out of phase (see Chap. 13). This drift could account for the damping seen in our experiments. Planned experimental work will focus on investigating the noise within our clock. For the time being, we assume that this damping can be seen within individual cells.

17.5.2 Developing an Explicit Model for the Synthetic Genetic Clock

Another approach to understanding the synthetic genetic clock is to attempt to explicitly model all interactions known to involve the clock components. A first attempt at this is presented in Fig. 17.10 and the following figures. This model is of

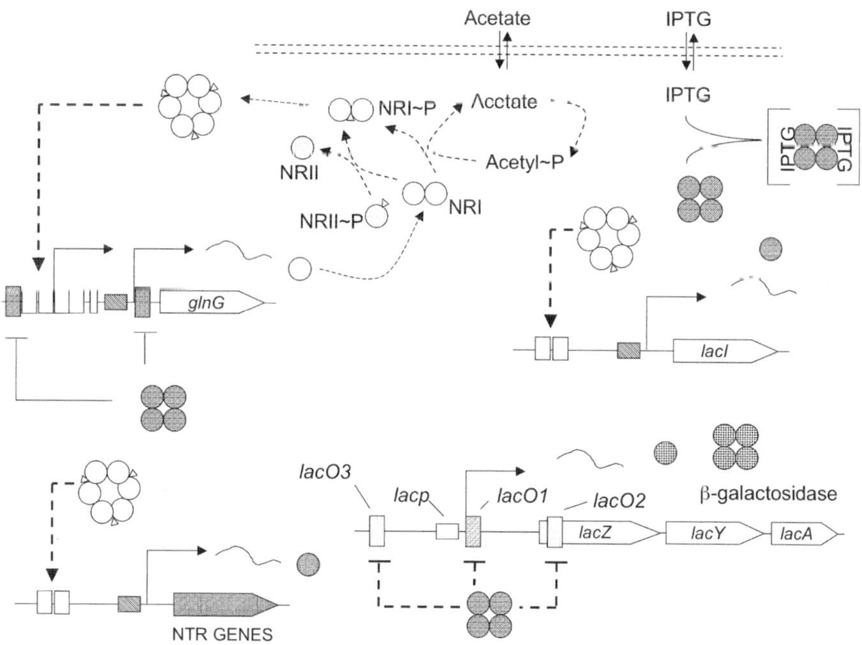

Fig. 17.10 An explicit model for clock function. Symbols are consistent with Fig. 17.1. Additional symbols: *double-dashed* line represents the cell membrane, which acetate and IPTG must cross to

course sensitive to gaps in our knowledge, which hopefully do not include any principle components essential for function. Interactions were included whether we thought them relevant or not. For example, acetyl phosphate (acetyl~P) is known to be able to phosphorylate the activator NRI in *E. coli* (Feng et al. 1992); therefore, this was included in the model along with the role of this reaction in providing acetate, which is exported from the cell and thus may be sensed by other cells. Whether or not acetyl~P plays a significant role in clock function remains to be determined. In one case, we did not model individual genes, but rather summarized a family of genes with a single model species. Specifically, all nitrogen-regulated genes and gene products are summarized as a single gene and protein in our model (depicted at lower left in Fig. 17.10). For the discussion that follows, we use a parameter set in the simulations that results in oscillator function similar to that seen in experiments. Specifically, we constructed a bacterial strain similar to our NC77 clock strain except that it lacks a functional copy of the *lacY* gene, and we attempted to fit the model to an experiment using this strain. The *lacY* mutant clock strain was specifically constructed to simplify the modeling.

The definitions of all parameters are presented in Table 17.1 and the parameter values that resulted from our preliminary efforts to fit the data to the model are presented in Table 17.2. We note that these results represent our initial efforts at the so-called "inverse problem" where data are fit to an explicit model and that some of the parameters obtained from the fitting are probably in error. For example, while the protein–DNA interactions show Kd in the nanomolar range, as expected, and the transcription rates are not unreasonable, certain other parameters are not within expected ranges (Table 17.2). The mRNA rates of decay appear to be much too low (Table 17.2); and the translation rates of the proteins showed far greater variation than expected and seem too slow (Table 17.2). It is certainly unreasonable to expect less than one protein translation event per generation for the proteins participating in function of our genetic oscillator (Table 17.2). Furthermore, the maximal growth rate appears to be much higher than expected or is feasible (Table 17.2). Thus, further efforts are required to directly determine one or more parameters and thus better constrain the fitting.

An interesting feature of the simulation of clock function was that external acetate was predicted to oscillate due to its coupling to the activator, which also oscillated (Fig. 17.11). It is not known at present whether acetyl~P has any role in

Fig. 17.10 (Continued) exert their effects intracellularly, *small triangle* on NRII or NRI depicts phosphoryl group. (It is assumed that each NRI dimer is phosphorylated on one site to allow assembly of the hexamer.) Acetate is assumed to have a constant rate of conversion to acetyl~P within the cell. Acetyl~P can directly transfer its phosphoryl group to NRI, forming NRI~P. NRI~P is also formed by transfer of phosphoryl groups from NRII~P. NRII binds ATP and phosphorylates itself, forming NRII~P. For clarity of the figure, the autophosphorylation of NRII is not depicted. The complex of the gratuitous inducer IPTG and the LacI repressor is shown within square brackets (*top right*); it is assumed that this complex contains two IPTG molecules bound to the repressor tetramer. At *bottom left*, the *ntr* genes of the host cell are depicted as if they were a single gene

Table 17.1 Definition of parameters, with additional symbols for the modeling figures

Parameter	Definition
k_{aA}	Association of activator for activator module enhancer
k_{-aA}	Dissociation of activator from activator module enhancer
k_{aR}	Association of repressor to the activator module proximal site
k_{-aR}	Dissociation of repressor from the activator module proximal site
k_{gA}	Association of activator to enhancer of nitrogen regulated genes, summarized G
k_{-gA}	Dissociation of activator from G enhancer
k_{rA}	Association of activator to the repressor module enhancer
k_{-rA}	Dissociation of activator from the repressor module enhancer
k_{zR}	Association of repressor with the *lacZYA* promoter
k_{-zR}	Dissociation of repressor from *lacZYA*
k_{sR}	Association of repressor to the "sink" site
k_{-sR}	Dissociation of repressor from the sink site
pa	Basal transcription rate from the activator module
Pa	Activated transcription rate from the activator module
Pg	Transcription rate of G
Pr	Transcription rate of the repressor gene
Pz	Transcription rate of *lacZYA*
ga	Decay rate of activator mRNA
gg	Decay rate of g mRNA
gr	Decay rate of repressor mRNA
gz	Decay rate of *lacZYA* mRNA
PA	Rate of translation of activator
PG	Rate of translation of G
PR	Rate of translation of repressor
PZ	Rate of translation of β-galactosidase
k_{a2}	Rate of association of activator monomers into the dimer
k_{a6p}	Rate of association of phosphorylated activator dimers into the hexamer
k_{-a6p}	Dissociation of the activator hexamer
k_{acp}	Rate of phosphorylation of activator by acetyl phosphate
k_{nr2}	Rate of phosphorylation of activator by NRII
k_{r4}	Rate of tetramerization of repressor
k_{z4}	Rate of tetramerization of β-galactosidase
Di	Rate of diffusion of IPTG into cells
De	Rate of diffusion of IPTG from the cells
k_{iR}	Rate of association of IPTG with the repressor
k_{-iR}	Rate of dissociation of IPTG from repressor
di	Rate of diffusion of acetate into cells
de	Rate of diffusion of acetate from cells
k_{acpac}	Rate of association of acetyl phosphate with activator
k_{-acpac}	Rate of dissociation of acetyl phosphate from activator
Pacp	Rate of production of acetyl phosphate
G	Maximal growth rate
$k_{translation}$	Metabolic costs of producing proteins
$k_{transcription}$	Metabolic costs of producing transcripts

(*continued*)

Table 17.1 (Continued)

Parameter	Definition
Additional symbols for the modeling figures	
A	Activator monomer
A2	Activator dimer
A2P	Phosphorylated activator dimer
A6P	Phosphorylated activator hexamer
aco	External acetate
aci	Cytoplasmic acetate
acp	Acetyl phosphate
Sa	Activator module unoccupied by repressor and activator
SaA	Activator module bound to activator
SaR	Activator module bound to repressor
SaAR	Activator module bound to both activator and repressor
Ss	Unoccupied "sink" site
SsR	"Sink" site occupied by repressor
Ma	mRNA for activator
Mr	mRNA for repressor
Sr	Repressor module unoccupied by activator
SrA	Repressor module bound by activator
Mr	Repressor mRNA
R	Repressor monomer
R4	Repressor tetramer
IR	IPTG bound to repressor
Ic	Cytoplasmic IPTG
Io	Extracellular IPTG
Sz	*lacZYA* control region unoccupied by repressor
SzR	*lacZYA* control region bound by repressor
Mz	*lacZYA* mRNA
Z	β-Galactosidase monomer
Z4	β-Galactosidase tetramer
Sg	Ntr promoters unoccupied by activator (composite)
SgA	Ntr promoters bound by activator (composite)
Mg	Ntr mRNA (composite)
G	Ntr gene products (composite)

clock function, but the model shows how it (or another factor that is coupled in a similar way to the activator) could play the role of intracellular signaling species, coordinating the functions of cells in the population (see Chap. 13).

Modeling the activator module (Fig. 17.12) suggested that a strong oscillation in activator protein and mRNA could be observed even under conditions where the activator module was occupied by both activator and repressor almost all of the time. Further, our simulation predicted that the distal operator or "sink" was essentially occupied all of the time. By contrast, the repressor module was predicted to be frequently unoccupied by activator (Fig. 17.13). Note that the transcription and translation machinery essentially serve as an amplifier; converting

Table 17.2 Parameter values from fitting experimental data to the explicit model

Activity/rate	Parameter	Value
Protein–DNA interaction	k_{aA}	22.8 (μM h)$^{-1}$
	k_{-aA}	0.06 h^{-1}
	k_{aR}	25.3 (μM h)$^{-1}$
	k_{-aR}	0.1 h^{-1}
	k_{gA}	0.49 (μM h)$^{-1}$
	k_{-gA}	1.14 h^{-1}
	k_{rA}	0.99 (μM h)$^{-1}$
	k_{-rA}	0.23 h^{-1}
	k_{zR}	18.3 (μM h)$^{-1}$
	k_{-zR}	0.07 h^{-1}
	k_{sR}	27.4 (μM h)$^{-1}$
	k_{-sR}	0.01 h^{-1}
Transcription rates	pa	0.42 h^{-1}
	Pa	8.99 h^{-1}
	Pg	54.5 h^{-1}
	Pr	0.66 h^{-1}
	Pz	3.54 h^{-1}
mRNA degradation rates	γa	0.04 h^{-1}
	γg	14.3 h^{-1}
	γr	0.01 h^{-1}
	γz	0.01 h^{-1}
Translation rates	PA	21.0 h^{-1}
	PG	0.02 h^{-1}
	PR	0.81 h^{-1}
	PZ	17.7 h^{-1}
Complex formation	k_{a2}	0.22 (μM h)$^{-1}$
	k_{a6p}	2.64 (μM2 h)$^{-1}$
	k_{-a6p}	2.54 h^{-1}
	k_{acp}	0.03 (μM h)$^{-1}$
	k_{nr2}	0.06 h^{-1}
	k_{r4}	16.0 (μM3 h)$^{-1}$
	k_{z4}	0.73 (μM3 h)$^{-1}$
Small molecules	Di	0.07 h^{-1}
	De	0.01 h^{-1}
	k_{iR}	9.36 (mM h)$^{-1}$
	k_{-iR}	1.38 h^{-1}
	di	0.1 h^{-1}
	de	46.9 h^{-1}
	k_{acpac}	0.44 h^{-1}
	k_{-acpac}	1.47 h^{-1}
	Pacp	0.05 mM h^{-1}
Growth rate	Γ	159.0 h^{-1}
	$k_{translation}$	20.2 h μM^{-1}
	$k_{transcription}$	0.39 h μM^{-1}

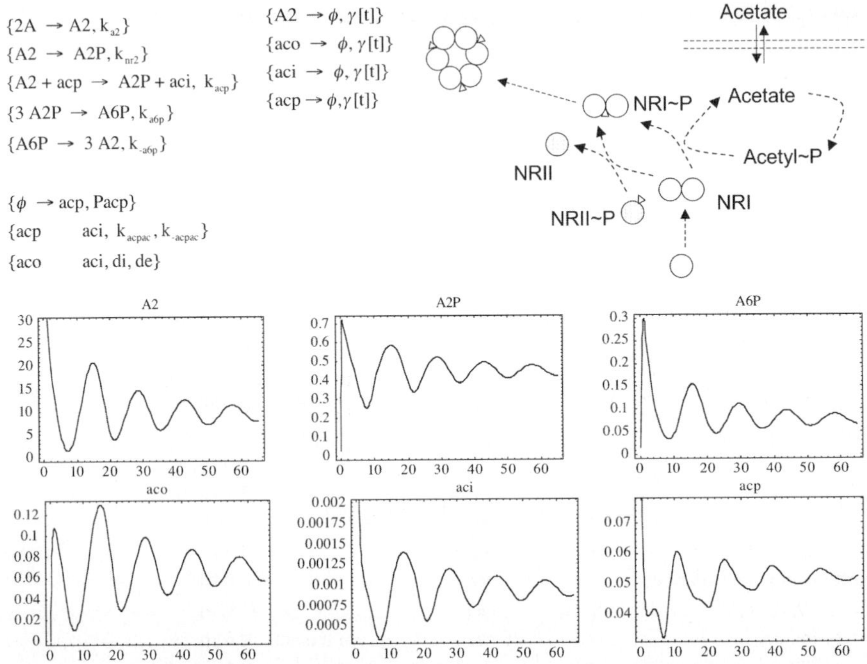

Fig. 17.11 Modeling the interconversions of activator and acetate. In this section of the model, we describe the phosphorylation of NRII and NRI, and the dephosphorylation of acetyl~P by NRI. We also model the diffusion of acetate between cells. All symbols are as in Fig. 17.10, and all parameters are as described in Table 17.1. The simulations showing the dynamical properties of each modeled species have the amplitude of the modeled species as their *y*-axis and time (h) as their *x*-axis. *A2* Dimeric form of the activator, *A2P* phosphorylated form of the dimeric activator, *A6P* active, hexameric form of the activator, *aco* acetate outside of the cells, *aci* acetate within the cell, *acp* acetyl~phosphate within the cell. Note that the model predicts that acetate outside of the cells oscillates as the clock oscillates, providing a potential means for cell–cell communication and coherent activity

the low-amplitude oscillation of bound activator at the repressor module into high-amplitude oscillations in repressor mRNA, repressor subunit, and repressor tetramer concentrations (Fig. 17.13). This was particularly helped by the non-linear tetramer formation. Similarly, low-amplitude oscillations in the extent of occupancy of the operators of the *lacZ* operon by repressor are converted into high-amplitude oscillations in the level of *lacZ* mRNA and protein (Fig. 17.14). Finally, the cellular *ntr* genes are predicted to act in a similar manner to that of the repressor module, with predictable effects on the cellular growth rate (Fig. 17.15). These effects were modeled by assuming that the production of unnecessary transcripts and proteins imposes a metabolic burden on the cell (Fig. 17.16). More specifically, when the repressor module is down-regulated, production of unnecessary β-galactosidase and galactoside permease slows the

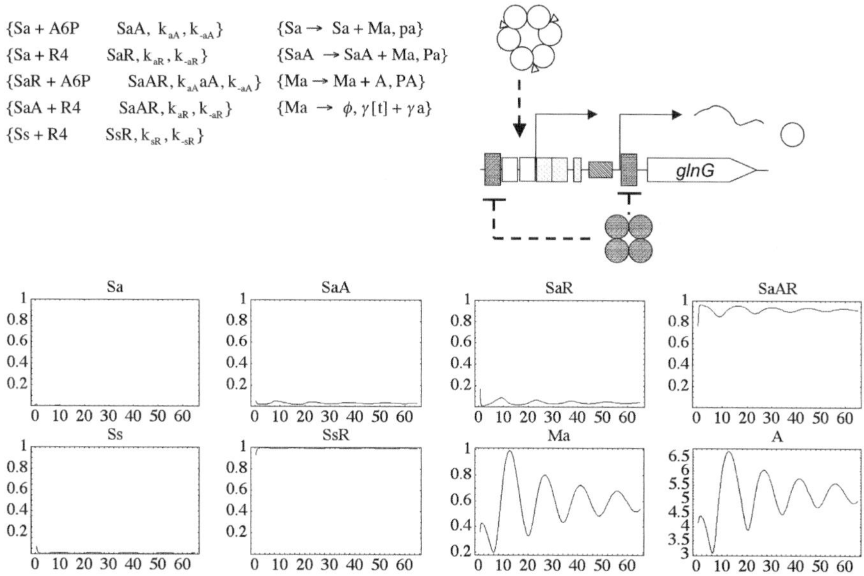

Fig. 17.12 Modeling the behavior of the activator module. In this section of the model, we describe the activation and repression of transcription from the activator module. All symbols are as in Fig. 17.10, and all parameters are as described in Table 17.1. In our model, there is no restriction blocking the simultaneous binding of activator and repressor to their binding sites within the activator module. We assumed that the distal operator functions as a "sink", and that the binding of repressor at the distal operator does not directly influence any other interaction. *Sa* Unbound activator module (which essentially never appears), *SaA* module with only activator bound, *SaR* module with just repressor bound at the proximal operator, *SaAR* activator module with both activator bound at the enhancer and repressor bound at the proximal site (which our model predicts is the major species at all times), *Ss* unoccupied distal operator, *SsR* occupied distal operator, *Ma* mRNA of the activator module (which oscillates), *A* activator protein subunit

growth of the cells. Similarly, when the activator module is producing activator, production of unnecessary GS and other nitrogen-regulated gene products should slow the growth rate (these nitrogen-regulated gene products are summarized as "G"). In contrast, rapid growth rate leads to rapid decrease in the concentrations of the proteins, due to their dilution by cell division. The overall effect is that gene products "Z" and "G" and cell growth are mutually inhibitory, with the consequence that cell growth fluctuates during the functioning of the genetic clock.

17.6 Discussion

While our synthetic genetic clock is clearly imperfect in that it produces damped oscillations, the dramatic nature of these oscillations and their reproducibility render it a system worthy of further study. As shown in the figures and before

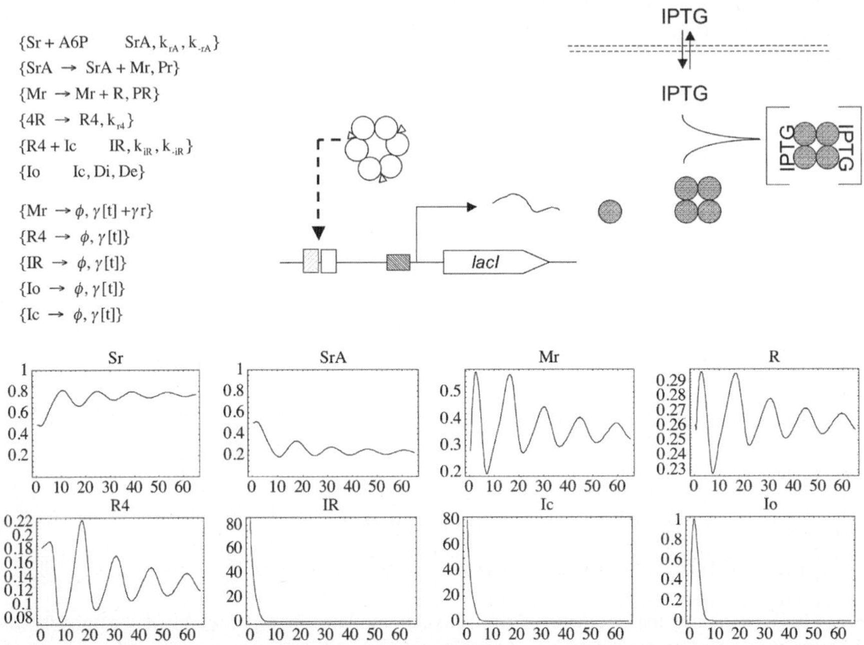

Fig. 17.13 Modeling the behavior of the repressor module. In this section of the model, we describe the activation of the repressor module, production of repressor, and inactivation of repressor by the gratuitous inducer IPTG, which diffuses into the cell. All symbols are as in Fig. 17.10, all parameters are as described in Fig. 17.8, and the graphs of modeled species are as in Fig. 17.11. In our model, there is no basal expression of the repressor module in the absence of activator. This assumption is supported by experimental studies of the glnK promoter (Atkinson et al. 2002). *Sr* Unoccupied repressor module (which is a common species), *SrA* repressor module bound by activator, *Mr* mRNA of the repressor module, *R* repressor subunit, *R4* repressor tetramer, *IR* repressor tetramer bound to IPTG (which we assume is innocuous), *Ic* IPTG within the cell, *Io* IPTG outside of the cell. Note that IPTG was washed away from the cells at the beginning of the clock experiments; the levels detected here in our simulations represents the IPTG that was bound by repressor within the cell and thus was slowly diluted out in the first few hours of the clock experiment

(Atkinson et al. 2003), we observe four or more clear waves of *lacZ* expression with amplitude of the waves on the order of hundreds of Miller units of β-galactosidase activity, in experiments lasting upwards of 60 h. Thus, there is nothing subtle about the phenomenon. The period of oscillations in our experiments is on the order of 10 generations, which means that cells inherit information on the status of their clock as they grow and divide. Since we can control the growth rate of the cells by altering the nutritional richness of the growth medium and the temperature of incubation, the real-time period of the synthetic genetic oscillator can be varied from about 8 h to about 30 h (data not shown). In the experiments shown here, we only examined the genetic oscillators in bacterial populations, with about 10^{11} cells maintained at

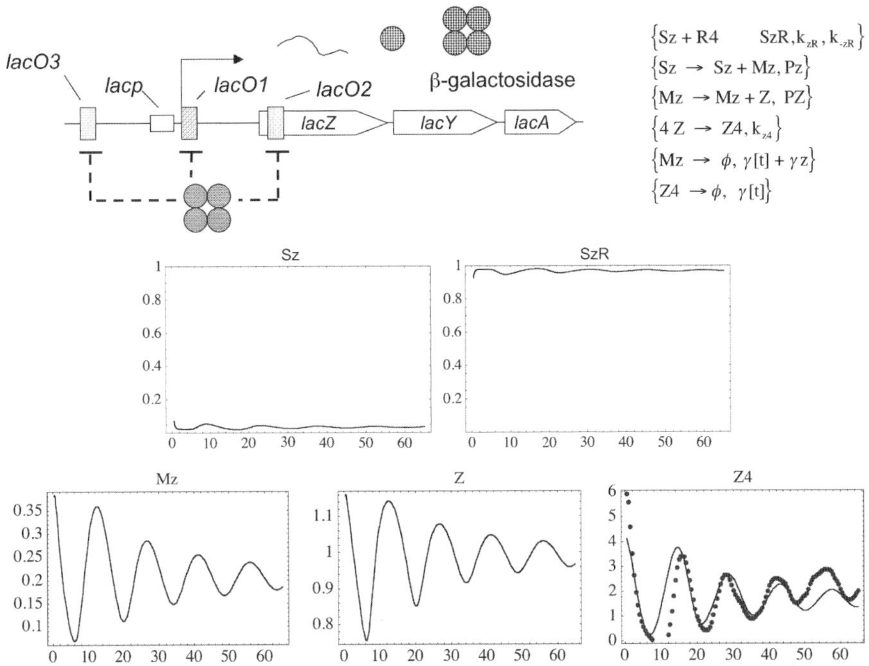

Fig. 17.14 Modeling the oscillations in *lacZ* expression. In this section of the model, we describe the transcription of the chromosomal *lacZYA* operon, which serves as a reporter of repressor concentration in our clock experiments. All symbols are as in Fig. 17.10, all parameters are as described in Table 17.1, and the graphs of modeled species are as in Fig. 17.11. *Sz*, *SzR lacZ* promoter unbound or bound by repressor, respectively. *Mz*, *Z*, and *Z4* mRNA, subunit, and tetramer for β-galactosidase (*lacZ* product), respectively. The data points (•) are an experimental run using the *lacY* mutant clock strain

a constant optical density in a reactor and growing with a doubling time of approximately 1 h.

We observed that alteration of the operators of the activator module resulted in a clock with somewhat reduced damping of oscillations relative to the starting system. Ideally, modifications of parameters may be identified that result in yet further reduction in the damping of oscillations. One possible means of "tuning" repression may be to use a strain that lacks the galactoside permease encoded by *lacY* and to conduct the experiments in the presence of various concentrations of IPTG to partially inactivate repressor to various extents. For such work, the suitable clock strain has already been constructed and shown, in the absence of IPTG, to produce damped oscillations similar to those obtained in the *lacY*$^{+}$ background (Fig. 17.14).

Interestingly, our data suggest that the distal operator of the activator module plays an important role in clock function, by reducing repression. That is, the distal operator does not seem to participate in a repression DNA loop (Oehler et al. 1990),

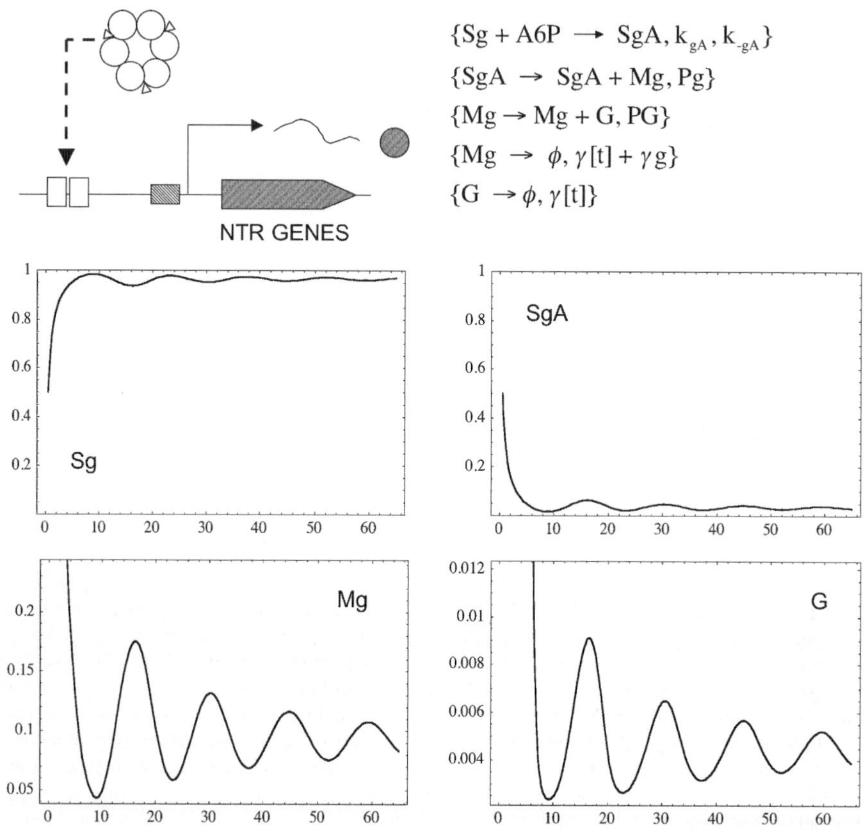

Fig. 17.15 Modeling the expression of nitrogen-regulated genes under the control of the genetic clock. In this section of the model, we consider all of the nitrogen-regulated genes as if they were a single, composite gene with average properties. All symbols are as in Fig. 17.10, all parameters are as described in Table 17.1, and the graphs of modeled species are as in Fig. 17.11. *Sg* Unoccupied promoter, *SgA* promoter bound by activator, *Mg* mRNA, *G* gene product protein subunit. Not surprisingly, the modeled gene behaves similarly to the repressor module (Fig. 17.13), as no additional regulatory events were included in the model. This is a simplification, as in intact cells the nitrogen regulated genes that are activated by NRI~P are in some cases subjected to additional independent controls

as intended. One hypothesis to explain these observations is that the distal operator serves as a "sink" for repressor, and by so doing, increases the kinetic order of repression of the module. A phenomenon of "pseudo-cooperativity" has been previously noted in a theoretical study of genetic autoregulation by repression (Goodwin. 1965), although the issue seems to have never been focused upon experimentally. If the distal operator site works by increasing the kinetic order of repression of the activator module, this would increase the stability of oscillations (Atkinson et al. 2003). Of course, a variety of experiments are underway, to investigate whether the

B

$$\gamma(t) = \frac{\Gamma}{1 + k_{transcription} \sum_i p_i S_i(t) + k_{translation} \sum_i P_i M(t)}$$

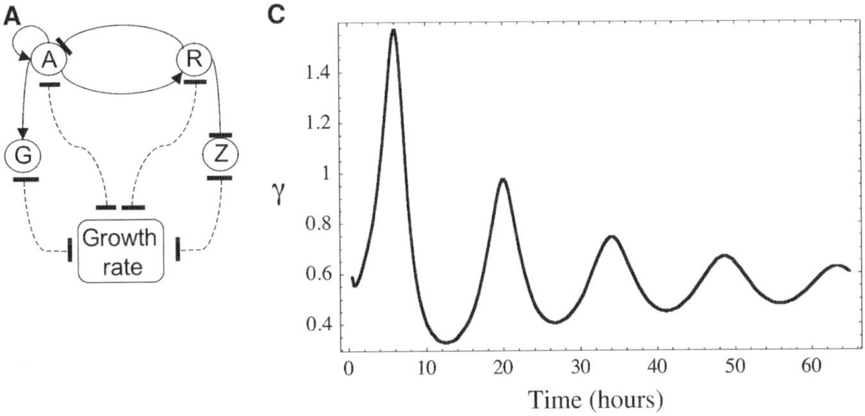

Fig. 17.16 Simulation of the effect of the genetic clock on the cellular growth rate. **A** Circuit diagram showing how various species are interconnected to each other and to cellular growth rate. *A* Hexameric activator, *R* tetrameric repressor, *Z* products of the *lacZYA* operon, *G* products of nitrogen regulated genes that are activated by NRI~P. *Solid lines* Direct regulatory interactions, *dashed lines* indirect regulatory interactions. **B** Simple model for the growth rate of the cells, $\gamma(t)$, as a function of the respective costs of producing transcripts and proteins. Γ Maximal growth rate, $k_{transcription}$ and $k_{translation}$ cost of producing transcripts and proteins, respectively. **C** Simulation of the model, showing that growth rate is expected to oscillate as a consequence of the functioning of the synthetic genetic clock

distal operator can work from more distant locations and whether multiple tandem copies of the distal operator work even better.

Our synthetic genetic clock seemed to be remarkably robust to some variations in parameters, yet appeared to be very sensitive to other changes. For example, although we observed that NC77 was our best clock, there were a few other combinations of operators within the activator module that also produced reasonably good clocks (Figs. 17.3, 17.4). Thus, while there is an optimal level of repression, oscillatory function was robust to some variation in repression. In contrast, our clock seemed to be very sensitive to changes in the number of chromosomal binding sites for activator and repressor; specifically, decreasing the number of repressor binding sites and increasing the number of activator binding sites seemed to have a dramatic effect on clock function (Fig. 17.7). Similarly, simply removing the distal operator of the activator module, or even converting it to the weak *lacO3* sequence when the proximal operator was *lacO1*, had a dramatic deleterious effect on clock function (Fig. 17.4). We are still working to explain these observations.

A significant drawback to the experimental system described in this paper is the labor-intensive aspect of the clock experiments when performed with standard laboratory instruments, as was done for the experiments shown here. The main problem is that the growth of the cells was affected by the synthetic genetic clock, such that it was not possible to maintain a culture under continuous conditions (including optical density) without continuous adjustment of the nutrient pump of the fermentor. That is, a turbidostat is required to study the system. Except as noted for the NC77 clock and a few others, the experiments shown here have mainly been performed just once each. Since expression of the Ntr-regulated *nac* gene is known to affect cell growth (Blauwkamp and Ninfa 2002), we constructed a mutant strain of *E. coli* in which the enhancer of the *nac* gene was mutated. This strain was used as the host cells for the clock experiments shown in Figs. 17.3 and 17.4, and indeed, the variations in cell growth rate as the clock produced damped oscillation of *lacZ* expression was reduced relative to the strain with normal control of *nac*, and it was easier to maintain the culture turbidity by manual control of the nutrient pump using this host cell background. But, apparently *nac* is not the only gene contributing to the effect, as even the *nac* mutated strain shows significant variation in growth rate as the clock functions. Because of this bottleneck, we recently described a homemade turbidostat that can be assembled from commercially available parts and controlled by computer (Ninfa et al. 2007). Results with the automated system have shown that the damped oscillations obtained for a given genetic clock are highly reproducible when experiments are repeated (N. Perry, personal communication). We expect the automated turbidostat to allow us to confirm the results shown here and extend our study of the sensitivity of the genetic clock to changes in various parameters and experimental conditions. We are also working on improved fluorescent reporters for the system that may allow continuous real-time observations of oscillations in populations. Hopefully, the experimental methods currently used here will soon be obsolete.

Theoretical studies with toy models, both here and elsewhere (Del Vecchio 2007), show that the circuit topology we used is capable of producing stable oscillations. Indeed, depending on parameters, a wide variety of oscillatory systems were identified, ranging from relaxation type oscillators to sinusoidal type oscillators. Thus, the "oscillation generator" is likely to be contained within the features forming the idealized models. But, this does not mean that interactions not contained within these models are without important effects. In our more explicit model, where we attempt to include as much information as possible, we could see that cells had the potential to signal to each other by virtue of clock effects on the extracellular concentration of acetate. Whether this in fact does contribute to the maintenance of phasing in the population is currently under investigation. The rise and fall of extracellular acetate that is predicted by the model may have a role in favoring rhythm in the population.

Acknowledgements This work was supported by grant GM063642 from the NIH-NIGMS. We thank Henry Wu, Pricilla Prior, Ritesh Senapati, and Grace Song for technical assistance with the clock experiments.

References

Atkinson MR, Blauwkamp TA, Bondarenko V, Studitsky V, Ninfa AJ (2002a) Activation of the *glnA*, *glnK*, and *nac* promoters as *Escherichia coli* undergoes the transition from nitrogen-excess growth to nitrogen starvation. J Bacteriol 184:5358–5363

Atkinson MR, Pattaramanon N, Ninfa AJ (2002b) Governor of the *glnA* promoter of *Escherichia coli*. Mol Microbiol 46:1247–1257

Atkinson MR, Savageau MA, Meyers J, Ninfa AJ (2003) Development of a genetic circuitry exhibiting toggle switch or oscillatory behavior in *Escherichia coli*. Cell 113:597–607

Blauwkamp TA, Ninfa AJ (2002) Nac-mediated repression of the *serA* promoter of *Escherichia coli*. Mol Microbiol 45:351–363

Conrad E, Mayo AE, Ninfa AJ, Forger DB (2008) Rate constants rather than biochemical mechanism determine behavior of genetic clocks. J R Soc Interface 1:9–15

Del Vecchio D (2007) Design and analysis of an activator–repressor clock in *Escherichia coli*. Proc Am Control Conf 2007:1589–1594

Feng J, Atkinson MR, McCleary W, Stock JB, Wanner BL, Ninfa AJ (1992) Role of phosphorylated metabolic intermediates in the regulation of glutamine synthetase synthesis in *Escherichia coli*. J Bacteriol 174:6061–6070

Goodwin BC (1965) Oscillatory behavior in enzymatic control processes. Adv Enzyme Regul 3:425–438

Kondo T, Mori T, Lebedeva NV, Aoki S, Ishiura M, Golden SS (1997) Circadian rhythms in rapidly dividing cyanobacteria. Science 275:224–227

Ninfa AJ, Magasanik B (1986) Covalent modification of the *glnG* product, NR_I, by the *glnL* product, NR_{II}, regulates the transcription of the *glnALG* operon in *Escherichia coli*. Proc Natl Acad Sci USA 83:5909–5913

Ninfa AJ, Reitzer LJ, Magasanik B (1987) Initiation of transcription at the bacterial *glnAp2* promoter by purified *Escherichia coli* components is facilitated by enhancers. Cell 50:1039–1046

Ninfa AJ, Selinsky S, Perry N, Atkins S, Song QX, Mayo A, Arps D, Woolf P, Atkinson MR (2007) Using two component systems and other bacterial regulatory factors for the fabrication of synthetic genetic devices. Methods Enzymol 422:488–512

Oehler S, Eismann ER, Kramer H, Muller-Hill B (1990) The three operators of the *lac* operon cooperate in repression. EMBO J 9:973–979

Reitzer LJ, Magasanik B (1985) Expression of *glnA* in *Escherichia coli* is regulated at tandem promoters. Proc Natl Acad Sci USA 82:1979–1983

Sadler JR, Sasmor H, Betz JL (1983) A perfectly symmetrical lac operator binds the lac repressor very tightly. Proc Natl Acad Sci USA 80:6785–6789

Yu D, Ellis HM, Lee EC, Jenkins NA, Copeland NG, Court DL (2000). An efficient recombination system for chromosome engineering in *Escherichia coli*. Proc Natl Acad Sci USA 97:5978–5983

Index

Printing: Krips bv, Meppel, The Netherlands
Binding: Stürtz, Würzburg, Germany